Smart Strategies and Technologies for Sustainability and Biodiversity in Herbaceous and Horticultural Crops

Smart Strategies and Technologies for Sustainability and Biodiversity in Herbaceous and Horticultural Crops

Editors

Christian Frasconi
Marco Fontanelli
Daniele Antichi

Basel • Beijing • Wuhan • Barcelona • Belgrade • Novi Sad • Cluj • Manchester

Editors
Christian Frasconi
Department of Agriculture,
Food and Environment
University of Pisa
Pisa
Italy

Marco Fontanelli
Department of Agriculture,
Food and Environment
University of Pisa
Pisa
Italy

Daniele Antichi
Department of Agriculture,
Food and Environment
University of Pisa
Pisa
Italy

Editorial Office
MDPI
St. Alban-Anlage 66
4052 Basel, Switzerland

This is a reprint of articles from the Special Issue published online in the open access journal *Agronomy* (ISSN 2073-4395) (available at: www.mdpi.com/journal/agronomy/special_issues/R93414033U).

For citation purposes, cite each article independently as indicated on the article page online and as indicated below:

Lastname, A.A.; Lastname, B.B. Article Title. *Journal Name* **Year**, *Volume Number*, Page Range.

ISBN 978-3-7258-0582-2 (Hbk)
ISBN 978-3-7258-0581-5 (PDF)
doi.org/10.3390/books978-3-7258-0581-5

Contents

About the Editors

Christian Frasconi

Christian Frasconi is associate professor of farm mechanization and farm machinery at the Department of Agriculture, Food and Environment of the University of Pisa, Italy. His main research topics are farm mechanization and farm machinery, precision agriculture and robotics, conservation agriculture, non-chemical weed control, and machines for turfgrass and landscape management.

Marco Fontanelli

Marco Fontanelli is associate professor of farm mechanization and farm machinery at the Department of Agriculture, Food and Environment of the University of Pisa, Italy. His main research topics are farm mechanization and farm machinery, precision agriculture and robotics, conservation agriculture, non-chemical weed control, and machines for turfgrass and landscape management.

Daniele Antichi

Daniele Antichi is associate professor of agronomy at the Department of Agriculture, Food and Environment of the University of Pisa, Italy. His main research topics are organic farming, conservation agriculture, non-chemical weed control, and cover crop management.

Editorial

Smart Strategies and Technologies for Sustainability and Biodiversity in Herbaceous and Horticultural Crops

Christian Frasconi, Marco Fontanelli * and Daniele Antichi

Department of Agriculture, Food and Environment (DAFE), University of Pisa, 56126 Pisa, Italy; christian.frasconi@unipi.it (C.F.); daniele.antichi@unipi.it (D.A.)
* Correspondence: marco.fontanelli@unipi.it; Tel.: +39-050-2218922

Citation: Frasconi, C.; Fontanelli, M.; Antichi, D. Smart Strategies and Technologies for Sustainability and Biodiversity in Herbaceous and Horticultural Crops. *Agronomy* **2024**, *14*, 528. https://doi.org/10.3390/agronomy14030528

Received: 19 February 2024
Accepted: 1 March 2024
Published: 4 March 2024

Current trends in modern farming systems are moving in the direction of technical solutions for improving the sustainability and biodiversity of agroecosystems. Innovative agronomical strategies and new technologies can help farmers to reduce or eliminate chemical inputs, preserve soil and water quality, decrease exhaust and greenhouse gas emissions, prevent pollution, and lower energy demand. Sustainable management also aims to enhance biodiversity in order to lengthen the "life" of agroecosystems. Herbaceous and horticultural crops are dominant crops that can contribute to achieving this goal.

This Special Issue presents contributions regarding innovative technologies, machines, and strategies for the sustainable management of herbaceous and horticultural crops, including applications in organic farming systems, conservation agriculture, integrated or non-chemical weed and pest control, cover crops and intercropping use, precision and digital farming technologies, and robotic technologies for sustainability. Moreover, this Special Issue concerns conservation agriculture, organic agriculture, cover crops, intercropping, integrated/non-chemical weed and pest control, and precision and digital agriculture for sustainability. These Special Issue is a collection of 12 papers: 5 on tree crops/orchards and 7 on herbaceous crops.

Starting on the topic of tree crops, Ran et al. [1] used local thermal damage as an innovative technique for jujube trees. After creating an appropriate heat transfer model, the authors used experimental results to show that heating the bark at a certain temperature for an appropriate time can effectively improve yield and quality. Compared with traditional girdling techniques, this method had a smaller impact on the health of jujube trees and did not cause permanent wounds.

The following three papers address the topic of weed control under tree rows. Lisek [2] studied the effect of five methods of in-row weed management on the composition and diversity of weed species in a plum orchard. *Festuca rubra* L. ssp. was sown as a cover crop. The obtained results indicated that the flora developing in the control plot, tilled and mowed plots, and plot sprayed with post-emergence herbicides showed greater potential to provide ecosystem services than the flora of mulched plots. Gagliardi et al. [3] studied an innovative method for weed control under tree rows in alley cropping farming systems; they did so using a mower, which they modified by replacing blades with chains (as they are more resistant to stones and pruning materials). No major differences emerged with the standard blade mower. The setting with a high working speed and high rotation speed resulted in a compromise, obtaining a weed biomass reduction of 59.6%. Peruzzi et al. [4] ran a two-year trial on a vineyard; they aimed to evaluate the effect of cover crops managed with an autonomous mower on *E. canadensis* weeds under a trellis. This combination induced a reduction of between 61 and 84% in *E. canadensis* when compared to conventional management techniques.

Roškarič et al. [5] studied a "0-pesticide residue" spray program in a Mediterranean vineyard scenario. The innovative strategy was compared to the standard integrated grape protection program. The level of infection of leaves and grapes by fungal pathogens did not significantly increase. The amount and quality of yield were not decreased significantly, but a small financial loss occurred. The zero-pesticide system enabled a significant decrease in the concentration of pesticide residues in wine (ranging from 20% to 99%).

Concerning herbaceous crops, two studies investigated the use of conservation agricultural practices on tomatoes in the Mediterranean area [6,7]. The first, from Abou Chehade et al., evaluated the impact of different cover crops and tillage systems on organic tomato processing. Rye and a mixture were the most productive and the best weed suppressor. Squarrose clover stimulated tomato growth regardless of tillage system. The results suggested that legume cover crops could be the key to developing feasible organic vegetable no-tilling systems [6]. Gagliardi et al. confirmed that organically processed tomato plants can successfully grow within a conservation farming system [7]. Benetti et al. studied the role of cover crop mulch in mitigating soil compaction [8]. Rye was chosen as a winter cover crop, and four different mulch treatments were compared within different traffic intensities. The results highlight that the cover crop maintains a lower soil penetration resistance during compaction events, helping subsequent field operations. Furthermore, roller crimpers and cover crop termination by flail mowers impact soil's capacity differently due to different soil moisture contents.

Clemente et al. and Rossi et al. carried out two interesting trials on industrial crops [9,10]. The spectral response of camelina under different regimes of N was studied in order to optimize fertilization. Positive and significant correlations were observed among several vegetation indices (obtained through UAV flights and seed yield) as well as between green canopy cover fractions and leaf N concentrations. Overall, these findings demonstrate the feasibility of utilizing remote sensing techniques from UAVs for predicting seed yield in camelina [9]. Linseed organic cultivation was also studied using different agronomic parameters. Generally, linseed showed good agronomic traits that make it suitable for introduction into organic systems. Autumn sowing coupled with milder and wetter conditions seemed to be more favorable for linseed cultivation, while the superior genotypes were Kaolin > Szafir > Galaad [10].

Two remaining articles describe the use of breeding techniques to enhance the agronomic performances of vegetable crops [11,12]. Karbarz et al. tested microalgae *Planktochlorella nurekis* clones obtained by co-treatment with colchicine and cytochalasin on four plant species (lettuce, wheat, broccoli, and radish) to check their potential use as biostimulators in agriculture. Eleven clone extracts showed both a stimulating and inhibitory effect on tested plants (depending on the concentration, plant species, and algal clone); thus, they could potentially be used as biostimulators or natural herbicides [11]. Li et al. assembled a mitochondrial genome map of *O. javanica* (watercress, a very popular vegetable in Asia) using the Illumina and Nanopore sequencing platforms. The results showed that watercress is closely related to species of *Bupleurum*, *Apium*, *Angelica*, and *Daucus*, providing a valuable resource for the study of the molecular breeding [12].

This collection of articles shows that many strategies can effectively enhance sustainability and biodiversity in herbaceous and horticultural crops within different farming scenarios. These goals can be achieved using innovative technologies developed to carry out specific tasks such as weed control, pest control, cover crop management, crop monitoring, biostimulation, and breeding.

Conflicts of Interest: The authors declare no conflict of interest.

References

1. Ran, J.; Zhang, J.; Wang, X.; Liu, Y.; Hu, C.; Xing, J.; Sun, B. Research on Jujube-Fruit-Yield-Increasing Technology Based on Local Thermal Damage of Jujube Bark. *Agronomy* **2023**, *13*, 2551. [CrossRef]
2. Lisek, J. Diversity of Summer Weed Communities in Response to Different Plum Orchard Floor Management in Row. *Agronomy* **2023**, *13*, 1421. [CrossRef]
3. Gagliardi, L.; Fontanelli, M.; Frasconi, C.; Sportelli, M.; Antichi, D.; Tramacere, L.G.; Rallo, G.; Peruzzi, A.; Raffaelli, M. Assessment of a Chain Mower Performance for Weed Control under Tree Rows in an Alley Cropping Farming System. *Agronomy* **2022**, *12*, 2785. [CrossRef]

4. Peruzzi, A.; Gagliardi, L.; Fontanelli, M.; Frasconi, C.; Raffaelli, M.; Sportelli, M. Continuous Mowing for *Erigeron canadensis* L. Control in Vineyards. *Agronomy* **2023**, *13*, 409. [CrossRef]

5. Roškarič, M.; Paušič, A.; Valdhuber, J.; Lešnik, M.; Pulko, B. Development of a "0-Pesticide Residue" Grape and Wine Production System for Standard Disease-Susceptible Varieties. *Agronomy* **2023**, *13*, 586. [CrossRef]

6. Abou Chehade, L.; Antichi, D.; Frasconi, C.; Sbrana, M.; Tramacere, L.G.; Mazzoncini, M.; Peruzzi, A. Legume Cover Crop Alleviates the Negative Impact of No-Till on Tomato Productivity in a Mediterranean Organic Cropping System. *Agronomy* **2023**, *13*, 2027. [CrossRef]

7. Gagliardi, L.; Sportelli, M.; Fontanelli, M.; Sbrana, M.; Luglio, S.M.; Raffaelli, M.; Peruzzi, A. Effects of Conservation Agriculture Practices on Tomato Yield and Economic Performance. *Agronomy* **2023**, *13*, 1704. [CrossRef]

8. Benetti, M.; Liu, K.; Guerrini, L.; Gasparini, F.; Peruzzi, A.; Sartori, L. How Much Impact Has the Cover Crop Mulch in Mitigating Soil Compaction?—A Field Study in North Italy. *Agronomy* **2023**, *13*, 686. [CrossRef]

9. Clemente, C.; Ercolini, L.; Rossi, A.; Foschi, L.; Grossi, N.; Angelini, L.G.; Tavarini, S.; Silvestri, N. Spectral Response of Camelina (*Camelina sativa* (L.) Crantz) to Different Nitrogen Fertilization Regimes under Mediterranean Conditions. *Agronomy* **2023**, *13*, 1539. [CrossRef]

10. Rossi, A.; Clemente, C.; Tavarini, S.; Angelini, L.G. Variety and Sowing Date Affect Seed Yield and Chemical Composition of Linseed Grown under Organic Production System in a Semiarid Mediterranean Environment. *Agronomy* **2023**, *13*, 45. [CrossRef]

11. Karbarz, M.; Piziak, M.; Żuczek, J.; Duda, M. Influence of Microalgae *Planktochlorella nurekis* Clones on Seed Germination. *Agronomy* **2023**, *13*, 9. [CrossRef]

12. Li, X.; Han, Q.; Li, M.; Luo, Q.; Zhu, S.; Zheng, Y.; Tan, G. Complete Mitochondrial Genome Sequence, Characteristics, and Phylogenetic Analysis of *Oenanthe javanica*. *Agronomy* **2023**, *13*, 2103. [CrossRef]

Article

Research on Jujube-Fruit-Yield-Increasing Technology Based on Local Thermal Damage of Jujube Bark

Junhui Ran [1], Jiajia Zhang [1], Xufeng Wang [1], Yuanjie Liu [1,*], Can Hu [1,*], Jianfei Xing [1] and Beibei Sun [2]

[1]   College of Mechanical and Electrical Engineering, Tarim University, Alar 843300, China; 120200005@taru.edu.cn (J.R.); 10757222245@stumail.taru.edu.cn (J.Z.); 119990012@taru.edu.cn (X.W.); 120200012@taru.edu.cn (J.X.)

[2]   Instrumental Analysis Center of Tarim University, Alar 843300, China; 120219002@taru.edu.cn

*   Correspondence: 120050023@taru.edu.cn (Y.L.); 120140004@taru.edu.cn (C.H.)

Abstract: Girdling is an important means of improving the yield and quality of jujube trees, but this measure can easily cause injury, or even death, to jujube trees. A technology for increasing yield and improving quality, based on local thermal damage of jujube bark, is proposed to address a series of issues in current jujube-tree-girdling technology. First, we measured the thermophysical parameters of jujube bark and established a heat-transfer model for jujube bark. Then, in order to investigate the impact of local thermal damage on jujube-tree yield and fruit quality, local heating experiments were conducted on jujube-tree bark, using the heat-transfer model. The experimental results indicated that heating the jujube bark at a certain temperature for an appropriate time can effectively improve the yield and quality of jujube fruit. Compared with traditional girdling techniques, this method has less impact on the health of jujube trees and does not form permanent wounds on them. The research results provide new ideas for exploring sustainable yield-increase methods for fruit trees.

Keywords: fruit trees; thermal damage; heat-transfer model; bark; girdling; yield; jujube-fruit quality

Citation: Ran, J.; Zhang, J.; Wang, X.; Liu, Y.; Hu, C.; Xing, J.; Sun, B. Research on Jujube-Fruit-Yield-Increasing Technology Based on Local Thermal Damage of Jujube Bark. *Agronomy* **2023**, *13*, 2551. https://doi.org/10.3390/agronomy13102551

Academic Editors: Christian Frasconi, Marco Fontanelli, Daniele Antich

Received: 4 September 2023
Revised: 28 September 2023
Accepted: 29 September 2023
Published: 3 October 2023

1. Introduction

Girdling is an important method for increasing the yield and improving the fruit quality of fruit trees, and this technology has been widely applied in various regions of China [1]. Practice has shown that girdling can promote the accumulation of sugar in fruits, improve skin color, and promote fruit ripening. Numerous research results have shown that many fruit trees, such as jujube, citrus, apple, cherry, and blueberry trees, can improve the quality and the yield of their fruits, to a certain extent, through appropriate girdling [1,2]. Girdling usually refers to the use of tools to girdle the trunk or the main branches of a fruit tree in a reasonable manner, inhibiting the transportation of nutrients generated by the leaves through the phloem to the roots and allocating as many nutrients as possible to the fruit, in order to improve the yield and the quality of the fruit tree [3].

However, fruit tree girdling also has numerous defects. First, the most direct impact of girdling on trees is physical damage to them. An unhealed girdling wound increases the risk of insects and pathogens invading the tree, making it susceptible to pests and diseases. Second, continuous girdling for many years can affect the distribution balance of the photosynthetic products in the tree, resulting in insufficient supply of root nutrients. This leads to weakened tree vigor and undermines the sustainable productivity of an orchard. More seriously, according to statistics from farmers and from our team over the years, the annual continuous girdling of jujube trees in jujube orchards, for about 7 years, leads directly to about 8% of the trees dying each year thereafter [4]. In addition, years of continuous girdling have led to a decrease in the quality of fruits, resulting in reduced sugar content and a sour taste. Related studies have shown that after girdling until the formation of new phloem tissue, continuous girdling treatment affects both the water transport in the xylem and the normal physiological activities of the fruit trees [5]. This, in turn, leads to a weakened physiological function of the roots and a reduced ability to absorb various ions from the soil. At the same time, the accumulation of cytokinins in the roots inhibits the development of lateral root elongation [6]. In an experiment, Schepper [7] found that a decrease in the sap flow rate can be observed shortly after girdling. In addition, some other studies suggested that long-term girdling hinders the synthesis and transportation of endogenous hormones in trees, limits overall plant-nutrient absorption, and hinders the growth of branches, roots, and trunks, ultimately leading to a decrease in tree vitality [8]. In short, girdling has a significant destructive effect on the regulation of tree bodies, and this damage accumulates with increases in girdling frequency over time.

In order to eliminate these negative effects caused by traditional fruit-tree girdling, researchers have made continuous attempts at new methods related to girdling. First, plastic or metal wires have been used to ligate fruit trees to inhibit the transport of nutrients to the roots during the fruiting period. However, when using plastic rolling strips, it is easy for the strips to relax under the action of the preload, so the effect is not as good as that of iron wire. Moreover, both methods are cumbersome to operate, and ultimately the plastic bag or the metal wire needs to be removed, which requires a large amount of manual investment. In addition, researchers have also attempted to use the method of locally heating the fruit bark at high temperatures to achieve the goal of inhibiting the transportation of photosynthetic products from the phloem to the roots during specific periods. Research has shown that heating the phloem of peach trees at a certain temperature for a certain period of time reduces the ability of the phloem to transport starch toward the root, resulting in a better inhibitory effect on the nutritional growth of the peach trees and promoting their reproductive ability [9]. However, there is very little existing research on these methods, and no quantitative analysis has been conducted on the impact of locally heating the fruit bark on fruit-tree yield and fruit quality; there is also no relevant heat-transfer model available to guide specific heating experiments.

Unlike girdling, quantitative heating does not require peeling the bark, thereby avoiding the numerous adverse effects caused by exposure of the wood. Additionally, unlike the tightening methods such as the use of iron wire, quantitative heating does not require the subsequent removal of a rolling strip, making the process simpler. In addition, theoretically speaking, by establishing a heat-transfer model for bark, reasonable control of the heating temperature and the time can be achieved to accurately control the degree of bark-cell death during heating. This is expected to improve the yield and the quality of fruit trees without causing significant damage to the tree body. Therefore, the use of this method to replace the traditional girdling method in achieving the goal of increasing fruit-tree yield may be comparatively ideal. By heating the main branches (with the diameter of 15–20 mm) of jujube trees with a 1 mm diameter electric wire at 200 °C for 1 min, our team studied the effect of this treatment on jujube tree yield, in 2021. The preliminary research results indicated that the jujube fruit yield of jujube tree branches treated with this method was significantly higher than that of jujube tree branches untreated. However, this study did not consider that the heating time of bark with different thicknesses should vary, and quantitative research was not conducted on the heating time of bark with different thicknesses. Therefore, in order to further explore the impact of this method on jujube yield and establish a feasible jujube bark heating model, further research is needed.

This study proposes the hypothesis that heating the bark of jujube trees can cause heat damage, making it impossible for nutrients to be transported through the phloem to the roots for a period of time, promoting jujube-fruit growth and, ultimately, achieving the goal of increasing yield and improving jujube-fruit quality. In order to verify the correctness of this hypothesis, we carried out the following work. First, we measured the thermophysical properties and other parameters of jujube bark. Then, we established a heat-transfer model for jujube bark to provide a basis for decision making on the heating temperatures and times for jujube bark. Finally, based on the constructed heat-transfer model of jujube bark, jujube-bark-heating experiments were conducted to study the specific impact of the local heating of jujube bark on the quality and the yield of jujube fruit. Through this study, the feasibility of using thermal-damage-based methods to increase jujube yield and improve jujube-fruit quality was explored, and theoretical guidance and basic data are provided for the design of specialized heating devices in the future.

2. Materials and Methods

2.1. Materials and Instruments

The jujube trees used in the experiment are located in a jujube garden in Alar City, Xinjiang. The jujube trees are 8 years old and the Hui jujube variety, as shown in Figure 1. In addition, all experiments were conducted during the jujube-tree-girdling period, on 18 June 2022. A CD-15APX digital display vernier caliper produced by Mitutoyo (with an error of ±0.02 m) was used to measure jujube-bark size parameters. An LU-920SERIES thermostat produced by Anthone Electronice Co., Ltd. (Xiamen, China) with an error of ±0.5 °C was used to control the heating temperature during the experiment. The outdoor power supply was the B7 type produced by Shenzhen Chuangwei Energy Technology Co., Ltd (Shenzhen, China). A WRNK191 thermocouple with 0.5 mm probe diameter produced by Shanghai Shenji Instrument Co., Ltd (Shanghai, China) was used to detect temperature. The heater had a 60 × 60 mm infrared arc with a built-in K-type thermocouple and heating wire material of Nice0Cr20; it was produced by Taizhou Dayi Electric Heating Appliance Co., Ltd (Taizhou, China). Furthermore, in order to measure the yield and quality of jujube fruits, the jujube samples from the trees with different treatments were picked when jujube was in the full maturity stage, on 28 September 2022.

Figure 1. Jujube garden for the experiments.

2.2. Determination of Physical Parameters of Jujube Bark

The thickness of bark has a greater impact on the heat-transfer performance of bark compared to factors such as moisture content, density, and surface texture structure. Therefore, the trunk diameter can be used as an important indicator of the sensitivity of the corresponding phloem to heating. The bark of a jujube trunk is very thick, with a rough surface and wide and deep longitudinal cracks, which is not conducive to heat conduction. The bark of the main branch is relatively thin, with a narrow and shallow surface texture structure. Therefore, the main branch is chosen to reduce energy loss during the heating process and improve heating efficiency. In order to provide basic parameters for the construction of a heat-transfer model for jujube bark, some physical and heat-transfer-related parameters of jujube bark were measured.

Firstly, it is necessary to measure the thermal conductivity of jujube bark. Although the thermal conductivity of bark varies with the age of the tree, the moisture content of the bark, density, and other factors, it is generally considered constant in transient heat-transfer models [10]. In addition, due to its convenient use and relatively high accuracy, the Wenger formula [11] is often used to estimate the thermal-conductivity determination of tree bark. Therefore, the Wenger formula was also used in this study to calculate the thermal conductivity of jujube bark, as shown in Equation (1). The density of each jujube-bark sample was measured by the mass displacement of water method. Then, the density value was substituted into this equation to obtain its thermal conductivity.

$$k = (4.684\rho + 0.076) \times 10^{-4} \tag{1}$$

where k is the thermal conductivity coefficient, $\mathrm{Wm^{-1}/K}$; ρ is the fresh jujube-bark density, $\mathrm{g/cm^{-3}}$.

Then, Equation (2) was used to calculate the specific heat capacity of jujube bark. Bark samples with volume of $2 \mathrm{~cm^{-3}}$ were collected using a sharp knife and their initial weight was immediately measured. Then, the samples were placed in a drying oven for 48 h at $60\,^\circ\mathrm{C}$ to obtain their dry mass. The data obtained through the above experiment were used to calculate the ratio of moisture mass to dry mass of bark (M).

$$\begin{cases} c = 4186.8 \left[0.264 + 0.00116T + \frac{Mc_w}{419 + \Delta_c} \right] \\ c_w = 3.8 + 130 \div (371.85 - T) \end{cases} \tag{2}$$

where c is the specific heat capacity of jujube bark, $\mathrm{J~kg^{-1}/K}$; M is the ratio of moisture mass to dry mass of bark,%; c_w is the specific heat capacity of water, $\mathrm{kJ~kg^{-1}/K}$; T is the temperature, $^\circ\mathrm{C}$; Δ_c is an empirical correction for the effect of moisture on heat capacity, $\mathrm{cal/(g\,^\circ C)}$; Δ_c is taken as 0.305 M for $M \le 27\%$ or 0.0832 M for $M > 27\%$.

The thermal diffusion coefficient of bark is the rate at which heat is transferred from the outer surface of the bark to the inner surface through a given thickness of bark. It is a measure of the insulation capacity of tree bark and an important parameter in the heat-transfer model. Finally, based on the above measurement data, the thermal diffusion coefficient of jujube bark is calculated by Equation (3).

$$\begin{cases} \alpha = \frac{k}{\rho c} \\ \rho = \rho_b (1 + M_c) \end{cases} \tag{3}$$

where α is the thermal diffusion coefficient, $\mathrm{cm^2/min}$; k is the thermal conductivity, $\mathrm{W~m^{-1}/K}$; ρ is the density of fresh jujube bark, $\mathrm{kg/m^3}$; c is the specific heat capacity, $\mathrm{J~kg^{-1}/K}$; ρ_b is the density of dried jujube bark, $\mathrm{kg/m^3}$; M_c is the moisture content of the bark, %.

2.3. Construction of Heat-Transfer Model for Jujube Bark

The heater is applied to the bark, and energy is mainly transmitted to the interior of the bark through thermal conduction [12]. High temperature causes damage to the phloem cells of jujube bark, leading to a decrease in the conductivity of the phloem, which can inhibit the normal transportation of photosynthetic products through the bark to the roots. Jujube bark is a complex biological material with a complex structure, being a typical anisotropic material. Therefore, its actual heat-transfer characteristics are also very complex. In order to establish a heat-transfer model that is convenient for practical application, it is necessary to make idealized assumptions about jujube bark. Peterson and Ryan [13] and Mantgem and Schwartz [14] established a bark heat-transfer model based on the standard one-dimensional heat-transfer equation. When using this method to describe the radial transfer of heat along the jujube-tree bark from the outer surface to the inner surface, it is assumed that the bark is a semi-infinite solid. The bark is not actually a semi-infinite object, and it is theoretically imperfect when viewed as a semi-infinite object. However, a large amount of practice has shown that the heat-transfer model of bark constructed by treating it as a semi-infinite object has practical value and significance when studying bark heat transfer. Therefore, this model is widely used for bark heat-transfer studies because of its convenient practical application [13] and this model was adopted in this study. Assuming that the outer surface temperature of the jujube bark reaches the temperature of the heater instantly when the heater contacts the jujube tree, then at a specific heating temperature, heat diffuses through the outer surface of the bark to the inner surface, so that the time required for the inner surface of the jujube bark to reach a specific temperature meets Equation (4).

$$\begin{cases} \frac{T_i - T_1}{T_0 - T_1} = erf(\eta) = \frac{2}{\sqrt{\pi}} \int_0^\eta e^{-\eta^2} d\eta \\ \eta = \frac{x}{2\sqrt{\alpha\tau}} \end{cases} \tag{4}$$

where T_i is the set temperature on the inner surface of the bark, °C; T_0 is the environmental temperature of the bark surface, °C; T_1 is the heater temperature, °C; x is the thickness of the bark, cm; α is the thermal diffusivity of the bark, cm^2/min; τ is the time required for the inner surface temperature of the bark to reach the set temperature, min; $erf(\eta)$ is the Gaussian error integral function.

It is generally believed that 60 °C is the lethal temperature for plant tissue cells, and the temperature of 60 °C for more than 1 min was used in some studies as a reference standard for plant-cell death [12]. However, studies also have shown that even within the range of 45–60 °C, plant cells still die as long as the temperature duration is long enough [15,16]. In this study, we assume that the death of jujube-bark cells occurs at a critical lethal temperature of 60 °C [17]. To ensure that the formation layer is not damaged, heating needs to be stopped quickly when the internal surface temperature of the jujube bark reaches 60 °C. Therefore, in the heating model, the internal surface temperature T_i of the bark is set to 60 °C. T_0 is the ambient temperature, and the heat-transfer model is not sensitive to this parameter. This study was conducted at an ambient temperature of approximately 25 °C. The thermal diffusion coefficient of jujube bark measured in the experiment varied slightly with the temperature of the heater, jujube-bark density, moisture content, and so on; but there were not significant differences. Therefore, in this study, the average value of the thermal diffusion coefficients of jujube-bark samples was used as the final α, which was taken as 0.057 cm^2/min.

The heating temperature T_1 of the heater was set to 200, 300, and 400 °C, respectively. The relationship between the thickness x of jujube bark and the heating time required for the inner surface of jujube bark to reach the set temperature can be obtained through Equation (4), as shown in Equation (5).

$$\begin{cases} \tau_2 = 5.42x^2 \\ \tau_3 = 3.77x^2 \\ \tau_4 = 3.10x^2 \end{cases} \tag{5}$$

where τ_2, τ_3, and τ_4 are the time required for the internal surface temperature of jujube bark to reach 60 °C at heating temperatures of 200, 300, and 400 °C, respectively, min; x is the thickness of jujube tree bark, cm.

2.4. Model Construction for the Relationship between Jujube Bark and Corresponding Diameter

From the results of the above heat-transfer-model construction, it can be seen that the heating time required for the inner surface of jujube bark to reach a certain temperature under a certain heat source temperature is a function of the bark thickness. Therefore, it is necessary to know the thickness of jujube bark when heating it. However, due to the complexity of the composition and structure of bark, there is currently no effective non-destructive rapid measurement method or related instrument for bark thickness. However, research has shown a correlation between bark and its corresponding diameter [18]. The research results indicated that by establishing a mathematical model of the relationship between tree diameter and bark thickness, bark thickness can be indirectly measured by measuring the diameter of the tree [19]. Therefore, this study also indirectly estimated the thickness of jujube bark in the heat-transfer model through this method. In order to construct a model for the relationship between jujube bark and corresponding diameter, 50 healthy jujube trees were randomly selected in the jujube orchard to measure the diameter of main branches and corresponding bark-thickness data. When measuring the diameter of jujube-tree branches and the corresponding thickness of jujube-tree bark, firstly, the bark with a width of 10 mm was removed from the randomly selected main branches of the jujube tree. Then, one location at the girdled opening of each branch was randomly selected to measure its diameter and corresponding bark thickness. After obtaining the data of jujube-bark thickness and corresponding diameter, the optimal mathematical model that jujube-bark thickness and corresponding diameter fulfilled was analyzed.

After obtaining the mathematical model, the accuracy of the model was tested by comparing the actual diameter of jujube-tree branches and the corresponding thickness of jujube bark with the predicted results of the model. Therefore, 30 main branch samples from different jujube trees were selected to verify the reliability of the model. After the accuracy of the bark thickness calculated by the model met the requirements, and the thickness of jujube-tree branches during the heating experiment based on the heating model was ultimately calculated using this mathematical model.

2.5. Accuracy Test of Heat-Transfer Model for Jujube Bark and Experimental Design of Two Different Treatments of Jujube-Tree Heating and Girdling

The accuracy testing of the jujube-bark heat-transfer model included two stages. In the first stage, experiments were conducted using measured jujube bark thickness data to verify the performance of the heat-transfer model itself. During the experiment, 30 jujube trees were randomly selected, and a main branch was randomly selected from each jujube tree for heating experiments. The experimental method is shown in Figure 2a. Firstly, the jujube-tree bark was girdled with a width of 8 mm, the thickness of it was measured, and a thermocouple inserted through the girdled part at the boundary between the jujube bark and the xylem. Then, to ensure the efficient heat transfer of the heater, the heater was tightly pressed on the jujube bark to start heating. Meanwhile, the theoretical and actual values of the heating time were recorded when the inner surface of the jujube bark reached 60 °C, respectively. In the second stage, after the jujube-bark thickness prediction model was substituted into the heat-transfer model, experiments on the heat-transfer model were conducted based on the predicted thickness data of the jujube bark. During the experiment, 30 jujube trees were randomly selected, and a main branch was randomly selected from each jujube tree for heating experiments. The heating and actual temperature measurement methods were consistent with the first-stage experimental method. The difference was that at this stage, the diameter of the jujube tree was a direct variable, and the thickness of the jujube-tree bark was predicted through the diameter.

(a)   (b)

Figure 2. Experimental methods for heating jujube bark: (**a**) experimental method for initial heat transfer model; (**b**) heating experimental method for final heat transfer model with the function of the bark thickness prediction.

Then, based on the prediction model of jujube-bark thickness, the heat-transfer model was used for heating experiments. One hectare of healthy jujube trees was randomly selected in the jujube garden and the main branches were heated. The test method is shown in Figure 2b, two heaters were used and tightly pressed on the jujube bark to simultaneously heat each jujube branch. The other jujube trees were all treated with traditional girdling as a control. Then, in the later stage of the experiment, jujube trees from different treatments were observed, and the conditions of trees from heating treatment and wound-healing conditions of the trees after girdling treatment were statistically analyzed.

2.6. Methods for Measuring Jujube Yield and Quality

To investigate the differences in the effects of heating and traditional girdling on the quality and yield of jujube fruit, the jujube-fruit yield and main quality indicators were measured during the jujube fruit harvest period under two different treatment methods. The indexes of jujube-fruit nutrients that were tested included the contents of soluble solids, water, total acid, protein, vitamin C, total sugar, reducing sugar, and sugar acid ratio. Moreover, the physical quality index of jujube tested in this study was single-fruit weight.

The national standard GB/T 5009.7-2008 was referenced to measure the total sugar content. In addition, the content of soluble solids was determined by a handheld refractometer [20]. During the soluble solids measurement, 15 jujubes were selected from each sample, and their juice was extracted by a juicer. Then, the juice was filtered by three layers of gauze, and an appropriate amount of fruit juice was taken to determine the content of soluble solids by an LH-B55 handheld refractometer (Luheng Environmental Technology Cable Co., Ltd., Shanghai, China). The content of vitamin C was determined by 2, 6-dichloroindophenol titration [21]. Total acid content was determined by the acid–base-titration method [22]. The content of reducing sugar was determined by ferin colorimetry. The moisture content of jujube was determined by normal atmospheric temperature and constant pressure drying method. The moisture content test samples of jujube were cut into 3 mm thick slices and dried in an oven at 105 °C to a constant mass. Then, the moisture content was measured by Equation (6).

$$M = \frac{W_1 - W_0}{W_1} \times 100\% \tag{6}$$

where M is the moisture content of jujube, %; W_1 is the weight of jujube before drying, g; and W_0 is the weight of jujube after drying, g.

In addition, the crude-protein content was determined by the Kjeldahl method. In the experiment of protein determination, the core of jujube was removed and jujube flesh was ground into powder after drying. A 0.300 g powder sample was taken, poured into a dry digestive tube, shaken well after 0.2 g copper sulfate and 6 g potassium sulfate were added, then 20 mL of concentrated sulfuric acid was added. It was heated at a low temperature until all the contents were carbonized, and the foaming in the digestive tube stopped; it was then heated at a low temperature to keep slightly boiling until the liquid was blue-green and clear. Then, heating continued for 30 min before cooling. Lastly, water was added up to a volume of 100 mL, which was the digestive liquid of the sample. According to the above method, copper sulfate, potassium sulfate, and concentrated sulfuric acid of the same amount as the sample used for digestion were taken for digestion. After cooling, it was diluted to 100 mL, which was the blank digestive solution. In total, 10 mL of sample digestion solution was taken in the digestive tube, put into a Kjeldahl nitrogen determinator for distillation for 7 min, then the liquid was titrated. At the same time, the 10.0 mL blank digestion solution was distilled according to the above method. All determinations were performed in triplicate. Finally, the protein content was calculated by Equation (7).

$$Y = \frac{c \times (V_1 - V_0) \times 14 \times F}{m \times 1000} \times 100\% \tag{7}$$

where Y is mass fraction of protein, %; c is concentration of hydrochloric acid standard solution, mol/L; V_0 is the volume of standard solution consumed in blank titration, mL; V_1 is the volume of standard solution consumed in sample titration, mL; 14 is the millimolar mass of nitrogen, g/mmol; F is the conversion coefficient of nitrogen to protein, 6.25; and m is sample weight, g.

2.7. Statistical Analysis

The means and standard deviations of the data related to this study were calculated by Microsoft Excel 2007 software. Bar graphs about nutrients in jujube and the significance of the data were analyzed by Origin 2022 (Origin Lab Corporation, Northampton, MA, USA). In addition, the non-parametric test method based on a Kruskal–Wallis test was used to analyze the significance of the difference between the predicted results of the model for jujube-tree-bark-thickness prediction and the actual measurement results. Additionally, the parameter test based on one-way ANOVA method was used to test the significance of the differences in the effects of jujube-tree-bark-girdling treatment and heating treatment on the yield and the quality of jujube of corresponding jujube trees. When performing the one-way ANOVA, a normality test of the experimental data was carried out according to Shapiro–Wilk, and the homogeneity test of variance was Levene's test.

3. Results

3.1. Construction of Jujube-Bark-Thickness-Prediction Model and Accuracy Test Results

The actual measurement results and linear fitting results of jujube-branch diameter and corresponding bark thickness are shown in Figure 3a. The linear fitting equation for the diameter of jujube branches and the corresponding bark thickness is shown in Equation (8). The determination coefficient of the fitting equation was 0.79, indicating a relatively high goodness of fit of the model.

$$\begin{cases} x = -0.516 + 0.073d \\ R^2 = 0.79 \end{cases} \tag{8}$$

where x is the thickness of the jujube-tree bark, mm; d is the diameter of the jujube tree, mm; R^2 is the determination coefficient.

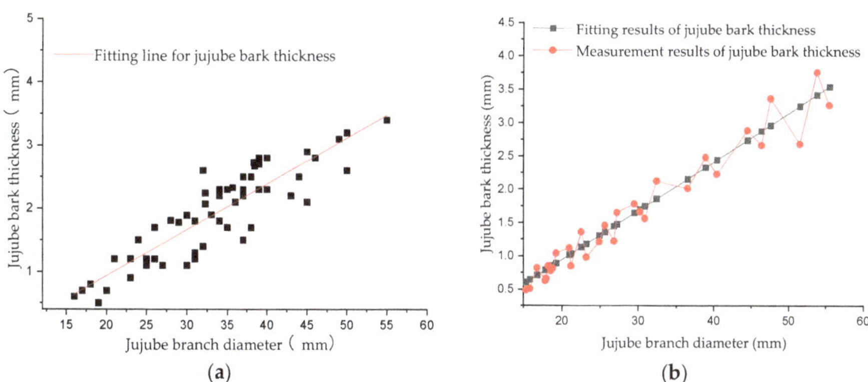

Figure 3. Construction and accuracy test results of the jujube-tree-bark-thickness-prediction model: (**a**) actual measurement results and linear fitting results of jujube tree branch diameter and corresponding bark thickness; (**b**) the thickness of jujube tree bark obtained through model calculation and the corresponding actual measurement results.

The thickness of jujube-tree bark obtained through model calculation and the corresponding actual measurement results are shown in Figure 3b. The minimum error of bark thickness obtained through the two methods was 2.07%, and the maximum error was 26.13%. In addition, the non-parametric test results based on a Kruskal–Wallis test showed that there was no significant difference ($p = 0.93$) between the actual measurement results of jujube-bark thickness and the results calculated through the model at the 0.05 level.

3.2. Construction Results of Heat-Transfer Model for Jujube Bark

3.2.1. Construction of Initial Jujube-Bark Heat-Transfer Model and Accuracy Test Results

During the preliminary heating experiment of jujube bark, it was found that there was a significant pyrolysis phenomenon on the surface of jujube bark at temperatures of 300 and 400 °C, respectively. During the heating process, visible gas was generated from the jujube bark, and after the heating was completed, black charcoal was produced on the surface of the jujube bark. This indicated that heating the bark within this temperature range caused significant damage, which was not conducive to the health of jujube bark. However, under heating conditions of 200 °C, the above phenomenon did not occur. Therefore, in this study, the temperature for all subsequent experiments was set to 200 °C. Figure 4a shows the typical heat transfer curve of jujube bark with a thickness of 2.12 mm under 200 °C heating conditions. From Figure 4a, it was shown that when the temperature of the inner surface of jujube bark reached 60 °C and heating was stopped, the temperature of the inner surface of jujube bark would rapidly decrease. When the temperature dropped to about 40 °C, its rate of decline significantly decreased. Figure 4b shows the heat transfer prediction of the initial jujube-bark heat-transfer model and experimental results. In addition, the parameter test results based on one-way ANOVA showed that there was no significant difference ($p = 0.59$) between the heat transfer experimental results of jujube-tree bark and the prediction results of the initial bark heat-transfer model, at the 0.05 level.

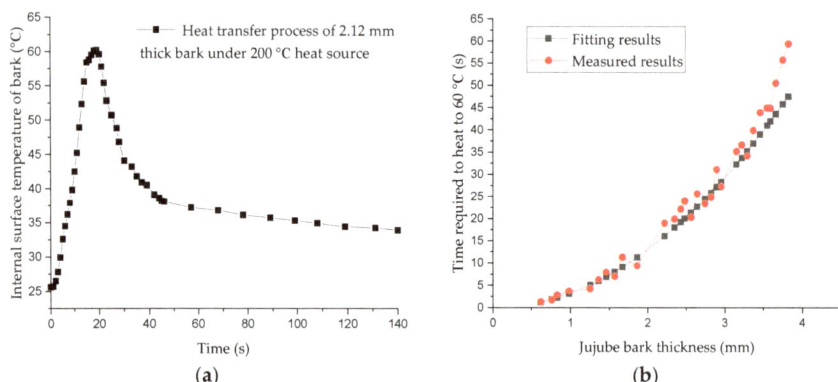

Figure 4. Construction of initial heat-transfer model for jujube bark and accuracy test results: (**a**) typical heat-transfer curve of jujube bark; (**b**) the heat-transfer prediction and experimental verification results of the initial jujube-bark heat-transfer model.

3.2.2. Construction and Accuracy Test Results of Heat-Transfer Model for Jujube Bark with Thickness Prediction

After incorporating the jujube-bark thickness prediction equation into the initial jujube-bark heat-transfer model, the relationship between jujube-branch diameter and the time required for heat transfer was obtained, as shown in Equation (9).

$$\tau'_2 = 325.20(-0.0516 + 0.0073d)^2 \tag{9}$$

where τ'_2 is the time required for the internal surface temperature of jujube bark to reach 60 °C when the heating temperature is 200 °C, s; d is the diameter of the jujube-tree branch, mm.

Figure 5 shows the predicted heating time of jujube bark by the heat-transfer model with bark-thickness-prediction function and actual measurement results. The parameter test results based on one-way ANOVA showed that there was no significant difference ($p = 0.66$) between predicted results by the heat-transfer model and actual measurement results, at the 0.05 level.

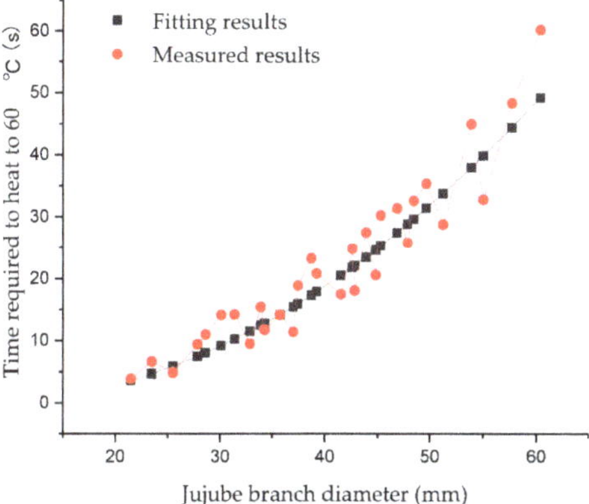

Figure 5. The heat-transfer prediction and experimental verification results of the heat-transfer model with bark-thickness-prediction function.

3.3. Experimental Results of the Effect of Heating-and-Girdling Treatments on the Yield and Quality of Jujube Fruit

3.3.1. Experimental Results of the Effects of Two Different Treatments on Jujube Trees and Fruit Yield

The experimental results of the effects of two different treatments on jujube-fruit yield and single-fruit weight are shown in Figure 6a,b, respectively. The statistical results of the healing of jujube-tree-girdling wounds after girdling treatment showed that 14.5% of jujube trees had incomplete healing of jujube-tree-girdling wounds, and 8.6% of jujube trees were dead. Meanwhile, the heated jujube trees showed no visible wounds or tree death. In addition, the single-fruit-weight test results showed that the weight of the single jujube obtained from both treatments was within the range of 9–13 g. The average single-fruit weight of fresh jujube from trees treated with heating was 11.23 g, while that from the jujube trees treated with girdling was 11.18 g. The analysis of variance results showed that the two treatments had no significant impact on the weight of the single fruit ($p = 0.94$). In terms of yield, the yield per hectare of jujube trees treated with heating and girdling was 19.87 tons and 20.76 tons, respectively. The analysis of variance results showed that the two treatments had a significant impact on jujube-fruit yield ($p = 0.02$).

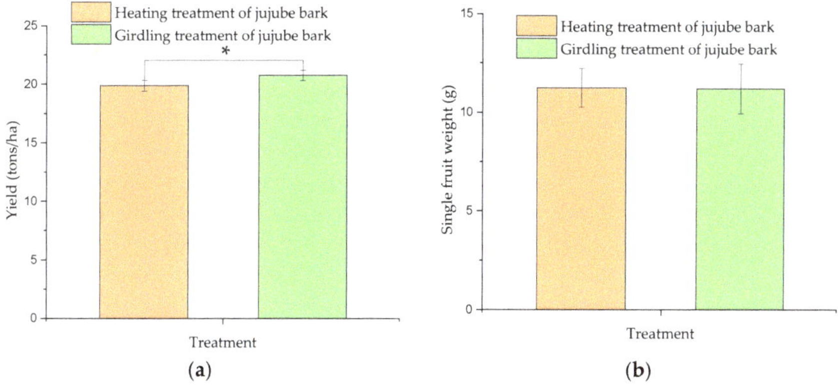

Figure 6. Experimental results of the effects of two different treatments on jujube-fruit yield and single-fruit weight: (**a**) effects of different treatments on jujube-fruit yield, (**b**) effects of different treatments on the single-fruit weight; * represents a significant difference (p values ≤ 0.05).

3.3.2. Experimental Results of the Effects of Two Different Treatments on the Quality of Jujube Fruit

The experimental results of the effects of two different treatments on Vc and protein content of jujube fruit are shown in Figure 7a,b, respectively. The average Vc content of fresh jujube fruits treated with heating and girdling was 3.302 mg/g and 3.384 mg/g, respectively. The analysis of variance results showed that the two treatments had no significant impact on the Vc content of jujube fruit ($p = 0.29$). The average protein content of fresh jujube fruits treated with heating and girdling was 2.23% and 2.31%, respectively. The analysis of variance results showed that the two treatments had no significant effect on the protein content of jujube fruit ($p = 0.14$). The experimental results of the effects of two different treatments on the content of reducing sugars and soluble solids in jujube fruit are shown in Figure 7c,d, respectively. The average reducing sugar content of fresh jujube fruits treated with heating and girdling was 16.06% and 16.89%, respectively. The analysis of variance results showed that the two treatments had no significant effect on the reducing sugar content of jujube fruit ($p = 0.24$). The average soluble solids content of fresh jujube fruits treated with heating and girdling was 47.36% and 48.08%, respectively. The analysis of variance results showed that the two treatments had no significant effect on the average soluble solids content of jujube fruit ($p = 0.12$). The experimental results of the effects of two different treatments on the water and acid content of jujube fruits are shown in Figure 7e,f, respectively. The average moisture content of fresh jujube fruits treated with heating and girdling was 56.44% and 53.26%, respectively. The analysis of variance results showed that the two treatments had a significant impact on the moisture content of jujube fruit ($p = 0.03$). The average content of fresh jujube acid of jujube trees treated with heating and girdling was 0.21% and 0.20%, respectively. The analysis of variance results showed that the two treatments had no significant impact on the average acid content of jujube fruit ($p = 0.28$). The experimental results of the effects of two different treatments on the sugar acid ratio and total sugar content of jujube fruit are shown in Figure 7g,h, respectively. The average sugar acid ratio of fresh jujube fruits treated with heating and girdling was 189.72% and 196.46%, respectively. The analysis of variance results showed that the two treatments had no significant effect on the sugar acid ratio of jujube fruit ($p = 0.41$). The average total sugar content of fresh jujube fruits treated with heating and girdling was 39.37% and 39.65%, respectively. The analysis of variance results showed that the two treatments had no significant effect on the average total sugar content of jujube fruit ($p = 0.80$).

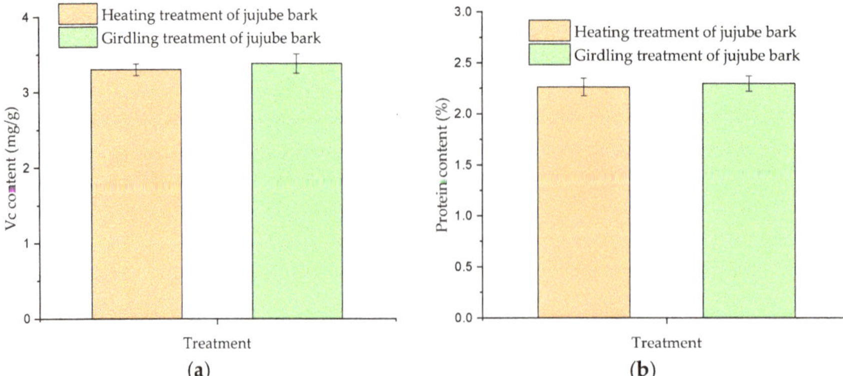

Figure 7. *Cont.*

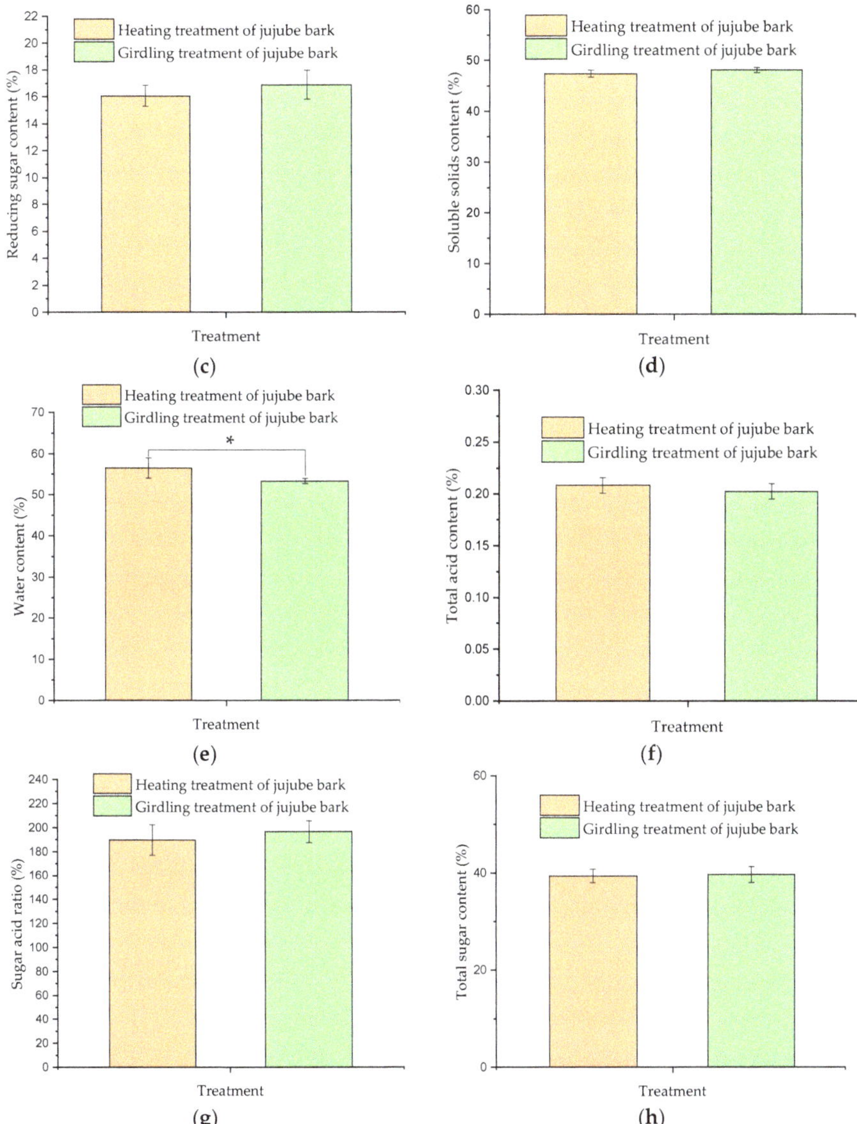

Figure 7. Experimental results of the effects of two different treatments on the quality of jujube fruit: (**a**) comparison of vitamin C content; (**b**) comparison of protein content; (**c**) comparison of reducing sugar content; (**d**) comparison of soluble solids content; (**e**) comparison of water content; (**f**) comparison of total acid content; (**g**) comparison of sugar acid ratio; (**h**) comparison of total sugar content; * represents a significant difference (*p* values $\leq$ 0.05).

4. Discussion

4.1. Prediction Model of Jujube-Bark Thickness Based on Jujube-Branch Diameter

Predicting the thickness of jujube bark is the foundation for precise heating of jujube bark. Therefore, in this study, a mathematical model of the diameter of jujube branches and the corresponding jujube-bark thickness was established, and the reliability of using this model to indirectly predict jujube-bark thickness was verified through experiments.

The fitting results of tree bark thickness indicated that there was a high linear positive correlation between branch bark thickness and its diameter, and the fitting model had a high degree of goodness of fit. The experimental analysis results indicated that there was no significant difference between the predicted results of jujube-bark thickness using this model and the actual measured data. This result was basically consistent with the relevant research conclusions on predicting bark thickness of other types of trees [23]. This indicated that the model is expected to become a simple and feasible method for predicting bark thickness before obtaining better methods for detecting it. Of course, we hope to explore faster and more accurate non-destructive testing methods for bark thickness, which is also one of the areas that we need to focus on in the future.

4.2. Heat-Transfer Characteristics and Model of Jujube Bark

Preliminary heating experiments on jujube bark have shown that temperatures above 300 °C can cause significant pyrolysis of jujube bark, which is consistent with existing wood pyrolysis experiments [24]. Due to the fact that high-temperature pyrolysis can cause numerous pores on the surface of wood [25], if high-temperature heating causes severe pyrolysis of jujube bark, it will directly damage the structure of jujube bark. In addition, the porous surface can easily provide breeding locations and habitat for insects, which will cause secondary damage to jujube trees. Therefore, this study conducted heating experiments on jujube bark at a temperature of 200 °C. The heating experiment of jujube-tree bark at a temperature of 200 °C showed that when the internal surface temperature of the main branch of the jujube tree reached 60 °C and heating was stopped, the time in which the internal surface temperature of the jujube-tree bark remained above 45 °C was approximately within the range of 20 to 40 s. Therefore, this heating condition is not sufficient to kill the cambium and the phloem cells near the inner surface of the jujube bark. However, when the inner surface temperature of jujube bark reached 60 °C, other phloem cells near the outer surface of the bark were all at temperatures higher than 60 °C. Therefore, overall, heating the jujube bark under this condition will not cause the death of all bark cells, let alone the death of cambium cells, but may cause the death of phloem cells near the outer surface of the bark. Therefore, this may damage the function of jujube bark to some extent, but it will not completely prevent the bark from transporting nutrients to the roots. That is to say, under this heating condition, it is expected that the bark will continuously transport a portion of the nutrients generated by photosynthesis to the root system, while maintaining the health of the tree to a greater extent and increasing yield. Therefore, as long as the parameters of thermal injury of jujube bark are selected appropriately, it can be used many times and not cause damage to the tree body. This is of great significance for promoting the sustainable development of jujube planting.

As a complex biological material [26], jujube bark has extremely complex structures and thermophysical properties in various aspects [27,28]. Therefore, in order to establish a feasible and convenient heat-transfer model, this study considered jujube bark as a semi-infinite solid, and then a jujube-bark heat-transfer model was established based on the standard one-dimensional heat-transfer equation. By solving the model, the initial heat-transfer equation of jujube bark was obtained. The time required for heat transfer in jujube bark under specific temperature parameters was a quadratic function of jujube-bark thickness. The analysis of variance on the experimental results showed that there was no significant difference between the predicted heat-transfer time of the model and the actual measured time. This indicated that the model is meaningful to be used to predict the heat-transfer time of jujube bark, and the idealization of jujube bark as a semi-infinite object did not significantly affect the effectiveness of the heat-transfer model. Then, the prediction model of jujube-bark thickness was incorporated into the initial heat transfer model to obtain a quadratic power function of heat-transfer time with respect to the diameter of jujube branches. The final experimental results also indicated that there was no significant difference between the predicted heat-transfer time by the model and the actual measured results. Therefore, using this model to predict the bark thickness and heat-transfer time of jujube branches simultaneously has certain reliability in guiding jujube-bark-heating experiments.

4.3. Effect of Heating Treatment on the Yield and Quality of Jujube Fruit

Through observation and statistics, it was found that 14.5% of jujube trees treated with traditional girdling methods did not fully heal, and 8.6% of jujube trees died. However, the heated jujube trees showed no visible wounds or tree death. Furthermore, there was no significant difference in single-fruit weight between two different treatments of jujube trees. Although there were significant differences in yield, the yield of jujube fruit obtained by heating treatment was only 4.28% lower than that obtained by girdling treatment, while it significantly increased compared with the yield of jujube trees without any treatment [29]. Because if the jujube tree is not subjected to any treatment such as no girdling, the ju-jube fruit will shed a large amount during the growth stage, resulting in extremely low yield. This result validates the correctness of the hypothesis proposed in this study, indicating that heating the jujube-tree bark can also increase production, and there is basically no permanent damage to the jujube tree. The detection and analysis of the main quality indicators of jujube fruit showed that there was no significant difference in the quality of jujube fruit between the two treatments, indicating that heating treatment also improved the quality of jujube fruit compared to untreated jujube trees.

As for the slight decrease in jujube-fruit yield caused by heating treatment compared to girdling treatment, there are many possible reasons. For example, it may be due to the fact that there is still a portion of jujube-tree bark around the heating treatment area that has not lost its function, allowing nutrients to be transported normally to the roots through this portion of the bark. Alternatively, the phloem that temporarily loses its function after heating can restore its ability to transport nutrients in a relatively short period of time. These reasons may to some extent lead to a relative decrease in the fruit-setting rate of jujube trees [1], thereby leading to a relative decrease in yield. Overall, heating treatment can greatly avoid damage to jujube trees while improving their yield and quality. This has positive significance for the healthy and sustainable development of jujube planting. In the future, quantitative research can be conducted on the relationship between the degree of damage to the bark and phloem of jujube trees and indicators such as jujube-fruit yield and quality; through the correction and optimization of heating model parameters, further improvements in the jujube-fruit yield of jujube trees are expected under heating treatment. In addition, in the research and development of heating-related equipment, automatic jujube bark thickness detection and automatic decision making of heating time can be studied to improve the automation and intelligence of heating equipment.

5. Conclusions

This study aimed to address a series of negative issues such as jujube-tree death caused by traditional methods of increasing yield through jujube-tree girdling. A method of heating the jujube-tree bark was proposed to cause local thermal damage to inhibit the transportation of photosynthetic products to the roots, thereby achieving the same goal of increasing yield as girdling treatment. The established heat-transfer model was used to predict the heating time required for jujube bark with different temperature parameters and thicknesses, and this was used to guide the development of heating experiments. The experimental results showed that heating treatment can increase yield and improve the quality of jujube fruit, and there is no significant damage to jujube trees. Therefore, as long as the degree of thermal damage to jujube bark is accurately controlled, this method can be applied for a long time without affecting the health of jujube trees. The research results provide innovative ideas for increasing fruit-tree production and are of great significance for the sustainable development of the fruit-tree-planting industry. Especially for fruit trees such as peach trees that may experience serious problems such as gum flow after girdling, adopting this method to increase fruit-tree yield and reduce fruit-tree damage is of great practical significance for the sustainable and healthy development of the fruit-tree-planting industry.

Author Contributions: Resources, X.W. and J.R.; data curation, C.H.; writing—original draft preparation, J.R.; writing—review and editing, Y.L.; visualization, J.Z., J.X. and B.S.; supervision, X.W.; project administration, Y.L. All authors have read and agreed to the published version of the manuscript.

Funding: This research was funded by the Scientific and Technological Planning Projects of First Division Alar City, Xinjiang Construction Corps, grant number 2022ZB03; the Innovative Research Project of the Tarim University President Foundation, grant number TDZKSS202112; Tarim University "Three Districts" Science and Technology Talent Support Program Project, grant number BT1720230017G.

Data Availability Statement: The data presented in this study are available on request from the corresponding author.

Acknowledgments: The authors acknowledge the assistance of Instrumental Analysis Center of Tarim University and Alar Wanda Agricultural Machinery Co., Ltd.

Conflicts of Interest: The authors declare no conflict of interest.

References

1. Xu, H.B.; Feng, J.C.; Fan, L.L.; Yang, H.Q.; Fan, W. Research progress on girdling mechanism and non-invasive girdling technology of pomiculture. *J. Henan Agric. Sci.* **2020**, *49*, 1–9.
2. Khandaker, M.M.; Hossain, A.S.; Osman, N. Application of girdling for improved fruit retention, yield and fruit quality in Syzygium samarangense under fifield conditions. *Int. J. Agric. Biol.* **2011**, *13*, 18–24.
3. Quentin, A.G.; Close, D.C.; Hennen, L.M. Down-regulation of photosynthesis following girdling, but contrasting effects on fruit set and retention, in two sweet cherry cultivars. *Plant Physiol. BioChem.* **2013**, *73*, 359–367. [CrossRef] [PubMed]
4. Ran, J.; Guo, W.; Hu, C.; Wang, X.; Li, P. Adverse effects of long-term continuous girdling of jujube tree on the quality of jujube fruit and tree health. *Agriculture* **2022**, *12*, 922. [CrossRef]
5. Tombesi, S.; Dayk, R.; Johnson, R.S. Vigour reduction in girdled peach trees is related to lower midday stem water potentials. *Funct. Plant Biol.* **2014**, *41*, 1336–1341. [CrossRef] [PubMed]
6. Wan, X.C.; Landhäusser, S.M.; Zwiazek, J.J.; Lieffers, V.J. Signals controlling root suckering and adventitious shoot formation in aspen (*Populus tremuloides*). *Tree Physiol.* **2006**, *26*, 681–687. [CrossRef]
7. Schepper, V.; Steppe, K. Tree girdling: A tool to improve our understanding of coupled sugar and water transport. *Acta Hortic.* **2013**, *990*, 313–320. [CrossRef]
8. Choi, S.T.; Park, D.S.; Kang, S.M.; Song, W.D. Effect of different girdling dates on tree growth, fruit characteristics and reserve accumulation in a late-maturing persimmon. *Sci. Hortic.* **2010**, *126*, 152–155. [CrossRef]
9. Onguso, J.M.; Mizutani, F.; Hossain, A.B.M.S. Effects of partial ringing and heating of trunk on shoot growth and fruit quality of peach trees. *Bot. Bull. Acad. Sin.* **2004**, *45*, 301–306.
10. Martin, R.E. Thermal properties of bark. *For. Prod. J.* **1963**, *13*, 419–426.
11. Hengst, G.E.; Dawson, J.O. Bark properties and fire resistance of selected tree species from the central hardwood region of North America. *Can. J. For. Res.* **1994**, *24*, 688–696. [CrossRef]
12. Dickinson, M.B.; Johnson, E.A. Temperature-dependent rate models of vascular cambium cell mortality. *Can. J. For. Res.* **2004**, *34*, 546–559. [CrossRef]
13. Peterson, D.L.; Ryan, K.C. Modeling post-fire conifer mortality for longrange planning. *Environ. Manag.* **1986**, *10*, 797–808. [CrossRef]
14. Mantgem, P.V.; Schwartz, M. Bark heat resistance of small trees in Californian mixed conifer forests: Testing some model assumptions. *For. Ecol. Manag.* **2003**, *178*, 341–352. [CrossRef]
15. Bova, A.S.; Dickinson, M.B. Linking surface-fire behaviour, stem heating, and tissue necrosis. *Can. J. For. Res.* **2005**, *35*, 814–822. [CrossRef]
16. VanderWeide, B.L.; Hartnett, D.C. Fire resistance of tree species explains historical gallery forest community composition. *For. Ecol. Manag.* **2011**, *26*, 1530–1538. [CrossRef]
17. Gutsell, S.L.; Johnson, E.A. How fire scars are formed: Coupling a disturbance process to its ecological effect. *Can. J. For. Res.* **1996**, *26*, 166–174. [CrossRef]
18. Cellini, J.M.; Galarza, M.; Burns, S.L.; Martinez-Pastur, G.J.; Lencinas, M.V. Equations of bark thickness and volume profiles at different heights with easymeasurement variables. *For. Syst.* **2012**, *21*, 23–30.
19. Kıtıkıdou, K.; Papageorgiou, A.; Milios, E.; Stampoulidis, A. A bark thickness model for Pinus halepensis in kassandra, Chalkidiki (Northern Greece). *Silva Balc.* **2014**, *15*, 47–55.
20. Wu, C.Y.; Li, Z.X.; Li, X.G.; Zeng, L.P. Determination of soluble solids in three tropical fruits. *Chin. J. Trop. Agric.* **2018**, *38*, 69–72.

21. Liu, K.P.; Huang, C.H.; Leng, J.H.; Chen, K.; Yan, Y.P.; Gu, Q.Q.; Xu, X.B. Principal component analysis and comprehensive evaluation of the fruit quality of 'Jinkui' kiwifruit. *J. Fruit Sci.* **2012**, *9*, 867–871.
22. Wang, H.F.; Shao, X.F. *Fruit and Vegetable Storage and Processing Experimental Guidance*; Science Press: Beijing, China, 2012; pp. 56–58.
23. Gordon, A.D. Estimating bark thickness of Pinus radiata. *N. Z. J. For. Sci.* **1983**, *13*, 340–345.
24. Jones, J.L.; Webb, B.W.; Jimenez, D.; Reardon, J.; Butler, B. Development of an advanced one-dimensional stem heating model for application in surface fires. *Can. J. For. Res.* **2004**, *34*, 20–30. [CrossRef]
25. Gu, X.L.; Ma, X.; Li, L.X.; Liu, C.; Cheng, K.H. Pyrolysis of poplar wood sawdust by TG-FTIR and Py-GC/MS. *J. Anal. Appl. Pyrolysis* **2013**, *102*, 16–23. [CrossRef]
26. Hamelinck, C.N.; Van, H.G.; Faaij, A.P.C. Ethanol from lignocellulosic biomass: Techno-economic performance in short, middle- and long-term. *Biomass Bioenergy* **2005**, *28*, 384–410. [CrossRef]
27. Frangi, A.; Fontana, M. Thermal expansion of wood and timber-concrete composite members under ISO-fire exposure. *Fire Saf. Sci.* **2003**, *7*, 1111–1122. [CrossRef]
28. Liu, R.; Cao, J.; Peng, Y. Influences of wood components on the property of wood-Plastic composites. *Chem. Ind. Eng. Prog.* **2014**, *33*, 2072–2083.
29. Ran, J.; Hu, C.; Zhang, F.; Wang, X.; Li, P. Effect of Micro- and Macro-Mechanical Characteristics of Jujube Bark on Jujube Girdling Quality. *Agriculture* **2022**, *12*, 278. [CrossRef]

Article

Diversity of Summer Weed Communities in Response to Different Plum Orchard Floor Management in-Row

Jerzy Lisek

The National Institute of Horticultural Research, Konstytucji 3 Maja 1/3 Str., 96-100 Skierniewice, Poland;
jerzy.lisek@inhort.pl

Abstract: The effect of five methods of in-row weed management on the species composition and diversity of summer weed communities in a plum orchard was evaluated. Different methods of orchard floor management (OFM) were implemented for seven consecutive years from 2009 to 2015. *Festuca rubra* L. ssp. *rubra*–rhizomatous perennial grass was sown as a cover crop in the alleys of the orchard, in the tree planting year. In the seventh year of OFM implementation, the treatments were ranked according to the decreasing value of the Shannon–Wiener floristic diversity index as follows: tillage, post-emergence herbicides spraying, mowing, mulch, and weedy control. The highest value of Simpson dominance index was found in the control treatment. In plots with such treatments as control, mowing, tillage, and mulch, the dominant species was *F. rubra*. This meant that the rhizomatous cover crop from the alleys penetrated and affected the in-row flora. Vegetation of mulched plots was characterized by low value of density and soil cover. The obtained results indicated that the flora developing in the control, sprayed with post-emergence herbicides, tilled and mowed plots had greater potential to provide ecosystem services, than the flora of mulched plots.

Keywords: spontaneous flora; Shannon–Wiener diversity index; Simpson dominance index; *Prunus domestica* L.; *Festuca rubra* L. ssp. *rubra*

Citation: Lisek, J. Diversity of Summer Weed Communities in Response to Different Plum Orchard Floor Management in-Row. *Agronomy* **2023**, *13*, 1421. https://doi.org/10.3390/agronomy13051421

Academic Editors: Christian Frasconi, Marco Fontanelli and Daniele Antichi

Received: 25 April 2023
Revised: 16 May 2023
Accepted: 19 May 2023
Published: 21 May 2023

1. Introduction

The area of orchards in the European Union, including pome and stone fruits, citrus and olives, is approximately 6 million hectares [1]. These cultivated plants are of great economic, environmental, and landscape importance. Fruit trees require effective weed management, which is carried out by various methods, such as herbicide treatment, tillage, mowing [2], flame weeding, mulching [3], and cover crops, including living mulch [4–6]. Weeds pose threats to crops, such as competition for light, water, nutrients, allelopathic effect and increase in risk caused by diseases and pests, including rodents and spring frosts during the flowering of fruit trees [7,8]. Spontaneous vegetation also brings benefits, referred to as ecosystem services, such as increasing biodiversity, creating a habitat for beneficial organisms, providing food for bees and other pollinators, protecting the soil from erosion, salinity and mechanical compaction, reducing soil nutrient leaching, increasing the organic matter content in the soil, landscape, and ornamental functions [8–11]. Considering the environmental role of weeds, the European Union pursues a policy aimed at increasing the number of wild plants in the agricultural landscape [12]. Botanical diversity supports the efficient performance of ecosystem services by weed communities and is monitored in different time horizons, in arable [13], horticultural [14], and perennial industrial [15] crops. The comparison shows that in the conditions of the Indian Western Himalayas, weed species biodiversity in horticulture (apple and vegetable gardens) is greater than in arable crops [16]. A similar relationship between permanent crops (orchards and vineyards) and arable plants (cereals and row crops) was found on the Istrian peninsula in Croatia [14]. The diversity of weed species depends on environmental conditions, landscape structure and agricultural practices [17–19]. In arable crops, the main factor limiting the diversity of weeds and selecting a specific set of species is the intensification of agricultural measures, primarily tillage, the use of herbicides and mineral fertilizers [20–23]. The intensification of agricultural practices may lead to a reduction in the diversity of orchard and vineyard flora in the long term [14]. The effect of various orchard floor management (OFM) methods on weed diversity was evaluated in orchards with different tree species, such as apple [24], apricot [25], fig [26], almond [27], and olives [28,29]. Closely related to the subject of the present studies are also the results concerning the effects of weed management on the flora diversity obtained in vineyards [30–33], where, similarly to orchards, an alley cropping system is maintained.

The hypothesis of the research was that the method of OFM leads to changes in agro-phytocenoses, important both for the environment, due to the ecosystem services provided by the flora, and for the fruit grower, who must protect trees from weed disservices, taking into account practical possibilities and economic realities. The aim of the study was to evaluate the multi-year implementation of selected OFM methods on the diversity of weed communities in a plum orchard, including the dynamics of changes in successive years, to indicate the methods that best preserve the initial diversity and the weak points of the selected methods.

2. Materials and Methods

2.1. Experimental Site, Material and Design

The field experiment was conducted in the Experimental Orchard of the National Research Institute of Horticulture in Dąbrowice, Central Poland (51°55′ N, 20°06′ E, 145 m a.s.l.). Orchard soil type, with neutral reaction (pH 6.5 in potassium chloride) and 2.3% of organic matter, was classified as luvisol, according to the international soil classification system [34]. Prior to the establishment of the orchard, forecrops–cereals and mustard—were cultivated for three years, but earlier, the area had been used for intensive orchards for over 30 years. The experimental plots were the internal part of a large, uniform plum orchard unit with an area of 2 ha (about 2500 trees) to limit the edge effect. Plum trees grew inside the Experimental Orchard with an area of 70 ha. Around it were commercial orchards, farmland with arable crops and ruderal places—villages and roads. One-year-old plum trees of the cultivar 'Valjevka' (*Prunus domestica* L.), grafted on Myrobalan seedlings (*Prunus cerasifera* Ehrh. *var. divaricata* Ledeb.), were planted in the spring of 2008. The trees were spaced at 2 m in rows and 4 m between rows. Trees trained to the central leader spindle system were drip irrigated. Soil water potential was kept between 0 and -0.02 MPa at a depth of 0.2 m, according to the reading on the tensiometers. Mineral fertilization—nitrogen (N): 50 kg/ha in 2008 and 2009, 30 kg/ha in 2010–2014, 15 kg/ha in 2015 (as ammonium nitrate); phosphorus (P): 50 kg/ha in 2008 (triple superphosphate); potassium (K): 75 kg/ha in 2008, 50 kg/ha in 2010, 50 kg/ha in 2012 (potassium chloride) and plant protection (three fungicide treatments against brown rot disease and three insecticide treatments against plum sawfly and plum moth per year) were carried out according to current recommendation for commercial orchards. Perennial, rhizomatous grass *Festuca rubra* L. ssp. *rubra* (red fescue), cultivar 'Leo', hereinafter referred to as *F. rubra*, was sowed in the inter-rows in September 2008. For seven consecutive years, from 2009 to 2015, the following methods of orchard floor management (OFM) in-row trees were introduced:

1. Control with limited weeding (manual weeding in spring within a radius of 0.5 from the tree trunk);
2. Spraying with post-emergence herbicides (glyphosate—two treatments per year at the rate of 2.88 kg a.i./ha in May and in the second half of August; glufosinate ammonium—one treatment, 0.6 kg a.i./ha in mid-June);
3. Mulching with organic waste—cereal straw with 2-yers-old compost from plant wastes in a volume ratio of 2:1 (layer of about 10 cm, filled in every 2 years, which was enough to effectively reduce the emergence of weeds);

4. Tillage—mechanical soil cultivation with the use of rotary cultivators and hoe—three times from the beginning of May to August, on average every six weeks;
5. Weed mowing—3 times between May and September. Mowing reduces weed growth less than tillage and herbicides. The last mowing was carried out about two weeks after tillage and herbicide spraying to limit weed regrowth closer to the onset of winter. Strong weed infestation in autumn attracts rodents.

Treatments were applied in completely randomized blocks with 4 replications (blocks) and 5 trees on the plots (20 trees per treatment). The width of the plots was 2 m, and their area was 20 m^2.

The study was carried out in the conditions of the temperate climate, intermediate between maritime and continental. During the study period, the average air temperature was 8.6 °C. January was the coldest month (-2.5 °C), and July was the hottest (19.4 °C). Average annual precipitation was 496 mm and ranged between 316 mm (in 2015) and 680 mm (in 2010).

2.2. Measurements and Analyses

2.2.1. Phytosociological Relevés

Relevé in phytosociology is a sample site in which all the plant species are described and documented. Data on plant occurrence of particular weed species, their share and importance in soil cover, were collected between 20 July and 10 August each year from 2008 to 2015, according to the Braun-Blanquet method [35], modified to the plot experiment design. Both early season and late-season weeds were present at the time of the survey. The ecosystem services provided by the summer vegetation are particularly important for biodiversity. Although the flora in the experiment was assessed in three terms, one mid-summer survey was chosen for the presentation, which enabled a precise and clear interpretation of the results. Weeds were additionally divided into groups–monocotyledonous (grassy) or dicotyledonous (broad-leaved); short-lived (annual + biennial) and perennial. The relevés with an area of 20 m^2 coincided with the experimental plots, therefore they are referred to as plots and were located under the tree canopies. In the year of establishing the orchard, hereinafter referred to as the base year, 20 plots were recorded, 5 in each block. In the first year of OFM implementation, the full cycle of interventions has not yet been completed. Weed response to OFM was studied starting from the second year of treatment differentiation, i.e., from the third year after planting the trees. In each plot, four sample squares, each with an area of 1m^2 (16 squares per treatment) were randomly placed, at least 0.5 m from tree trunks. Plots data were calculated as the mean of 4 sample squares.

2.2.2. Phytosociological Stability (S)

S was expressed on a 5-point scale. The stability classes according to the percentage of sample squares in which a given species was found were as follows: 5—81–100%, 4—61–80%, 3—41–60%, 2—21–40%, 1—1–20%.

2.2.3. The Cover Factor (CF)

CF was determined according to the formula:

$$CF = \frac{\sum CP_i}{N} \times 100$$

where: CP_i—percentage cover by i-th plant species in the sample square in which i-th species occurs; N—total number of sample squares.

By adding the cover factors for the various short-lived species and the perennials, the percentage share of these two groups in covering the soil with weeds was determined.

2.2.4. Weed Infestation Rate

The following importance classes were distinguished [36]:
I—very high: S = 5 or 4, CF > 1000;

II—high: S = 5 or 4, CF = 501–1000 or S = 3, CF > 750;
III—moderate: S = 5 or 4, CF = 251–500 or S = 3, CF = 501–750;
IV—low: S = 5 or 4, CF = 51–250 or S = 3, CF = 251–500;
V—sporadic: other lower stability classes and cover factors.

2.2.5. Diversity of Weed Communities

Diversity was compared for the OFM methods with the following indices:

- Shannon–Wiener diversity index–H' [37]

$$H' = -\sum_{i=1}^{s} p_i \ln p_i,$$

- Simpson dominance index–D [38]

$$D = \sum_{i=1}^{s} (p_i)^2,$$

where: $p_i = n_i/N$, n_i—number of individuals of the i-th species; N—total number of individuals; s—number of species.

The Shannon–Wiener diversity index (H') increases with the number of weed species in the community and the degree of equalization in their numbers. H' values range from 0 to 5, usually ranging from 1.5 to 3.5, rarely above 4.5. The Simpson dominance index (D) takes on values in the range (0; 1>, and when the value reaches 1, it means there is no diversity (a single-species community). The data on the number of each species, necessary for the calculation of the indices, were collected from sample squares with an area of 1.0 m^2, as shown in Section 2.2.1. Weeds were identified by the author. Latin plant names follow the Polish national botanical Key book [39]

2.3. Statistical Analysis

Results concerning number of weed species in one plot, total number of weeds per m^2, the Shannon–Wiener diversity index (H') value, and the Simpson dominance index (D) value, were analyzed statistically using analysis of variance. The significance of the means was evaluated using Duncan's test at 5% level. Statistical analyses were performed using the package STATISTICA 10.0 (StatSoft Inc., Tulsa, OK, USA).

3. Results

To show the effect of multi-year OFM on the flora, data on species composition, density and weed cover were compiled for the base year and the seventh year of OFM implementation.

3.1. Weed Species Number, Density and Cover

The comparison of selected weed infestation parameters showed that in the base year, the total number of weed species (36), the average number of species in one plot (21.5), and weed density (188.5) were higher than in the plots of each treatment, determined in the seventh year of OFM implementation, i.e., eight years later (Table 1). The lowest values of the three mentioned parameters were characteristic of the mulched plots, where the total number of species reached 14, the mean number of species per plot was 7.75, and the mean density of weeds was 18.4 pcs/m^2. In the base year, weed cover was dominated by short-lived species, which accounted for 94.9% of the cover (Table 1). In the seventh year of OFM implementation, the share of short-lived weeds in weed infestation was highest in the herbicide plots (88%) and lowest in weed control plots (3.6%). In the last year of the study, the relative share of grasses in the weed cover was higher in all treatments than in the base year, and the largest share (84.8%) was in the control plots (Table 1).

Table 1. In-row weed flora in the base year (2008) and in the seventh year of OFM implementation (2015).

Feature	Base Year	Seventh Year of Implementation				
		Control	Herbicides	Mulch	Tillage	Mowing
Total number of weed species	36	20	21	14	22	19
Number of short-lived species	25	11	14	4	14	12
Number of perennial species	11	9	7	10	8	7
Number of broad-leaved species	32	16	18	12	17	17
Number of grassy species	4	4	3	3	5	2
Total CF	9692	9971	8433	1188	8890	9183
Total CF of short-lived weeds	9196	358	7425	272	3386	2040
Total CF of perennial weeds	496	9613	1008	916	5504	7143
Share of short-lived species in weed cover (%)	94.9	3.6	88.0	22.9	38.1	22.2
Share of perennial species in weed cover (%)	5.1	96.4	12.0	77.1	61.9	77.8
Share of broad-leaved species in weed cover (%)	85.8	14.6	72.0	26.6	58.1	30.5
Share of grassy species in weed cover (%)	13.1	84.8	26.5	60.9	41.9	69.5
Mean number of weed species in one plot	21.5 ± 2.52 [d]	11.25 ± 1.26 [b]	14.5 ± 1.91 [c]	7.75 ± 0.5 [a]	14.0 ± 0.82 [c]	12.25 ± 1.26 [b]
Weed density (pcs m^2)	188.5 ± 9.75 [d]	138.8 ± 6.67 [c]	121.9 ± 5.63 [b]	18.4 ± 2.01 [a]	121.6±7.69 [b]	143.7 ± 5.10 [c]

Means followed by the same letter do not differ significantly at p = 0.05. Values with the prefix ± represent standard deviation.

3.2. Weed Infestation Rate

In the base year, four dominant species of short-lived weeds were distinguished–*Chenopodium album*, *Capsella bursa-pastoris*, *Stellaria media*, and *Poa annua*, which were characterized by a very high infestation rate (class I) and one species–*Polygonum aviculare* occurring in II class with a high infestation rate (Table 2). After seven years of varied OFM, significant differences were found among the dominant species in the plots of individual treatments. The control plots were clearly dominated by two perennial species with a very high infestation rate, *Festuca rubra* ssp. *rubra* and *Epilobium ciliatum* (Table 3). Two short-lived species–*Stellaria media* and *Poa annua*—belonged to the I class of weed infestation in herbicide plots (Table 4). In the mulched plots, none of the species occurred at a very high infestation rate, and the dominant species turned out to be *F. rubra*, occurring in a II class infestation (Table 5). Three perennial species–*F. rubra*, *Rumex acetosella*, and *Elymus repens*—in class I of infestation, and weeds in a class II infestation included perennial–*Taraxacum officinale*—and short-lived–*Chenopodium album*, *Poa annua*, and *Stellaria media*—dominated the tilled plots (Table 6). *F. rubra* clearly dominated the mowed plots and 3 species–*Taraxacum officinale*, *Bromus hordeaceus*, and *Crepis biennis*–belonged to a class II infestation (Table 7). After 7 years of OFM implementation, the dominant species in the plots of all treatments, except for herbicides, was the *F. rubra*. Its occurrence was found under the tree canopy for the first time in the second year of OFM implementation, and from the fourth year it was more and more numerous, especially in control plots.

3.3. Shannon–Wiener Diversity Index (H')

After seven years of OFM implementation, only one treatment with soil tillage did not significantly differ in the value of the in-row flora H' index, compared to the baseline value, which was 2.310 (Table 8). For all other treatments, the value of this index was significantly lower than the baseline value. The sequence of treatments according to the decreasing value of H' was as follows: tillage (2.285), herbicides (1.770), mowing (1.536), mulch (1.338), and weedy control (0.817). This means that the last of the treatments was characterized by the smallest biodiversity of flora. Differences between all OFM variants in the seventh year of implementation were statistically significant. In individual years of the assessment, both the value of H' and the relationship between treatments regarding the value of H' changed. In the second year of OFM implementation, the H' value for all treatments, except for control, was significantly lower than the baseline value. In the third, fourth, and fifth year, the value of H' was significantly lower than the baseline value for all evaluated methods of OFM. In the sixth year of OFM implementation, the relationships between treatments were shaped in the same way as in the seventh year of implementing the five OFM methods.

Table 2. In-row weed flora in the base year (2008).

Species	Phytosociological Stability (S)	Cover Factor (CF)	Weed Infestation Rate (Class)	Mean Number of Weeds (Pcs m^2)
Short-lived				
Chenopodium album L.	5	2146	I	44.3
Capsella bursa-pastoris (L.) Medik.	5	1772	I	34.8
Stellaria media (L.) Vill.	5	1554	I	30.5
Poa annua L.(G)	5	1116	I	26.3
Polygonum aviculare L.	5	574	II	10.7
Senecio vulgaris L.	5	376	III	7.0
Polygonum persicaria L. (*P. maculosa* Gray)	4	224	IV	6.3
Matricaria maritima L. ssp. *inodora* (L.) Dostál	4	212	IV	4.3
Galium aparine L.	4	158	IV	3.0
Fallopia convolvulus (L.) A. Löve	4	148	IV	2.8
Veronica persica Poir.	4	145	IV	2.3
Echinochloa crus-galli (L.) P. Beauv. (G)	3	128	V	2.0
Veronica arvensis L.	2	106	V	1.5
Viola arvensis Murr.	2	101	V	1.3
Lamium purpureum L.	2	98	V	1.0
Geranium pusillum Burm. F. ex L.	2	82	V	1.0
Chamomilla suaveolens (Pursh) Rydb.	1	64	V	0.5
Amaranthus retroflexus L.	1	42	V	0.5
Conyza canadensis (L.) Cronq.	1	41	V	0.5
Crepis capillaris (L.) Wallr.	1	38	V	0.3
Crepis biennis L.	1	32	V	0.2
Vicia villosa Roth.	1	12	V	<0.05
Bromus hordeaceus L. (G)	1	10	V	<0.05
Erodium cicutarium (L.) L'Her.	1	9	V	<0.05
Atriplex patula L.	1	8	V	<0.05
Perennial				
Equisetum arvense L.	3	114	V	2.3
Taraxacum officinale F. H. Wigg.	3	102	V	2.0
Cerastium holosteoides Fr. em. Hyl.	3	76	V	1.3
Trifolium repens L.	2	44	V	0.5
Plantago major L.	1	42	V	0.5
Epilobium ciliatum Raf.	1	38	V	0.5
Convolvulus arvensis L.	1	26	V	0.2
Cirsium arvense (L.) Scop	1	18	V	0.1
Rumex acetosella L.	1	16	V	<0.05
Elymus repens (L.) Gould (G)	1	12	V	<0.05
Urtica dioica L.	1	8	V	<0.05

G—grassy species.

3.4. Simpson Dominance Index (D)

The baseline value of index D was 0.139. In the seventh year of OFM implementation, the lowest value of the indice, not significantly different from the baseline value, was obtained for treatment tillage (Table 9). The value of D for all other treatments was significantly higher than the baseline value. Treatments were ranked according to increasing D index values in the following order: tillage (0.123), herbicides (0.269), mulch (0.365), mowing (0.381), and control (0.681). Differences between mulch and mowing treatments were insignificant. Weedy control was the treatment with the most marked species dominance among synanthropic plants of tree understory, in the last year of the assessment. In the fifth, sixth, and seventh year of OFM implementation, the relationships in the significance of differences between treatments were the same.

Table 3. In-row weed flora in the seventh year of OFM implementation (2015)–control.

Species	Phytosociological Stability (S)	Cover Factor (CF)	Weed Infestation Rate (Class)	Mean Number of Weeds (Pcs m^2)
Short-lived				
Matricaria maritima L. ssp. *inodora* (L.) Dostál	4	85	IV	1.8
Galium aparine L.	4	62	IV	1.3
Crepis biennis L.	4	54	IV	1.0
Conyza canadensis (L.) Cronq.	4	48	V	0.8
Bromus hordeaceus L.(G)	3	30	V	0.5
Fallopia convolvulus (L.) A. Löve	3	22	V	0.3
Stellaria media (L.) Vill.	2	19	V	0.2
Chenopodium album L.	?	12	V	0.1
Echinochloa crus-galli (L.) P. Beauv. (G)	1	10	V	0.1
Geranium pusillum Burm. F. ex L.	1	9	V	0.1
Tragopogon pratensis L.	1	7	V	< 0.05
Perennial				
Festuca rubra L. ssp. rubra (G)	5	8452	I	115.8
Epilobium ciliatum Raf.	5	508	I	6.5
Rumex acetosella L.	5	316	III	4.3
Taraxacum officinale F. H. Wigg.	5	89	IV	1.8
Cerastium holosteoides Fr. em. Hyl.	5	82	IV	1.5
Equisetum arvense L.	4	64	IV	1.3
Convolvulus arvensis L.	3	55	V	0.8
Elymus repens (L.) Gould (G)	2	28	V	0.5
Artemisia vulgaris L.	1	19	V	<0.1

G—grassy species.

Table 4. In-row weed flora in the seventh year of OFM implementation (2015)—herbicides.

Species	Phytosociological Stability (S)	Cover Factor (CF)	Weed Infestation Rate (Class)	Mean Number of Weeds (Pcs·m^2)
Short-lived				
Stellaria media (L.) Vill.	5	4080	I	52.3
Poa annua L. (G)	5	1950	I	27.5
Lamium purpureum L.	5	375	III	6.8
Conyza canadensis (L.) Cronq.	5	280	III	4.8
Echinochloa crus-galli (L.) P. Beauv. (G)	5	212	IV	4.5
Chenopodium album L.	4	169	IV	3.3
Capsella bursa-pastoris (L.) Medik.	4	115	IV	2.5
Bromus hordeaceus L. (G)	3	76	V	1.5
Viola arvensis Murr.	3	59	V	1.0
Veronica arvensis L.	4	38	V	0.8
Polygonum aviculare L.	2	29	V	0.5
Fallopia convolvulus (L.) A. Löve	1	19	V	0.3
Geranium pusillum Burm. F. ex L.	1	15	V	0.3
Galium aparine L.	1	8	V	0.1
Perennial				
Taraxacum officinale F. H. Wigg.	5	460	III	7.3
Epilobium ciliatum Raf.	5	302	III	5.3
Equisetum arvense L.	5	124	IV	2.5
Cerastium holosteoides Fr. em. Hyl.	3	62	V	1.3
Trifolium repens L.	1	26	V	0.5
Rumex acetosella L.	1	18	V	<0.1
Urtica dioica L.	1	16	V	<0.1

G—grassy species.

Table 5. In-row weed flora in the seventh year of OFM implementation (2015)–mulch.

Species	Phytosociological Stability (S)	Cover Factor (CF)	Weed Infestation Rate (Class)	Mean Number of Weeds (Pcs·m^2)
Short-lived				
Galium aparine L.	5	152	IV	2.3
Atriplex patula L.	4	69	IV	1.0
Chenopodium album L.	4	42	V	0.5
Fallopia convolvulus (L.) A. Löve	3	9	V	0.1
Perennial				
Festuca rubra L. ssp. *rubra* (G)	5	640	II	10.5
Urtica dioica L.	4	115	IV	1.5
Elymus repens (L.) Gould (G)	4	84	IV	1.3
Convolvulus arvensis L.	4	18	V	0.3
Artemisia vulgaris L.	3	18	V	0.3
Epilobium ciliatum Raf.	2	15	V	0.2
Malva neglecta L.	2	8	V	0.1
Rumex crispus L.	1	8	V	0.1
Tanacetum vulgare L.	1	5	V	<0.1
Calamagrostis epigejos (L.) Roth (G)	1	5	V	<0.1

G—grassy species.

Table 6. In-row weed flora in the seventh year of OFM implementation (2015)–tillage.

Species	Phytosociological Stability (S)	Cover Factor (CF)	Weed Infestation Rate (Class)	Mean Number of Weeds (Pcs·m^2)
Short-lived				
Chenopodium album L.	5	829	II	13.0
Poa annua L. (G)	5	628	II	11.5
Stellaria media (L.) Vill.	5	514	II	9.3
Conyza canadensis (L.) Cronq.	5	382	III	5.1
Bromus hordeaceus L. (G)	4	246	IV	3.0
Crepis biennis L.	4	198	IV	2.8
Echinochloa crus-galli (L.) P. Beauv. (G)	4	154	IV	2.5
Polygonum persicaria L. (*P. maculosa* Gray)	3	110	V	1.8
Polygonum aviculare L.	3	94	V	1.3
Galium aparine L.	2	81	V	0.8
Fallopia convolvulus (L.) A. Löve	2	63	V	0.5
Capsella bursa-pastoris (L.) Medik.	1	42	V	0.5
Senecio vulgaris L	1	25	V	0.3
Geranium pusillum Burm. F. ex L.	1	20	V	0.3
Perennial				
Festuca rubra L. ssp. *rubra* (G)	5	1620	I	21.5
Rumex acetosella L.	5	1584	I	18.8
Elymus repens (L.) Gould (G)	5	1080	I	15.5
Taraxacum officinale F. H. Wigg.	5	572	II	5.3
Cerastium holosteoides Fr. em. Hyl.	4	346	III	4.5
Equisetum arvense L.	4	148	IV	1.8
Epilobium ciliatum Raf.	4	126	IV	1.5
Cirsium arvense (L.) Scop.	1	28	V	<0.1

G—grassy species.

Table 7. In-row weed flora in the seventh year of OFM implementation (2015)–mowing.

Species	Phytosociological Stability (S)	Cover Factor (CF)	Weed Infestation Rate (Class)	Mean Number of Weeds (Pcs·m^2)
Short-lived				
Bromus hordeaceus L. (G)	5	742	II	11.8
Crepis biennis L.	5	516	II	9.0
Conyza canadensis (L.) Cronq.	5	294	III	4.3
Matricaria maritima L. ssp. *inodora* (L.) Dostál	4	156	IV	3.0
Stellaria media (L.) Vill.	3	82	V	1.8
Polygonum aviculare L.	2	76	V	1.5
Geranium pusillum Burm. F. ex L.	2	54	V	1.0
Lamium purpureum L.	1	42	V	0.8
Chenopodium album L.	1	40	V	0.8
Vicia villosa Roth.	1	22	V	0.2
Crepis capillaris (L.) Wallr.	1	10	V	0.1
Tragopogon pratensis L.	1	6	V	<0.05
Perennial				
Festuca rubra L. ssp. *rubra* (G)	5	5640	I	84.0
Taraxacum officinale F. H. Wigg.	5	512	II	9.5
Cerastium holosteoides Fr. em. Hyl.	5	454	III	7.3
Rumex acetosella L.	4	280	III	4.5
Epilobium ciliatum Raf.	4	196	IV	3.3
Cirsium arvense (L.) Scop.	1	52	V	0.8
Trifolium repens L.	1	9	V	<0.1

G—grassy species.

Table 8. Shannon–Wiener in-row flora diversity index (H') in response to OFM.

Treatment	H' Value					
Base year	2.310 ± 0.094 [c]	2.310 ± 0.094 [c]	2.310 ± 0.094 [d]	2.310 ± 0.094 [d]	2.310 ± 0.094 [e]	2.310 ± 0.094 [e]
	Year of OFM implementation					
	2nd	**3rd**	**4th**	**5th**	**6th**	**7th**
Control	2.165 ± 0.049 [c]	1.747 ± 0.292 [b]	1.602 ± 0.147 [b]	1.277 ± 0.080 [a]	1.055 ± 0.072 [a]	0.817 ± 0.130 [a]
Herbicides	1.897 ± 0.057 [b]	1.820 ± 0.113 [b]	1.786 ± 0.160 [b]	1.781 ± 0.064 [b]	1.777 ± 0.110 [d]	1.770 ± 0.155 [d]
Mulch	1.620 ± 0.259 [a]	1.418 ± 0.293 [a]	1.360 ± 0.054 [a]	1.353 ± 0.049 [a]	1.343 ± 0.057 [b]	1.338 ± 0.119 [b]
Tillage	1.873 ± 0.092 [b]	1.928 ± 0.124 [b]	2.059 ± 0.175 [c]	2.140 ± 0.154 [c]	2.183 ± 0.131 [e]	2.285 ± 0.072 [e]
Mowing	1.869 ± 0.095 [b]	1.854 ± 0.059 [b]	1.709 ± 0.068 [b]	1.660 ± 0.045 [b]	1.616 ± 0.051 [c]	1.536 ± 0.077 [c]

Means within column followed by the same letter do not differ significantly at $p = 0.05$. Values with the prefix ± represent standard deviation.

Table 9. Simpson in-row flora dominance index (D) in response to OFM.

Treatment	D Value					
Base	0.139 ± 0.019 a	0.139 ± 0.019 a	0.139 ± 0.019 a	0.139 ± 0.019 a	0.139 ± 0.019 a	0.139 ± 0.019 a
	Year of OFM implementation					
	2nd	**3rd**	**4th**	**5th**	**6th**	**7th**
Control	0.173 ± 0.009 [ab]	0.226 ± 0.057 [b]	0.316 ± 0.075 [bc]	0.447 ± 0.045 [d]	0.569 ± 0.026 [d]	0.681 ± 0.064 [d]
Herbicides	0.213 ± 0.016 [bc]	0.203 ± 0.030 [ab]	0.250 ± 0.074 [b]	0.234 ± 0.033 [b]	0.240 ± 0.041 [b]	0.269 ± 0.026 [b]
Mulch	0.241 ± 0.051 [c]	0.372 ± 0.091 [c]	0.381 ± 0.030 [c]	0.329 ± 0.019 [c]	0.330 ± 0.020 [c]	0.365 ± 0.047 [c]
Tillage	0.208 ± 0.030 [bc]	0.215 ± 0.32 [ab]	0.171 ± 0.035 [a]	0.155 ± 0.030 [a]	0.145 ± 0.027 [a]	0.123 ± 0.010 [a]
Mowing	0.245 ± 0.022 [c]	0.195 ± 0.012 [ab]	0.295 ± 0.022 [b]	0.314 ± 0.035 [c]	0.329 ± 0.021 [c]	0.381 ± 0.022 [c]

Means within column followed by the same letter do not differ significantly at $p = 0.05$. Values with the prefix ± represent standard deviation.

4. Discussion

The results of the present study indicate that the diversity of weed communities was strongly differentiated depending on the OFM method used, which is consistent with previous reports on this topic in apple, apricot, fig, almond, and olives orchards [24–29] and vineyards [30–33]. As in the apple orchard, this should be justified by the diverse spectrum of weeds controlled by each of the OFM methods and the competitive ability of the dominant flora species [6]. In the last year of the study, i.e., in the seventh year of OFM implementation, regardless of the treatment, the density of weeds and the number of species in one plot were significantly lower than in the base year. The post-emergence herbicides application did not result in a drastic reduction in the number of weeds, and in the long term, even favored a greater diversity of flora compared to the control. The lack of pre-emergence activity of the herbicides allowed for a quick recovery of secondary and primary (in the spring of the next season) weed infestation. The large number of weeds in the herbicide plots also resulted from the rich seed bank in the soil. This bank was supplemented by weeds developed between herbicide treatments and by the wind, and they were mainly seeds of weeds from the Asteracea (*Taraxacum officinale*, *Conyza canadensis*) and Onagraceae (*Epilobium ciliatum*) families. Additionally, some perennial weeds regrow after spraying with herbicides. The greatest reduction in the weed number and cover was achieved after the use of mulch, which prevented the germination of weed seeds. The unmown vegetation in the control effectively stopped falling plum leaves. In combination with drying weed shoots, they formed a mulch that limited the germination and development of short-lived weeds. While weed infestation with short-lived species prevailed in the base year, perennial species prevailed in the plots of four treatments–control, mulch, tillage, and mowing–during the final assessment. Intensive weeding treatments act as a management filter, i.e., they effectively eliminate specific species or groups of weeds [10,30]. This creates conditions for the development of species that are difficult to eliminate and is particularly visible in the absence of weed control methods rotation over a long period of time [10,30]. The most numerous short-lived weeds in herbicide and tilled plots were *Poa annua* and *Stellaria media*–a species with a long period of emergence—and in tilled plots, *Chenopodium album*–a species characterized by a large number of seeds per plant and long persistence of seeds in the soil [40]. Creeping weeds dominated both in plots where the soil was not disturbed (control, mulch, and mowing) and in tilled plots, which indicated that cutting of rhizomes was conducive to their proliferation. In the seventh year of OFM implementation, the relative share of grasses in the weed cover on plots of all treatments increased compared to the baseline year. The present research confirmed the tendency to promote the development of grasses after spraying with post-emergence herbicides and mowing, noted in an almond orchard [27], but the grasses on the control plots developed even stronger. Grasses tolerated mowing much better than dicotyledonous plants, which results from their biology [27]. Glufosinate ammonium, one of the herbicides used in the present study (second spraying), was more effective at controlling dicotyledonous weeds than grasses. Mulch was not a sufficient barrier to the development of creeping weeds, among which *F. rubra* was dominant. The obtained results were related to the specific composition of the flora and the strong pressure of the cover plant from the orchard alleys. In order to interpret the present results, it is necessary to consider the data on the dominant flora species, especially their share in cover after seven years of OFM implementation. In the base year, *F. rubra* was sown in the orchard alleys as a cover plant to reduce weed infestation, soil erosion, and compaction, to ensure easy passage of machines and nutrient recirculation. *F. rubra* plants appeared in the experimental plots in the second year of OFM implementation, and from the fourth year, their dominance progressed rapidly. The highest dynamics of plant development of this species were found in the control plots. In mulched and tilled plots, *F. rubra* occurred mainly on their edges facing the alleys.

The final value of the flora Shannon–Wiener's diversity index (H') for treatments with active measures, ranged in present studies from 1.338 to 2.285, and for the control it was 0.817. Taking into account the results of other agro-phyteconoses obtained by many authors [13–16,24,25,31,33,41], it should be recognized that the evaluated OFM methods maintained the species diversity of the synanthropic flora at a satisfactory level. For comparison, the value of weed H' in cereals in southern Poland was 0.97–1.08 [13], and in perennial industrial plants in eastern Poland it ranged from 0.7 to 2.3, depending on the crop and season [15]. The relatively low diversity of flora in cereals resulted from the uniformity of habitat conditions and agriculture landscape, and the long-term dominance of cereals in the assessment sites [13]. Data from industrial plants referred to a variety of crop species, and the experiment place was surrounded by a diverse landscape [15]. The H' value for flora in apple orchards was 0.67–1.18 in South Korea [41], 3.325 in Indian Western Himalayas [16], and 2.264 for orchards and vineyards on the Istrian peninsula in Croatia [14]. According to the cited authors, the intensification of production in orchards, related to the increase in their area and the frequency of activities, resulted in a decrease in orchard vegetation diversity. Large differences in the value of H' result from different environmental conditions, agricultural structure, intensity of agricultural practices [13], and the date (season) of assessment [41]. It should be noted that the experimental orchard from the present research was located in the middle of a large complex with fruit crops, cultivated for many years with high intensity. Common species prevailed, and no species of high environmental value were found–endangered, rare, or endemic taxa—under the conditions of the present research, as well as in a homogenous olive-dominated landscape in South-East of Italy [29]. Valuable species may be more easily found if research is conducted at various locations across the region, close to semi-natural habitats such as South African vineyards [33]. Valuable plants are those species that especially promote functional agrobiodiversity, e.g., *Achillea millefolium* L., *Centaurea jacea* L., *Leucanthemum vulgare* Lam., *Lotus corniculatus* L., and *Trifolium pratense* L. [42].

The second indice characterizing the diversity of weed communities was Simpson dominance index (D). The highest value of D, among the evaluated treatments was characterized by control (0.681), which proves the strong dominance of a few species. For comparison, in other studies conducted in Poland, the D value was 0.17–0.22 in cereal crops [13] and 0.13–0.45 in perennial industrial crops [15]. In the present studies, the lowest value of D was obtained for tillage, which at the same time was characterized by the highest value of H'. With a similar number of weed species, the treatment with herbicides was characterized by a significantly lower H' value and a higher D value than tillage. This shows that the dominance of the most abundant species was more pronounced after herbicide application than after tillage. Herbicides had a stronger filtering effect, i.e., limiting the weed species diversity more strongly than tillage. The results of the present experiment, probably due to the specific species composition, did not confirm that herbicides and tillage are a stronger management filter than mowing [22]. The use of herbicides in the present study reduced the H' value compared to tillage, as in South African vineyards [33] and in an apricot orchard in Turkey [25], but did not reduce diversity compared to the weed control, which has been noted in an apricot orchard [25] and in vineyards located in southern France [30]. When discussing the reports of other authors, the duration of the study should be taken into account. In the case of apricot orchard and vineyards, these were results from 2 to 3 years [25,30]. The present results covered seven years of diversified OFM, so with each year the importance of succession decreased, and the importance of the weed management method increased. After the second year of OFM implementation in the plum orchard, herbicides significantly reduced the flora diversity compared to the control, as reported by other authors [25,30]. With each subsequent year of research, the diversity in herbicide plots, represented as H', slightly decreased.

Herbicides were the only effective management filter that completely eliminated *F. rubra* from weed cover in the tree rows. In Polish orchards, bunch-type grasses are usually sown in the inter-rows of the orchard, and post-emergence herbicides are used under the tree canopies. With such a model, the problem of cover plants penetrating tree understory practically does not occur. Under the conditions of the present experiment, the rhizomatous groundcover crop from the alleys significantly influenced the diversity of in-row flora. The development of *F. rubra* on experimental plots raises the question of whether this species can be used in orchards as a so-called service crop that is grown to reduce weed pressure and provide ecosystem services without generating serious disservices and impairing fruit production. An example of service plants in the tree understory are perennial groundcovers, also referred to as living mulches [6]. Service plants, in addition to intercrops and cover crops, may also include some weeds [10,30,32]. *F. rubra* turned out to be a very effective competitor to weeds in the present research, especially in the control plots. In contrast, *Festuca ovina*, used in the understory of apple trees as a live mulch, did not effectively limit the development of perennial weeds [6]. Cover crops, as an effective competitor, can reduce flora diversity [24,31]. Perennial *Duchesnea indica* (Andrews) Focke (*Potentilla indica* (Andrews) Th. Wolf) used as a cover crop in a Chinese apple orchard reduced the flora H' value by 53.8%, from approximately 1.8 to 0.9.within three years [24]. In Californian vineyards, the H' value for treatments of cover crops–oat and/or legumes—or cover crops + tillage was 1.2–1.4, while for the treatment of resident weeds + tillage, it was 1.8 [31]. For comparison, in the present study, this value for tillage was 2.285 in the seventh year of implementation. The effect of spreading red fescue from the alleys can be compared to the effects of invasive plants, e.g., *Xanthium strumarium* L., which reduced the diversity of weed communities [43].

Whether *F. rubra* is a strong competitor to trees remains an important question. A study conducted in parallel with the present one showed that plum trees grafted on 'Myrobalan' seedlings (*Prunus cerasifera* Ehrh. var. *divaricata* Ledeb.) growing on the control plots, dominated by *F. rubra*, were characterized by a similar cumulative yield and productivity index as on the plots of other treatments [44]. In-row flora, including *F. rubra*, did not worsen the nutritional status of plum trees in the parallel study either [45].

Fracchiolla et al. [27] found that vegetation mowing and post-emergence herbicides promoted in the almond orchard the growth of sufficiently diversified and balanced flora that could lead to potential ecological services. The results of the present research suggest that such services can also be carried out by flora in tilled and control plots. In plots sprayed with post-emergence herbicides, mowed and tilled, the spontaneous flora regrowth was quick after the intervention and was not disturbed in the period from mid-September to May of the following year. Numerous weeds, such as *Stellaria media*, *Taraxacum officinale*, *Galium aparine*, and *Cirsium arvense*, which are considered to be melliferous plants [15], were found in the plots of the mentioned treatments and in the control. Mulching with natural materials is a good way to reduce weed infestation in organic orchards, as it is not associated with soil disturbance and chemical residues [3]. In the present study, mulching generated sufficient weed diversity, but due to poor development and low soil cover, spontaneous flora could not provide ecosystem services as well as other treatments.

From a practical point of view, the present research confirmed, in accordance with previous reports, the insufficient effectiveness of controlling perennial creeping weeds (*Elymus repens*, *Rumex acetosella*) by soil tillage [46] and mulching [3].

The results of the present research indicate the potential possibility of using *F. rubra* ssp. *rubra* sown in orchard alleys as spontaneous, self-propagating living mulch that limits weed infestation. Such application should be supplemented with studies on the tolerance of fruit trees to the presence of *F. rubra* in the understory, monitoring the occurrence of rodents and combining with other OFM methods. Post-emergence herbicide application followed by flora mowing may be needed in young orchards to protect poorly growing trees from early and strong *F. rubra* competition.

5. Conclusions

The OFM in-row strongly influenced the floristic diversity of summer weed communities in the plum orchard. The values of the diversity indices and the relationships between the treatments regarding the values of these indices changed during the subsequent years of the assessment. After seven years of OFM implementation, the treatments were ranked according to the decreasing value of the Shannon–Wiener diversity index as follows: tillage, post-emergence herbicides spraying, mowing, mulch, and control. The highest value of Simpson dominance index was found in the control treatment. *F. rubra* ssp. *rubra* was the dominant species in plots with treatments such as control, mowing, tillage, and mulch. This rhizomatous perennial grass, which was sown in the alleys of the orchard in the year of its establishment, penetrated and influenced in-row flora. Mulch efficiently reduced weed infestation, and therefore the vegetation of mulched plots was characterized by a low value of density and soil cover. The obtained results indicated that the flora overgrowing the control, sprayed with post-emergence herbicides, tilled, and mowed plots had greater potential to provide ecosystem services than the flora of mulched plots.

Funding: The publication was financed (co financed) by the Ministry of Education and Science of Poland as part of the statutory activities of The National Institute of Horticultural Research, Project No. 3.2.1., "The response of synanthropic orchard flora and plum trees to chosen systems of weed control and soil management".

Data Availability Statement: Data are contained within the article.

Conflicts of Interest: The author declares no conflict of interest.

References

1. EUROSTAT. Available online: https://ec.europa.eu/eurostat/statistics-explained/index.php?title=Agricultural_production_-_orchards (accessed on 20 April 2023).
2. Mia, M.J.; Massetani, F.; Murri, G.; Facchi, J.; Monaci, E.; Amadio, L.; Neri, D. Integrated weed management in high density fruit orchards. *Agronomy* **2020**, *10*, 1492. [CrossRef]
3. Granatstein, D.; Andrews, P.; Groff, A. Productivity, economics, and fruit and soil quality of weed management systems in commercial organic orchards in Washington State, USA. *Org. Agr.* **2014**, *4*, 197–207. [CrossRef]
4. Mia, M.J.; Furmańczyk, E.M.; Golian, J.; Kwiatkowska, J.; Malusá, E.; Neri, D. Living mulch with selected herbs for soil management in organic apple orchards. *Horticulturae* **2021**, *7*, 59. [CrossRef]
5. Paušič, A.; Tojnko, S.; Lešnik, M. Permanent, undisturbed, in-row living mulch: A realistic option to replace glyphosate-dominated chemical weed control in intensive pear orchards. *Agric. Ecosyst. Environ.* **2021**, *318*, 107502. [CrossRef]
6. Żelazny, W.R.; Licznar-Małańczuk, M. Living mulch persistence in an apple orchard and its effect on the weed flora in temperate climatic conditions. *Weed Res.* **2022**, *62*, 85–99. [CrossRef]
7. Lipecki, J. Weeds in orchards–pros and contras. *J. Fuit Ornam. Plant Res.* **2006**, *14*, 13–18.
8. Yvoz, S.; Cordeau, S.; Ploteau, A.; Petit, S. A framework to estimate the contribution of weeds to the delivery of ecosystem (dis)services in agricultural landscapes. *Ecol. Indic.* **2021**, *132*, 108321. [CrossRef]
9. Blaix, C.; Moonen, A.C.; Dostatny, D.F.; Izquierdo, J.; Le Corff, J.; Morrison, J.; von Redwitz, C.; Schumacher, M.; Westerman, P.R. Quantification of regulating ecosystem services provided by weeds in annual cropping systems using a systematic map approach. *Weed Res.* **2018**, *58*, 151–164. [CrossRef]
10. MacLaren, C.; Storkey, J.; Menegat, A.; Metcalfe, H.; Dehnen-Schmutz, K. An ecological future for weed science to sustain crop production and the environment. A review. *Agron. Sustain. Dev.* **2020**, *40*, 24. [CrossRef]
11. Balfour, N.J.; Ratnieks, L.F.W. The disproportionate value of 'weeds' to pollinators and biodiversity. *J. Appl. Ecol.* **2022**, *59*, 1209–1218. [CrossRef]
12. European Commission. Common Agricultural Policy. Available online: https://agriculture.ec.europa.eu/common-agricultural-policy/income-support/greening_en (accessed on 11 May 2023).
13. Dąbkowska, T.; Grabowska-Orządała, M.; Łabza, T. The study of the transformation of segetal flora richness and diversity in selected habitats of southern Poland over a 20-year interval. *Acta Agrobot.* **2017**, *70*, 1712. [CrossRef]
14. Štefanić, E.; Kovačević, V.; Antunović, S. Decline of arable flora diversity in Istria (from the year 2005 to the year 2017). *Zbornik Veleučilišta Rijeci* **2018**, *6*, 385–398. [CrossRef]
15. Radzikowski, P.; Matyka, M.; Berbeć, A.K. Biodiversity of Weeds and Arthropods in Five Different Perennial Industrial Crops in Eastern Poland. *Agriculture* **2020**, *10*, 636. [CrossRef]
16. Haq, S.M.; Lone, F.A.; Kumar, M.; Calixto, E.S.; Waheed, M.; Casini, R.; Mahmoud, E.A.; Elansary, H.O. Phenology and diversity of weeds in the agriculture and horticulture cropping systems of Indian Western Himalayas: Understanding implications for agro-ecosystems. *Plants* **2023**, *12*, 1222. [CrossRef] [PubMed]
17. Baessler, C.; Klotz, S. Effects of changes in agricultural land-use on landscape structure and arable weed vegetation over the last 50 years. *Agric. Ecosyst. Environ.* **2006**, *115*, 43–50. [CrossRef]
18. Glemnitz, M.; Radics, L.; Hoffmann, J.; Czimber, G. Land use impacts on weed floras along a climate gradient from South to North Europe. *J. Plant Dis. Prot.* **2006**, *20*, 577–586.

19. Raselle, R.L.; Freitas, S.P.; Colombo, J.N.; Krause, R.M.; Barth, H.; Barth, H.T. Phytosociological survey of weeds in the grapevine. *Biosci. J.* **2022**, *38*, e38093. [CrossRef]
20. Hawes, C.; Squire, G.R.; Hallett, P.D.; Watson, C.A.; Young, M. Arable plant communities as indicators of farming practice. *Agric. Ecosyst. Environ.* **2010**, *138*, 17–26. [CrossRef]
21. Ryan, M.R.; Smith, R.G.; Mirsky, S.B.; Mortensen, D.A.; Seidel, R. Management filters and species traits: Weed community assembly in long-term organic and conventional systems. *Weed Sci.* **2010**, *58*, 265–277. [CrossRef]
22. Gaba, S.; Fried, G.; Kazakou, E. Agroecological weed control using a functional approach: A review of cropping systems diversity. *Agron. Sustain. Dev.* **2014**, *34*, 103–119. [CrossRef]
23. Schumacher, M.; Ohnmacht, S.; Rosenstein, R.; Gerhards, R. How management factors influence weed communities of cereals, their diversity and endangered weed species in Central Europe. *Agriculture* **2018**, *8*, 172. [CrossRef]
24. Meng, J.; Li, L.; Liu, H.; Li, Y.; Li, C.; Wu, G.; Yu, X.; Guo, L.; Cheng, D.; Muminov, M.A.; et al. Biodiversity management of organic orchard enhances both ecological and economic profitability. *PeerJ* **2016**, *4*, e2137. [CrossRef] [PubMed]
25. Karaman, Y.; Işik, D.; Tursun, N. The effect of cover crops on composition and diversity of weeds in an apricot orchard. *Plant Prot. Bull.* **2021**, *61*, 10–19. [CrossRef]
26. Costa, T.; Giacobbo, C.L.; Galon, L.; Forte, C.T.; Damis, R.; Tironi, S.P. Management of soil cover and its influence on phytosociology, physiology and fig production. *Comun. Sci.* **2020**, *11*, e3236. [CrossRef]
27. Fracchiolla, M.; Terzi, M.; Frabboni, L.; Caramia, D.; Lasorella, C.; De Giorgio, D.; Montemurro, P.; Cazzato, E. Influence of different soil management practices on ground-flora vegetation in an almond orchard. *Renew. Agric. Food Syst.* **2016**, *31*, 300–308. [CrossRef]
28. Solomou, A.D.; Sfougaris, A.I.; Kalburtji, K.L.; Nanos, G.D. Effects of organic farming on winter plant composition, cover and diversity in olive grove ecosystems in central Greece. *Commun. Soil Sci. Plan.* **2013**, *44*, 312–319. [CrossRef]
29. Terzi, M.; Barca, E.; Cazzato, E.; D'Amico, F.S.; Lasorella, C.; Fracchiolla, M. Effects of weed control practices on plant diversity in a homogenous olive-dominated landscape (South-East of Italy). *Plants* **2021**, *10*, 1090. [CrossRef]
30. Kazakou, E.; Fried, G.; Richarte, J.; Gimenez, O.; Violle, C.; Metay, A. A plant trait-based response-and-effect framework to assess vineyard inter-row soil management. *Bot. Lett.* **2016**, *163*, 373–388. [CrossRef]
31. Steenwerth, K.L.; Orellana-Calderón, A.; Hanifin, R.C.; Storm, C.; McElrone, A.J. Effects of various vineyard floor management techniques on weed community shifts and grapevine water relations. *Am. J. Enol. Vitic.* **2016**, *67*, 153–162. [CrossRef]
32. Garcia, L.; Celette, F.; Gary, C.; Ripoche, A.; Valdés-Gómez, H.; Metay, A. Management of service crops for the provision of ecosystem services in vineyards: A review. *Agric. Ecosyst. Environ.* **2018**, *251*, 158–170. [CrossRef]
33. MacLaren, C.; Bennett, J.; Dehnen-Schmutz, K. Management practices influence the competitive potential of weed communities and their value to biodiversity in South African vineyards. *Weed Res.* **2019**, *59*, 93–106. [CrossRef]
34. IUSS Working Group Wrb. *World Reference Base for Soil Resources 2014, Update 2015 International Soil Classification System for Naming Soils and Creating Legends for Soil Maps*; World Soil Resources Reports No. 106; FAO: Rome, Italy, 2015.
35. Braun-Blanquet, J. *Pflanzensoziologie*, 3rd ed.; Soringerverlag: Wien, Austria, 1964; 865p.
36. Wysocki, C.; Sikorski, P. *Fitosocjologia Stosowana*; Wyd. SGGW: Warszawa, Poland, 2002; 449p.
37. Shannon, C.E. A mathematical theory of communication. *Bell System Tech. J.* **1948**, *27*, 379–423, 623–656. [CrossRef]
38. Simpson, E.H. Measurement of diversity. *Nature* **1949**, *163*, 688. [CrossRef]
39. Mirek, Z.; Piękoś-Mirkowa, H.; Zając, A.; Zając, M. Flowering plants and pteridophytes of Poland: A checklist. In *Biodiversity of Poland*; Polish Academy of Sciences, W. Szafer Institute of Botany: Cracow, Poland, 2002; Volume 1, 441p.
40. Tymrakiewicz, W. *Atlas Chwastów*; PWRiL: Warszawa, Poland, 1976; 450p.
41. Woo, I.S.; Pyon, J.Y. Characterization of weed occurrence in apple orchards. *Korean J. Weed Sci.* **1988**, *8*, 164–168.
42. Pfiffner, L.; Cahenzli, F.; Steinemann, B.; Jamar, L.; Bjorn, M.C.; Porcel, M.; Tasin, M.; Telfser, J.; Kelderer, M.; Lisek, J.; et al. Design, implementation and management of perennial flower strips to promote functional agrobiodiversity in organic apple orchards: A pan-European study. *Agric. Ecosyst. Environ.* **2019**, *278*, 61–71. [CrossRef]
43. Qureshi, H.; Anwar, T.; Arshad, M.; Osunkoya, O.O.; Adkins, S.W. Impacts of *Xanthium strumarium* L. invasion on vascular plant diversity in Pothwar region (Pakistan). *Ann. Bot.* **2019**, *9*, 73–82. [CrossRef]
44. Lisek, J.; Buler, Z. Growth and yield of plum trees in response to in-row orchard floor management. *Turk. J. Agric. For.* **2018**, *42*, 97–102. [CrossRef]
45. Lisek, J.; Stępień, T. Macroelements concentration in plum tree leaves and soil in response to orchard floor management. *Acta Sci. Pol. Hortorum Cultus* **2021**, *20*, 115–124. [CrossRef]
46. Rabcewicz, J.; Białkowski, P. The efficiency of mechanical weed control in ecological apple production. *J. Res. Appl. Agric. Eng.* **2011**, *56*, 79–83.

Article

Assessment of a Chain Mower Performance for Weed Control under Tree Rows in an Alley Cropping Farming System

Lorenzo Gagliardi, Marco Fontanelli *, Christian Frasconi, Mino Sportelli, Daniele Antichi, Lorenzo Gabriele Tramacere, Giovanni Rallo, Andrea Peruzzi and Michele Raffaelli

Department of Agriculture, Food and Environment, University of Pisa, Via del Borghetto 80, 56124 Pisa, Italy
* Correspondence: marco.fontanelli@unipi.it

Abstract: In the area under tree rows of alley cropping systems, coarse plant material as well as pruning material or stones may be present, so the use of a mower equipped with chains as cutting a tool could be advantageous. A mower designed for under-row weed control in orchards, equipped with an automatic tree-skipping mechanism, was modified by replacing blades with chains with the aim of evaluating its performance in an alley cropping system. A first trial was carried out in an open field to preliminarily compare the chain mower with the version equipped with blades in relation to different settings of working speed (1.6 and 2.4 km·h^{-1}) and rotation speed of the cutting tool (1830 and 2500 rpm). Weed biomass reduction, weed cover reduction, weed height reduction, weed biomass regrowth, and clipping size were assessed. In a second trial, the performance of the mowers with different setting configurations was assessed in an alley cropping system under a more critical environmental condition for mowing, i.e., the presence of dew. Weed biomass reduction, weed cover reduction, weed height reduction, and the mowers' field capacity with different working speed settings were assessed. No major differences emerged between the mowers and the chain mower performance was comparable to that of the standard blade mower. The setting with the high working speed and high rotation speed of the cutting tool turns out to be the best compromise, obtaining a weed biomass reduction of 59.6%, a weed cover reduction of 40.9%, and a higher field capacity compared to the setting with the low working speed, with an increase of 47.9%.

Keywords: agroforestry; under-row weed control; automatic tree-skipping; mechanical weed control; no-till strategies

Citation: Gagliardi, L.; Fontanelli, M.; Frasconi, C.; Sportelli, M.; Antichi, D.; Tramacere, L.G.; Rallo, G.; Peruzzi, A.; Raffaelli, M. Assessment of a Chain Mower Performance for Weed Control under Tree Rows in an Alley Cropping Farming System. *Agronomy* **2022**, *12*, 2785. https://doi.org/10.3390/agronomy12112785

Academic Editor: Anestis Karkanis

Received: 15 September 2022
Accepted: 6 November 2022
Published: 9 November 2022

Publisher's Note: MDPI stays neutral with regard to jurisdictional claims in published maps and institutional affiliations.

1. Introduction

The ability of agriculture to fulfill the food demand of a growing world population is currently threatened by multiple factors, such as climate change, soil degradation, depletion, and pollution of water resources [1]. As environmental issues and ecological sustainability gain more relevance and the research for environmentally friendly practices becomes increasingly important [2,3], agroforestry strategies are being explored with growing interest as sustainable approaches to land use. Agroforestry essentially refers to land use systems that, through the intercropping of trees and/or shrubs along with crops and/or animals, diversify and support production for greater socioeconomic and environmental benefits for land users [4,5]. Alley cropping represents a common agroforestry practice that consists of growing arable crops in alleys between widely spaced rows of trees or shrubs. This practice proved to benefit the agricultural systems by increasing overall productivity and resilience and promoting the rehabilitation of degraded soils [6,7]. However, some constraints may be related to a potential decrease in arable crop yield due to competition with trees, higher management needs, and greater labor demand [8]. In particular, under-row weed control can represent a major issue. Weeds can compromise the success of new agroforestry planting establishment [9]. Indeed, the optimal growth and survival of many tree species grown in temperate agroforestry systems can be hindered by weed competition [9–11].

Schroeder and Naeem [12] observed that weed control positively influenced the annual height increment, total basal diameter, and height of the tested agroforestry tree species. Furthermore, the risk that some weeds spread from the under-row area towards the crop in the alley, leading to yield losses, should be considered [13–15]. To date, the most common methods used to manage under-row weeds in alley cropping systems are mulching, chemical applications, and mowing. Mulching can be achieved by applying straw or other recalcitrant and cheap plant material under the row or by applying plastic mulch [16,17]. However, generally, large-scale mulching is cost-intensive [18]. The use of living mulch, i.e., cover crops planted in main crop stands and maintained as a living ground cover for part of or the whole growing season, provides many useful ecosystem services. Nevertheless, the proliferation of vegetation can create a too-competitive environment for the establishment and growth of trees [19]. On the other hand, the application of chemicals (i.e., herbicides), despite their effectiveness, not only causes environmental pollution but can also damage trees, making it necessary to shield them before treatments [12,16]. Mowing represents another weed control method commonly adopted in alley cropping systems. This practice can effectively reduce the competitiveness and longevity of perennial weeds and prevent the production of seeds for many types of weeds avoiding their spread [20,21]. Chen et al. [22] found that mowing effectively regulates the root amount and depth of mowed plants with consequent improvement of soil moisture. Furthermore, this practice can promote soil conservation by maintaining a permanent sod that helps to protect the soil from erosion [23]. In the under-row area of alley cropping systems, mowing can be performed by operators with motorized mowers [24], string trimmers, or with tractor-mounted mowers [25]. The tractor-mounted flail mowers employed in these contexts perform a side shift movement through a hydraulic lateral transverse function or hydraulic arms, cutting weeds along the tree row and skipping trees [26,27]. However, the use of these machines for this purpose can be laborious, as the lateral movement occurs following the command of the operator. On the other hand, the use of machines with an automatic tree-skipping mechanism (by means of a feeler, for example, such as those for under-row weed control in orchards), would make the practice easier for the operator [23]. Indeed, according to this tree-skipping mechanism, the mowing implement, due to the presence of a feeler rod, can enter the row in the working position and automatically exit whenever the feeler perceives the pressure of tree trunks. Among mowers, the rotary impact types are increasingly used due to their simplicity in construction and low maintenance cost [28]. These mowers can be equipped with various types of cutting tools, and their performance is strictly related to cutting edge sharpness and cutting speed. Generally, as the rotation speed of the cutting tool increases and the working speed decreases, the effectiveness of the cutting tool increases [29–31]. Furthermore, for tools that perform impact cutting, the cutting action is also linked to the plant inertia that, by conferring the opposing force, allows tools to penetrate the plant material. Thick stalks have higher inertia than thinner stalks, thus requiring lower cutting speed to be penetrated by the cutting tool [29]. The environmental conditions in which mowing is performed can also influence its effectiveness. In the presence of moisture on the leaves' surface, wet clippings, tending to easily bunch up, can cause the clogging of the mowers, thus jeopardizing the mowing operation [32]. In forest scenarios, chain mowers are often employed for clearing operations [33–35]. Chain mowers can effectively clear coarse plant material and are suitable for use in areas with high stoniness since chains simply bend over these obstacles whereas blades become frequently damaged [34–36]. It is well-known that the longer a cutting tool lasts before wearing out and needing to be changed, the better it is for the user. The use of chain mowers could be advantageous for under-row management in alley cropping contexts since coarse plant material, tree residues (such as pruning materials), or stones may be present in this area. To the best of our knowledge, there has been no research testing a chain mower with an automatic tree-skipping mechanism for under-row weed control in alley cropping farming systems. In this research, a mower designed for under-row weed control in orchards was modified by replacing blades with chains with the aim of evaluating its performance in an alley cropping context. The weed control effect of the chain mower was evaluated in comparison with the commercial version equipped with blades. The performances of the two mower versions were assessed in relation to different settings of the working speed and rotation speed of the cutting tool, preliminarily in an open field and then in an alley cropping farming system.

2. Materials and Methods

Two trials were performed in 2022 at the Centre for Agri-environmental Research "Enrico Avanzi", San Piero a Grado (Pisa), Italy (43°40′48″ N, 10°20′49″ E, 1 m.a.s.l.) to assess the chain mower performance as a means of under-row weed control in alley cropping systems. The first trial was carried out in an open field managed as weedy fallow, with the aim of comparing the weed control effect of the two mower versions with different setting configurations to preliminarily identify any ineffective ones. The second trial was conducted in a real alley cropping farming system to compare the two mower versions' weed management performance in the under-row area. In this second trial, the mowers' performance with different setting configurations was also assessed in a more critical environmental condition for mowing, such as with the presence of dew, to identify the most suitable setting configuration.

2.1. The Mowing Machine

Mowing was carried out with a tractor-mounted under-row mower (Dondi, Bastia Umbra, Italy) designed for weed control in orchards (Figure 1). The machine is driven by the power take-off and is equipped with an independent hydraulic system. The mowing implement is represented by a horizontal disc case (Ø 0.40 m) and 4 axial cutting blades (Figure 2a,b) driven by a hydraulic motor. The chain mower was obtained by replacing each blade with grade 8 chains in tempered steel (EN 818-2) (Figure 2c). The technical characteristics of the two cutting tools are shown in Table 1.

(a)   (b)

Figure 1. Rear (**a**) and lateral (**b**) view of the under-row mower.

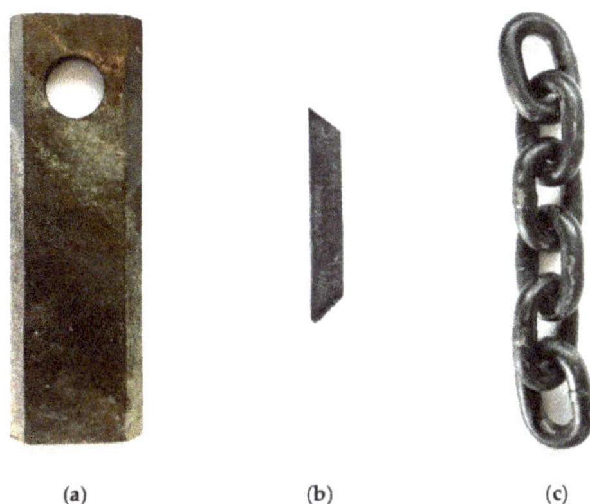

Figure 2. Different kinds of cutting tool: (**a**) blade; (**b**) cross section of the blade; (**c**) chain.

Table 1. Cutting tools' technical characteristics.

Parameters		Chain *	Blade *
Mass	g	94.3	397.6
Length	mm	118	151
Width	mm	21	51
Thickness **	mm	6	8

* Measurements refer to the single cutting tool. ** For the chain, the value corresponds to the wire diameter.

The cutting widths were 0.35 m and 0.32 m, respectively, for the blade and the chain mower versions. The mower is the one-sided type; thus, one pass on either side of the tree row is required for completing the row management. The presence of a horizontal feeler rod and a single-acting hydraulic cylinder allow the mowing disc to enter the row and automatically exit whenever the feeler perceives the pressure of tree trunks. The minimum distance between tree trunks to guarantee the optimum operation of the mower is 0.80 m. The disc case prevents cutting tools from damaging tree trunks. The mass of the machine is equal to 400 kg. During each trial, the mower was coupled with a Fiat DT 70-90 tractor (FiatAgri, Torino, Italy) powered by a 52.2 kW diesel engine.

2.2. Experiment Layout

In the first trial, the performances of the chain mower and the blade mower were compared in an open field with sandy loam soil and managed as weedy fallow, considering two working speeds (1.6 and 2.4 km·h^{-1}) and two rotation speeds of the cutting tool (1830 and 2500 rpm). The configuration settings of the machine with different working speeds and rotation speeds of the cutting tool were adjusted by identifying the appropriate combination of the range and speed gearbox of the employed tractor. The working speed was measured by recording the time taken by the tractor coupled with the mower to travel a known distance. The rotation speed of the cutting tool was detected with Extech 461920 Mini Laser Photo Tachometer Counter (Extech, Nashua, NH, USA). The tachometer employed measures in rpm in the range from 2 to 999,999 rpm, with a resolution of 0.1 rpm. During the first trial, mowing was performed with both mower versions on 5 February 2022. The adopted experimental design was a randomized complete block design with three replications. Each block was 10 m long. Both mower versions were tested with four different types of setting configurations each, according to the working and rotation speeds: (i) low working speed and low rotation speed of the cutting tool (Lws + Lrs); (ii) low working speed and high rotation speed of the cutting tool (Lws + Hrs); (iii) high working speed and low rotation speed of the cutting tool (Hws + Lrs); (iv) high working speed and high rotation speed of the cutting tool (Hws + Hrs). The tested setting configurations and the diagram describing the tested factors and relative levels in the first trial are shown in Table 2 and Figure 3, respectively.

Table 2. Main properties of the research concerning the tested setting configurations.

Settings		Working Speed (ws)		Rotation Speed (rs)
Lws + Lrs	km·h^{-1}	1.6	rpm	1830
Lws + Hrs	km·h^{-1}	1.6	rpm	2500
Hws + Lrs	km·h^{-1}	2.4	rpm	1830
Hws + Hrs	km·h^{-1}	2.4	rpm	2500

L—Low; H—high; ws—working speed; rs—rotation speed.

Trial 1 (three replications)

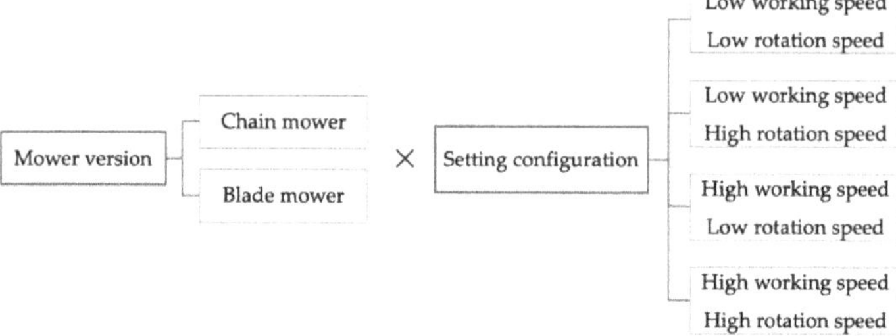

Figure 3. Diagram describing tested factors and relative levels in the first trial.

During the first trial, the most abundant weeds presents in the field when mowing was performed belonged to the genus *Lolium*. For this trial, weed biomass reduction, weed cover reduction, weed height reduction, weed biomass regrowth, and size of clippings produced were assessed.

The second trial was performed in an alley cropping system that consisted of a three-year rotation of annual grain crops planted between rows of poplar trees. The crop rotation included durum wheat (*Triticum turgidum* subsp. durum (Desf.) Husn.), sorghum (*Sorghum bicolor* L. Moench), and pigeon bean (*Vicia faba* L. var. minor Beck). The poplar trees along the rows were various clones of the hybrid poplar *Populus* × *canadensis* Moench planted in February 2019. Poplar trees provide high-value timber that is sold for plywood production. Three rows consisted of trees placed with 5 m of distance between each other and the distance between rows was 18 m. In this system, the arable crop is sowed between tree rows and a 2 m wide zone without crop was provided between tree rows and the arable crop. During this second trial, mowing was carried out on 11 February 2022 (Figure 4). Performing under-row weed control in this period allows the reduction of weed pressure and therefore competition with poplar trees at the time of their spring vegetative restart. The under-row weed control performance of the two mower versions with the above-mentioned setting configurations was evaluated both in the presence and in the absence of dew (Figure 5). To evaluate the mowers' performance in a critical environmental condition, mowing was tested at 9:30 a.m., according to the higher level of dew. Successively, at 12:00 a.m., a no-dew condition was identified and then tested. The times when mowing interventions were performed were chosen according to a preliminary dew monitoring followed in the two days preceding the trial. At this scope, leaf wetness sensors, described in Section 2.3, were used. Dew and no-dew mowing interventions lasted 50 min each. To ensure that both mower versions and their setting configurations acted under the same dew conditions, each mowing treatment was arranged so that the respective mowing alternately ran. During this second trial, a randomized complete 10 m long block design with three replications was adopted. In particular, each block was identified within the three spatial replicates of the field occupied by durum wheat in 2021/2022. Soil texture was sandy clay loam. Weed cover reduction, weed height reduction, weed biomass reduction, and field capacity of the two mower versions with different setting configurations were assessed. In the second trial, the majority of weeds encountered in the fields belonged to the genera *Lolium* and *Poa*.

(a)   (b)

Figure 4. Mowing performed on 11 February 2022: (**a**) front view of the working mowing disc; (**b**) rear view of the working mowing disc.

Trial 2 (three replications)

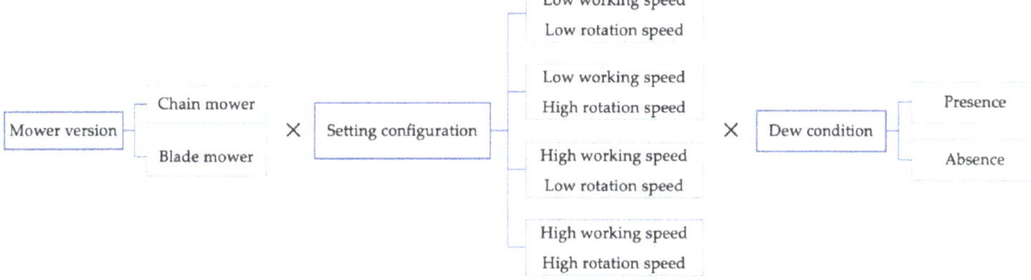

Figure 5. Diagram describing tested factors and relative levels in the second trial.

2.3. Data Collection

In each trial, weed cover, weed height, and weed biomass were determined before and four days after mowing treatments, according to Pergher et al. [23]. For the first trial, the measurements before treatment were carried out on 5 February 2022, just before the treatment was performed, while the measurements after treatment were executed on 9 February 2022. During the second trial, the surveys before treatment were carried out on 10 February 2022, i.e., the day before treatment, while the surveys after treatment were performed on 15 February 2022. Weed cover, which corresponds to the weed soil visual cover, was determined using a Nikon Coolpix 7600 (Nikon corporation, Tokyo, Japan) to shoot picture-task inside a square frame of 30 × 25 cm. Subsequently, pictures were analyzed with the Canopeo app [37], which provides the weed cover percentage by measuring the green pixel percentage of the 30 × 25 cm picture-task. Average weed population height was measured inside the square frame with a folding ruler. Three measurements for each replicate were carried out for weed cover and weed height before and after treatments. Weed biomass measurements were performed by cutting and collecting the live above-ground weed biomass present in the square frame of 30 × 25 cm. Total fresh biomass was then oven-dried at 100 °C for 3–4 days (until mass was constant) and dry biomass was then determined. One measurement per replicate of weed biomass was performed, before and after treatments. During the second trial, weed biomass, weed cover and weed height values were collected in the area between tree trunks.

For the parameters relating to weed biomass, weed cover, and weed height, the efficacy of both mower versions with each setting configuration was measured as the percentage of reduction (*R*) of initial values of the aforementioned parameters assessed on each plot according to the following Equation (1):

$$R(\%) = \frac{(x_b - x_a)}{x_b} \times 100 \tag{1}$$

where x_b and x_a are the entity of the parameter (in this case weed biomass, weed cover, or weed height) assessed before and after treatments, respectively.

During the first trial, clipping size, and weed biomass regrowth were also assessed. Clippings were collected immediately after the treatment and manually measured in length, whereas weed biomass regrowth was collected fifteen days after treatment in the properly marked same areas and the percentage increase in biomass was estimated.

The field capacity of both mower versions with different setting configurations of working speed was estimated in the alley cropping system considering the theoretical field time (i.e., the time the machine effectively operates at an optimum working speed and works over its full width of action) and the turning time.

In the second trial, weed leaf wetness was measured with the A-series leaf Wetness Sensor (Spectrum Technologies Inc., Aurora, IL, USA), a grid-like resistance-based sensor coupled to a datalogger LogBox-AA (Novus, Canoas, Brazil). LogChart-II software (Novus, Canoas, Brazil) was used to configure the datalogger and download and display data. The sensor estimates the wetness by measuring the resistance on the grid. Indeed, the sensor presents a circuit board with interlacing gold-plated fingers. Condensation on the device decreases the resistance between the fingers, which is measured as an analog signal in 16 voltage levels. Subsequently, the measurement was converted into a leaf wetness value through the application of the following Equation (2), such that if $\frac{V}{5} < 0.83$:

$$Leaf\ Wetness = -18.3 \times \left(\frac{V}{5}\right) + 15 \tag{2}$$

Instead, if $\frac{V}{5} > 0.83$, leaf wetness equals 0. The parameter *V* is the sensor voltage output ranking from 0 to 5 volts and 5 is the voltage supplied at the sensor. Leaf wetness value ranges from 0 to 15 and leaves are considered wet with values above 6. Dataloggers were set up to record measurements every 15 min. Two sensors were arranged in the field for two days before treatment and were removed after the last mowing treatment. Sensors were placed above the ground in order to mimic the condition of weed leaves as much as possible. In order to perform accurate measurements, at the time of installation, it was ensured that the sensors' surface were cleaned of conductive material, such as leaves, grass, dirt, and salts.

2.4. Statistical Analysis

Data were analyzed using the statistical software R (version 4.1.2.; R Foundation for Statistical Computing: Vienna, Austria) [38]. Normality distribution was evaluated using the Shapiro–Wilk test, while the Bartlett test was employed for the homoscedasticity. Arcsine or square root transformations of data were applied when needed to meet the normality assumption. For the first trial, two-way ANOVA was performed to assess the effect of mower version, setting configuration, and interactions between factors on weed biomass reduction, weed cover reduction, weed height reduction, weed biomass regrowth, and size of clippings.

For the second trial, three-way ANOVA was performed to assess the significance of different mower versions, setting configurations, dew conditions, and interactions between factors on weed biomass reduction, weed cover reduction, and weed height reduction. An LSD post hoc test at the 0.05 probability level was executed for both trials with the package "agricolae" [39]. The extension package "ggplot2" (Elegant Graphics for Data Analysis) was used to plot graphs.

3. Results

3.1. First Trial: Performances of the Two Mower Versions in the Open Field

3.1.1. ANOVA Analysis Results

Two-way ANOVA revealed a significant effect of setting configuration on weed biomass reduction ($p < 0.01$), whereas mower version and the interaction between mower version and setting configuration did not affect the parameter. Mower version had a significant effect on weed cover reduction ($p < 0.001$), while setting configuration and their interaction did not affect the parameter. Weed height reduction was not affected by the considered fixed factors. ANOVA highlighted a significant effect of both mower version and setting configuration on clipping size ($p < 0.001$ and $p < 0.05$, respectively), while their interaction did not affect the parameter. Weed biomass regrowth was not affected by the considered fixed factors. Results of two-way ANOVA are reported in Table 3.

Table 3. F-values and *p*-values estimations from two-way ANOVA.

Factors	WBR (%)		WCR (%)		WHR (%)		CS (cm)		WBRg (%)	
	F-Value	*p*-Value	F-Value	*p*-Value	F-Value	*p*-Value	F-Value	*p*-Value	F-Value	*p*-Value
Mower Version	0.043	0.839	20.280	***	0.068	0.795	12.555	***	0.605	0.450
Setting Configuration	6.585	**	0.888	0.453	2.446	0.073	2.763	*	0.355	0.787
Mower Version ´ Setting Configuration	0.454	0.718	0.293	0.830	2.140	0.105	1.681	0.174	0.468	0.709

*** $p < 0.001$; ** $p < 0.01$, * $p < 0.05$. WBR—Weed biomass reduction; WCR—weed cover reduction; WHR—weed height reduction; CS—clipping size; WBRg—weed biomass regrowth.

Since data are disputed as a % of the reduction of the parameters' initial values and not as absolute values, mean values of weed biomass, weed cover, and weed height measured before the mowing treatment on the experimental field are shown in Table 4.

Table 4. Parameter mean values recorded before the mowing treatments of the first trial (standard error).

	Nr. Observations		Values (SE)
Weed Biomass	24	g d.m.·m^{-2}	133.88 (7.78)
Weed Cover	72	%	45.7 (1.3)
Weed Height	72	cm	13.86 (0.28)

3.1.2. Differences between the Two Mower Versions' Performance

During the first trial, differences between chain and blade mower were only observed in terms of weed cover reduction and clipping size. The mower equipped with chains achieved the best results of weed cover reduction compared to the blade mower ($p < 0.05$), with back-transformed mean values of 92.6% and 84.3%, respectively. Concerning the clipping size, the chain mower produced larger clippings (back-transformed mean value of 8.66 cm) compared to the blade mower (back-transformed mean value of 6.95 cm) ($p < 0.05$).

3.1.3. Setting Configuration Effect on Mowing Performance

Both weed biomass reduction and clipping size were affected by each machine's setting configuration. In the first trial, averaged across mowing versions, the weed biomass reduction achieved by mowers with the Lws + Hrs setting was higher compared to mowers with Hws + Lrs and Lws + Lrs setting configurations (68.6% vs. 28.7% and 24.9%, respectively) ($p < 0.05$). No difference emerged between weed biomass reduction obtained by mowers with the Hws + Hrs setting and mowers with the other tested settings (Figure 6).

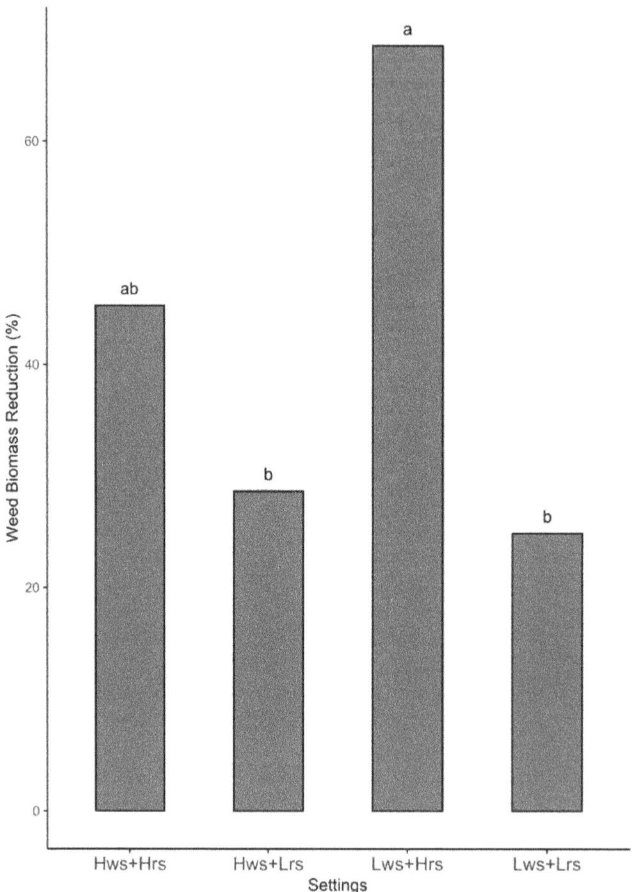

Figure 6. Effect of setting configurations on weed biomass reduction. Means denoted by different letters are significantly different at $p < 0.05$ (LSD test). Hws—High working speed; Lws—low working speed; Hrs—high rotation speed; Lrs—low rotation speed.

Mowers with the Lws + Hrs setting also produced smaller clippings (back-transformed mean value of 6.77 cm) compared to the other setting configurations (back-transformed mean values ranging from 7.93 to 8.31 cm) ($p < 0.05$). Results achieved with the other setting configurations were similar (Figure 7).

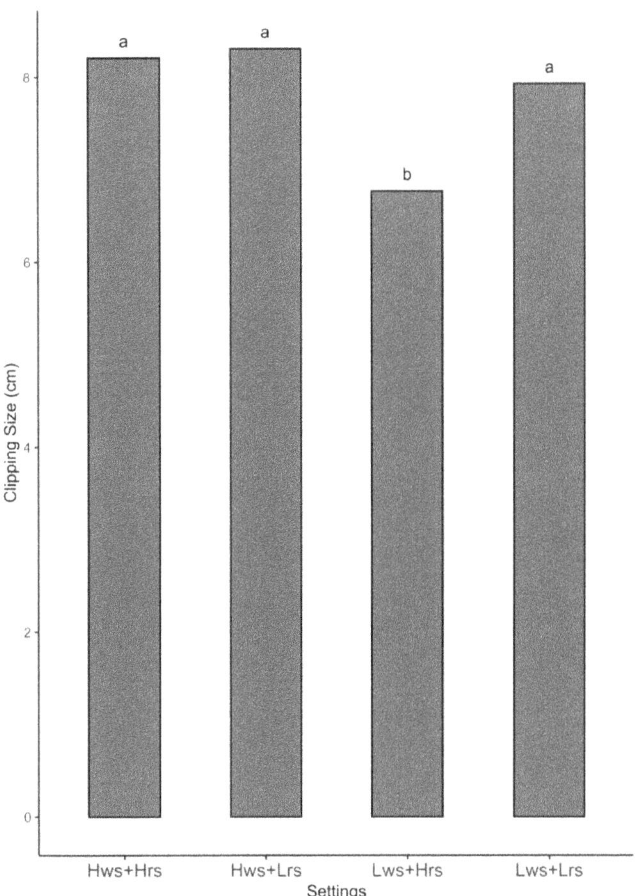

Figure 7. Effect of setting configuration on clippings size. Means denoted by different letters are significantly different at $p < 0.05$ (LSD test). Hws—High working speed; Lws—low working speed; Hrs—high rotation speed; Lrs—low rotation speed.

3.2. Second Trial: Performance of the Two Mower Versions in the Alley Cropping Farming System

Leaf wetness sensors measured an average leaf wetness of 6.35 during the mowing treatment carried out from 9:30 a.m. to 10:20 a.m. on 11 February 2022, thus evidencing the presence of dew on the leaves' surface. Sensors detected an average leaf wetness of 0 during the mowing treatment performed from 12:00 p.m. to 12:50 p.m. on the same day, pointing to the absence of dew. Figure 8 reports the trend of the average leaf wetness on the day of the mowing treatments measured by sensors.

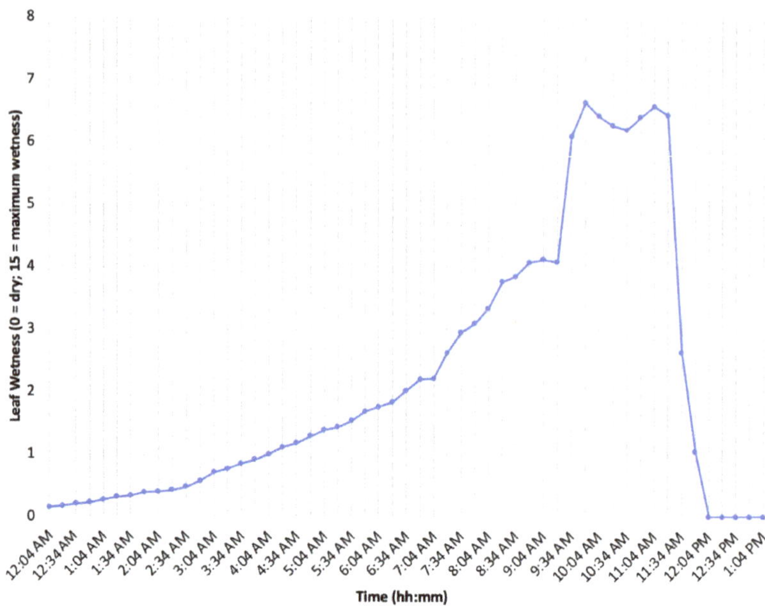

Figure 8. Trend of average leaf wetness values recorded on 11 February, i.e., the day the mowing treatments were performed.

3.2.1. ANOVA Analysis Results

Three-way ANOVA showed that both the setting configuration and the interaction between mower version and dew condition affected weed biomass reduction ($p < 0.05$ for both the fixed factors). Mower version, dew, and the other interactions did not affect the parameter. Weed cover reduction was affected by setting configuration ($p < 0.01$), dew condition ($p < 0.01$), and the interaction between mower version and dew condition ($p < 0.05$), while the mower version and the other interactions were not significant for this dependent variable. Weed height reduction was affected by the mower version ($p < 0.001$), the interaction between mower version and setting configuration ($p < 0.01$), and the interaction between mower version, setting configuration, and dew ($p < 0.05$). Results of three-way ANOVA are shown in Table 5.

Table 5. F-value and *p*-value estimations from three-way ANOVA.

Factors	WBR (%)		WCR (%)		WHR (%)	
	F-Value	***p*-Value**	**F-Value**	***p*-Value**	**F-Value**	***p*-Value**
Mower Version	1.461	0.236	2.159	0.144	16.832	***
Setting Configuration	4.077	*	4.566	**	1.169	0.324
Dew	2.760	0.107	7.042	**	0.580	0.448
Mower Version ´ Setting Configuration	0.523	0.670	0.514	0.673	5.297	**
Mower Version ´ Dew	4.564	*	5.258	*	1.337	0.250
Setting Configuration ´ Dew	2.027	0.131	1.543	0.207	1.264	0.290
Mower Version ´ Setting Configuration ´ Dew	0.305	0.822	2.054	0.110	2.897	*

*** $p < 0.001$; ** $p < 0.01$, * $p < 0.05$. WBR—Weed biomass reduction; WCR—weed cover reduction; WHR—weed height reduction.

Since data are disputed as a % of the reduction of the parameters' initial values also for the second trial, mean values of weed biomass, weed cover, and weed height measured in the alley cropping field (along the tree row) before the mowing treatments are shown in Table 6.

Table 6. Parameter mean values recorded before the mowing treatments of the second trial (standard error).

	Nr. Observations		Values (SE)
Weed Biomass	48	g d.m.·m^{-2}	205.49 (12.79)
Weed Cover	144	%	50.8 (1.4)
Weed Height	144	cm	15.54 (0.31)

3.2.2. Differences between the Two Mower Versions' Performance

During the second trial, the blade mower achieved a significantly higher reduction in weed height compared to the chain mower (46.9% vs. 36.3%, respectively) ($p < 0.05$). Regarding the effect of the mower version and setting configuration interaction on weed height reduction, the blade mower with the Hws + Hrs setting obtained the best result (57.2%) ($p < 0.05$). No difference emerged in the blade mower performance with Hws + Lrs, Lws + Hrs, and Lws + Lrs settings. The chain mower with the Lws + Hrs setting obtained a higher reduction in weed height compared to the machine with Hws + Hrs and Lws + Lrs settings (44.1% vs. 30.0% and 32.9%, respectively) ($p < 0.05$), and similar results were observed compared to both the same mower with the Hws + Lrs setting and the blade mower with Lws + Hrs, Lws + Lrs, Hws + Lrs settings. The chain mower with the Hws + Hrs setting resulted in a lower weed height reduction compared to the machine with the Lws + Hrs setting and the blade mower with different setting configurations ($p < 0.05$) (Figure 9).

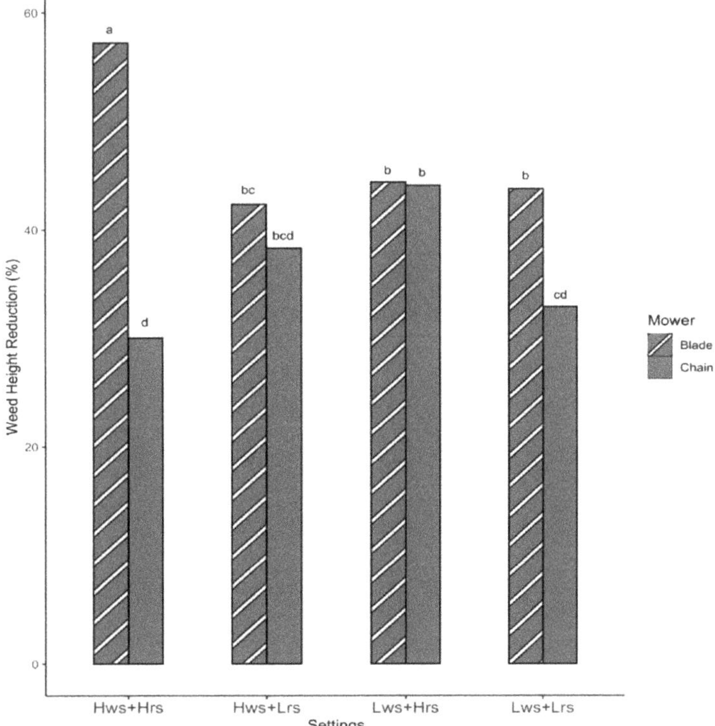

Figure 9. Effect of the interaction between mower version and setting configuration on weed height reduction. Means denoted by different letters are significantly different at $p < 0.05$ (LSD test). Hws—High working speed; Lws—low working speed; Hrs—high rotation speed; Lrs—low rotation speed.

3.2.3. Setting Configuration Effect on Mowing Performance

Mowers with the Lws + Lrs setting achieved a lower reduction in weed biomass compared to Lws + Hrs and Hws + Hrs settings (44.9% vs. 67.9% and 59.6%, respectively) ($p < 0.05$) while no differences emerged between these two settings. Mowers with the Hws + Lrs setting achieved results for weed biomass reduction that were found to be similar to mowers with the other tested settings (Figure 10).

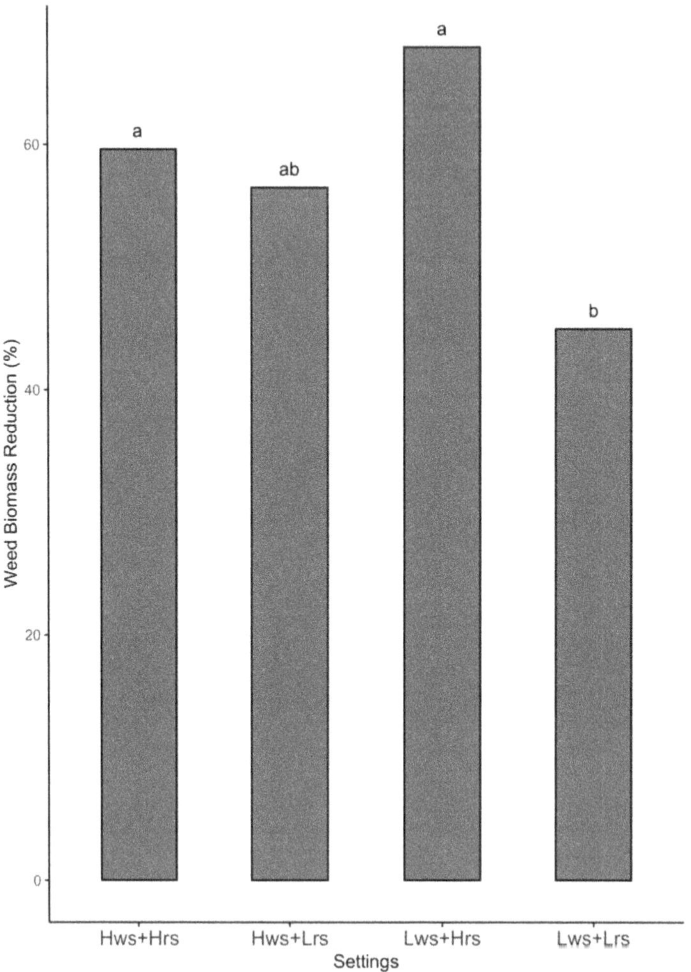

Figure 10. Effect of setting configuration on weed biomass reduction. Means denoted by different letters are significantly different at $p < 0.05$ (LSD test). Hws—High working speed; Lws—low working speed; Hrs—high rotation speed; Lrs—low rotation speed.

Concerning the weed cover reduction, mowers with the Lws + Lrs setting achieved a lower reduction in weed cover compared to mowers with Hws + Hrs and Lws + Hrs settings ($p < 0.05$), with back-transformed mean values of 26.8%, 40.9%, and 36.2%, respectively. No differences emerged in the comparison of mowers with Hws + Hrs and Lws + Hrs settings and between mowers with Lws + Lrs and Hws + Lrs settings (Figure 11).

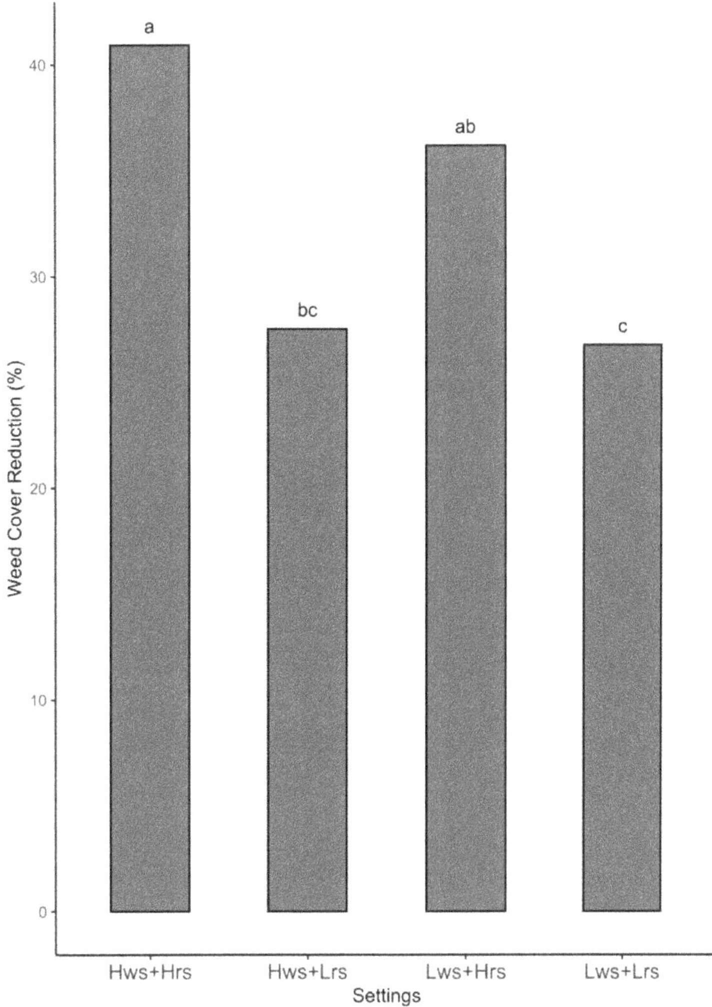

Figure 11. Effect of setting configuration on weed cover reduction. Means denoted by different letters are significantly different at $p < 0.05$ (LSD test). Hws—High working speed; Lws—low working speed; Hrs—high rotation speed; Lrs—low rotation speed.

3.2.4. Effect of Dew Conditions on Mowing Performances

The blade mower achieved a greater weed biomass reduction in the absence of dew compared to the presence of dew (69.0% vs. 51.1%), while no differences emerged in the comparison of the chain mower performance in different dew conditions. In the absence of dew, the blade mower achieved a higher weed biomass reduction compared to the chain mower (53.3%) ($p < 0.05$), while in presence of dew, no significant differences between the two mower versions emerged (Figure 12).

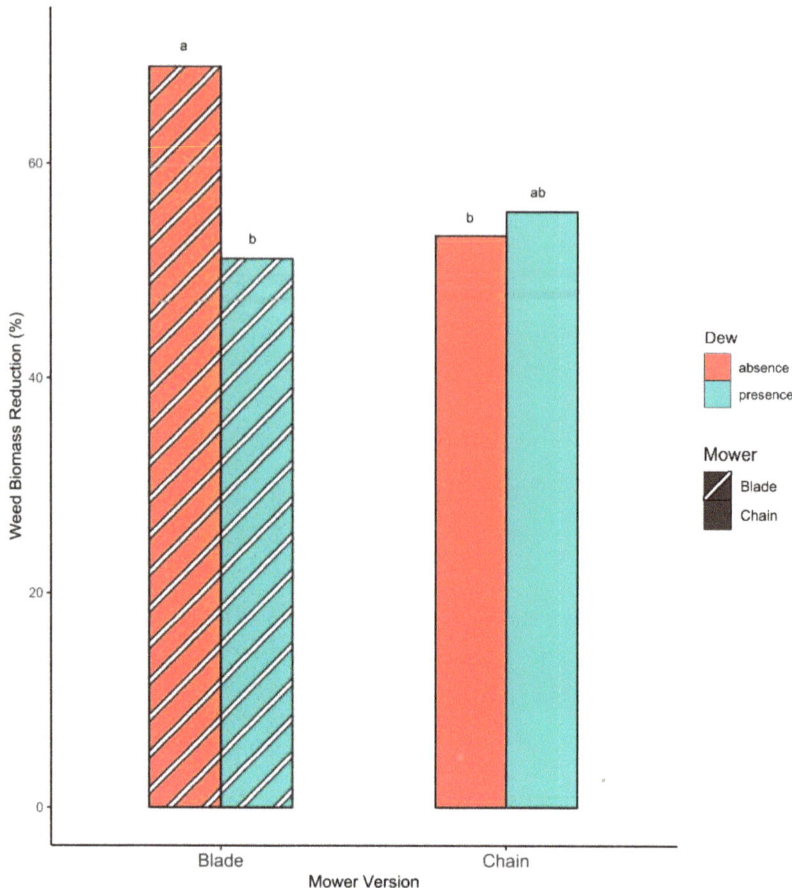

Figure 12. Effect of the interaction between mower version and dew condition on weed biomass reduction. Means denoted by different letters are significantly different at $p < 0.05$ (LSD test).

In general, mowers in the presence of dew obtained a lower reduction in weed cover compared to the performed treatment in the absence of dew ($p < 0.05$), with back-transformed mean values of 28.9% and 37.0%, respectively. The blade mower operating in the absence of dew resulted in a greater reduction in weed cover (back-transformed mean value of 43.0%) compared to the presence of dew (back-transformed mean value of 27.9%) and compared to the chain mower in both the absence and presence of dew (back-transformed mean values of 30.9% and 30.0%, respectively) ($p < 0.05$). The chain mower achieved similar results for this parameter in both dew conditions (Figure 13).

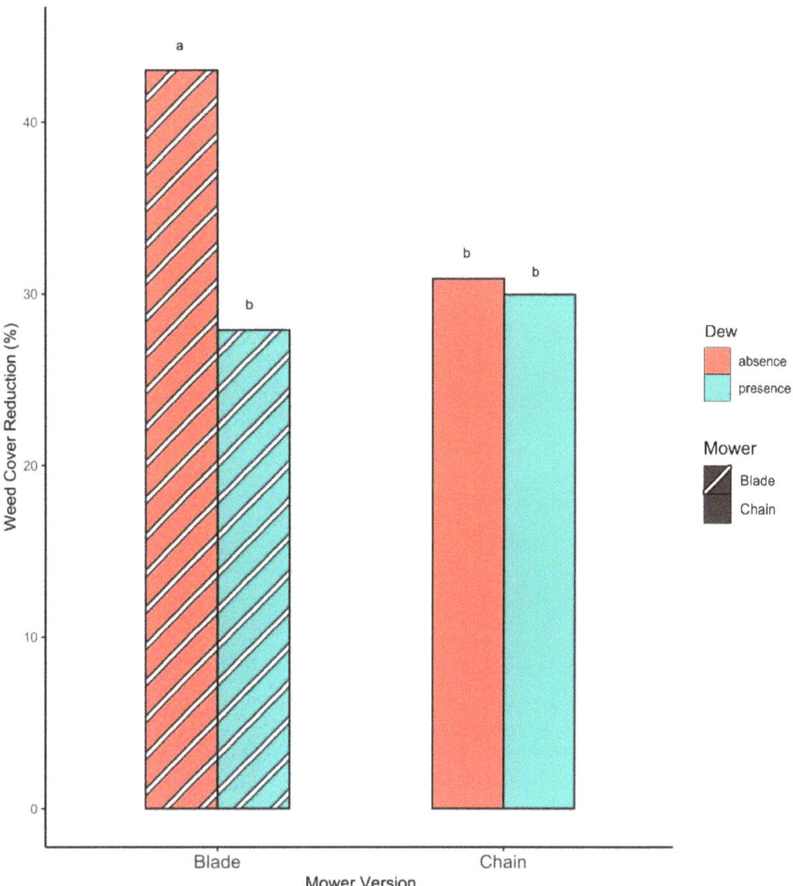

Figure 13. Effect of the interaction between mower version and dew conditions on weed cover reduction. Means denoted by different letters are significantly different at $p < 0.05$ (LSD test).

Dew did not affect weed height reduction within the individual combinations of mower version and setting, with the exception of the chain mower with the Lws + Hrs setting (Figure 14). The chain mower with the Lws + Hrs setting in the presence of dew obtained a higher weed height reduction (53.6%) compared to the absence of dew (34.6%) and compared to the same mower with Hws + Hrs and Lws + Lrs settings in both dew conditions (with weed height reduction of 27.0% and 33.1% for Hws + Hrs setting, and of 31.9% and 33.8% for Lws + Lrs setting in the absence and the presence of dew, respectively). No differences emerged between the chain mower with the Lws + Hrs setting in the presence of dew and the same mower with the Hws + Lrs setting in absence of dew, while the Hws + Lrs setting in the presence of dew achieved a lower weed height reduction (34.8%) ($p < 0.05$). The blade mower with the Hws + Hrs setting in the absence of dew achieved a higher weed height reduction (64.3%) compared to the same mower with the other settings in both dew conditions (ranging from 36.9% to 44.6% in the absence of dew and from 42.9% to 47.8% in the presence of dew) ($p < 0.05$). The blade mower with the Hws + Hrs setting in the absence of dew obtained a higher weed height reduction compared to the chain mower with different settings ($p < 0.05$), with the exception of the chain mower with the Lws + Hrs setting in presence of dew.

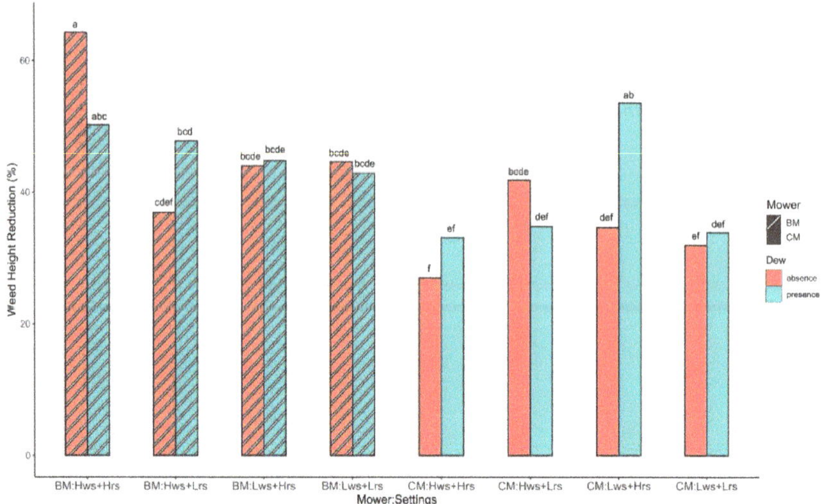

Figure 14. Effect of the interaction between mower version, setting configuration, and dew conditions on weed height reduction. Means denoted by different letters are significantly different at $p < 0.05$ (LSD test). BM—Blade mower; CM—chain mower; Hws—high working speed; Lws—low working speed; Hrs—high rotation speed; Lrs—low rotation speed.

3.3. Field Capacity with Different Working Speed Settings

Both mower versions' operative parameter values with the two different working speed settings are reported in Table 7.

Table 7. Estimated field capacity of the mowers with the two different working speeds.

Performance		Low Working Speed (Lws)	High Working Speed (Hws)
Working speed	$km \cdot h^{-1}$	1.60	2.40
Working width	m	9.00	9.00
Theoretical field capacity	$ha \cdot h^{-1}$	1.44	2.16
Theoretical field time *	$h \cdot ha^{-1}$	0.69	0.46
Total turning time *	$h \cdot ha^{-1}$	0.02	0.02
Field time **	$h \cdot ha^{-1}$	0.71	0.48
Field capacity **	$ha \cdot h^{-1}$	1.40	2.07

* Time required to carry out the under-row mowing treatment in a hypothetical area of 10,000 m^2 (30.00 m wide and 333.33 m long). ** Considering the theoretical field capacity and time and the turning time.

Comparing the mowers' field capacity with different working speed settings, it is possible to observe the higher field capacity of the mower with the high working speed setting with an increase of 47.9% compared to the low working speed setting.

4. Discussion

During the first trial, no major differences emerged between the two mower versions, except for the weed cover reduction, which was greater for the chain mower (92.6% vs. 84.3% for the chain mower and the blade mower, respectively) and the clipping size, which was lower for the blade mower (6.95 cm vs. 8.66 cm for blade mower and chain mower, respectively). The smaller size of clippings produced by the blade mower could be due to the greater sharpness of the cutting tool, which allowed for the cutting of the same plant material several times. In contrast, chains, having a flail action on weed [40], could tend to sweep and then drag the plant material without cutting it several times. Concerning the differences that emerged at the first trial between the tested setting configurations, mowers with the Lws + Hrs setting achieved a higher reduction in weed biomass than machines with Hws + Lrs and Lws + Lrs settings (68.6% vs. 28.7% and 24.9%, respectively). These results confirm those obtained by Yiljep et al. [29], such that as the cutting speed of the tool increases, the cutting efficiency also increases. Lei et al. [30] also stated that the rotation speed of the cutting tool is inversely related to the leakage rate, corresponding to the weight of leakage weeds and the weight of the weeds cut per unit area ratio. On the contrary, the working speed of the under-row mower is positively related to this cutting quality index. However, no significant differences emerged on weed biomass reduction, neither in the comparison between mowers with Hws + Hrs and Lws + Hrs settings, nor between mowers with Hws + Hrs, Hws + Lrs, and Lws + Lrs settings. Moreover, mowers with the Lws + Hrs setting configuration produced smaller clippings (6.77 cm) compared to mowers with other settings (ranging from 7.93 to 8.31 cm). The most likely explanation for this finding is that the low working speed and high rotation speed of the cutting tool allowed for the cutting of the same plant material several times. However, for mulching purposes, it is well-known that a smaller residue is more subject to wind displacement and tends to decompose faster than a coarser material [41].

During the second trial, no differences emerged between the two mower versions in terms of weed cover reduction as in the first trial. The two mowers stood out for the weed height reduction, which was higher for the blade mower compared to the chain mower (46.9% vs. 36.3%, respectively). The blade mower with the Hws + Hrs setting configuration achieved a better result for weed height reduction (57.2%) compared to the other tested settings (ranging from 42.4% to 44.4%). The chain mower with the Lws + Hrs setting obtained a higher weed height reduction (44.1%) compared to the same mower with Hws + Hrs and Lws + Lrs settings (30.0%, and 32.9%, respectively) but no differences emerged between Lws + Hrs and Hws + Lrs settings. The lower weed height reduction obtained by the chain mower may be due to the greater oscillatory movement of the cutting tool associated with the lower stiffness compared to the blade. Furthermore, this could also be due to the different cutting action of a dull cutting tool, such as a chain mower, which tends to mutilate the tip of weeds and therefore present a greater height compared to a sharp one, as other authors observed [42,43].

Mowers with Lws + Hrs and Hws + Hrs settings obtained a greater weed biomass reduction than mowers with the Lws + Lrs setting (67.9% and 59.6%, vs. 44.9%, respectively), while no significant differences emerged neither in the comparison between Lws + Hrs and Hws + Hrs settings, nor between Lws + Hrs, Hws + Hrs, and Hws + Lrs settings. However, results achieved by mowers with different tested settings are in line with results obtained by Pergher et al. [23] for an under-vine mower whose weed biomass reduction values ranged from 40.7% to 59.7%. Furthermore, mowers with Hws + Hrs and Lws + Hrs settings obtained better results than the Lws + Lrs setting for weed cover reduction (40.9% and 36.2% vs. 26.8%, respectively). Regarding the effect of dew on mowers' performances, the blade mower weed cover reduction obtained in the absence of dew (43.0%) was higher than in the presence of dew (27.9%), and compared to the chain mower, both in the absence of dew (30.9%) and in the presence of dew (30.0%). Dew did not affect weed height reduction within the individual combination of mower version and setting, except for the chain mower with the Lws + Hrs setting that achieved a higher weed height reduction in the presence of dew (53.6%) compared to in the absence of dew (34.6%). The blade mower with the Hws + Hrs setting in the absence of dew obtained a higher weed height reduction compared to the chain mower with different settings, with the exception of the chain mower with the Lws + Hrs setting in the presence of dew. Concerning the effect of dew conditions on the mowers' weed biomass reduction, the chain mower seems to be less affected by dew conditions than the blade mower, whose performance was worse in the presence of dew compared to the absence of dew (51.1% vs. 69.0%). No differences emerged between the two mower versions in the presence of dew while in the absence of dew, the blade mower achieved a higher weed biomass reduction than the chain mower (53.3%). However, no clear differences emerged between the two mower versions as the chain mower in the presence of dew achieved similar results of weed biomass reduction compared to the blade mower in the absence of dew.

Overall, no major differences emerged between the two tested mower versions and the chain mower performance was comparable to that of the blade mower, whose reliability is already established [23,30]. This occurred despite the lower sharpness of chains and the context in which the mowing treatments were performed. Indeed, during both trials, the predominant weed species were grasses in the vegetative phase, that is, with a prevalence of leaves over stalks and poorly lignified tissues. These weeds have a lighter inertia compared to weeds in advanced phenological stages with more lignified tissues and thick stalks, making the penetration of the cutting tools into the plant material more difficult [29]. Therefore, the chain mower proved to be a valid tool for under-row weed management in an alley cropping system. The use of a chain mower as a cutting tool works well under challenging conditions, such as clearing operations in forests, allowing the management of grass, shrubs, and coarse plant material. In addition, chains were found to be more versatile than blades as they can be used in the presence of stones where blades tend to become damaged, leading to premature wear and costly repair [33–36,44]. Moreover, blade regrinding can reduce the durability of costly knives [45].

The findings obtained from those trials highlighted that mowers set with the high rotation speed of the cutting tool (2500 rpm) tended to achieve the best results for the analyzed weed control parameters. This is in agreement with Kakahy et al. [46], who found that in passing from a rotation speed of the cutting tool of 1830 to 2553 rpm, despite a slight increase in power consumption of 15%, the effectiveness of mowing performance increased. It is desirable that in the presence of weeds in a more advanced phenological stage, that is, with more lignified tissues and thick stalks, the performance of mowers with both rotation speed settings would have been greater.

In regards to the working speed, the choice of the setting with the high working speed (2.4 km·h^{-1}) results in an increase in the mowers' field capacity of 47.9% compared to the setting with the low working speed. Although the chain mower with the Lws + Hrs setting achieved a slightly higher weed height reduction compared to the Hws + Hrs setting, the choice of the Hws setting involves greater efficiency by halving the time required for mowing. Furthermore, it is possible to state that with the higher working speed setting, the feeler has always worked correctly, allowing the mowing disc to enter the row and exit near the trees without damaging trunks, thus ensuring a proper functioning of the machine. Under-row weed management in alley cropping systems can be more efficiently accomplished by using this type of mower with automatic tree-skipping mechanisms than a tractor-mounted flail mower with an operator-controlled side shift or string trimmers that are commonly employed in these contexts. Lei et al. [30] found that the work efficiency of an obstacle avoiding mowers for under-row weed control in orchards is 4.44 times higher than an operator with a string trimmer and 20 times higher than manual weeding. Time-saving is crucial for users, involving lower operational costs and enabling managers to have time for other management operations.

In order to improve the efficiency of the chain mower, the cutting width could be increased. By adopting a mowing implement with a wider diameter instead of requiring a passage on each side to complete the tree row management, a single passage may be sufficient to guarantee satisfactory weed control.

5. Conclusions

In the present study, the use of the chain mower with automatic tree-skipping mechanism for under-row weed control in an alley cropping system obtained encouraging results. Indeed, the performance of the mower with chains was comparable to that of the blade mower, whose effectiveness is well-established. These findings highlight that the chain mower could be employed as a reliable means for under-row weed control in alley cropping systems, proving to be a valid alternative to the methods that are conventionally applied in these contexts. Concerning the tested settings, the setting with the high working speed and high rotation speed turns out to be the best compromise between effectiveness and efficiency, having obtained a satisfactory weed control (weed biomass reduction of 59.6% and weed cover reduction of 40.9%) and a higher field capacity compared to the setting with the low working speed, with an increase of 47.9%.

Nevertheless, further studies are needed to better understand the cutting mechanisms of chains in relation to different weed communities and weed phenological stages. In order to improve the performance of chain mowers as cutting tools, it would also be useful to test square section chains to evaluate any increase in cutting efficiency with a different cutting edge. Furthermore, since chains are suitable for their use in challenging scenarios, such as stony soils, the applicability of the chain mower for under-row management could be evaluated in vineyards, where the frequent soil-high stoniness hinders the equipment that is conventionally employed for this purpose.

Author Contributions: Conceptualization, M.R., M.F., C.F. and D.A.; methodology, M.F., C.F., L.G., G.R. and D.A.; software, M.F. and L.G.; validation, M.F., C.F., M.S., G.R., D.A. and L.G.T.; formal analysis, M.F.; investigation, L.G.; resources, M.R. and D.A.; data curation, L.G. and M.F.; writing—original draft preparation, L.G.; writing—review and editing, M.F., C.F., M.S., G.R., D.A. and L.G.T.; visualization, M.R. and A.P.; supervision, M.R. and M.F.; project administration, M.R. and M.F.; funding acquisition, M.R. and A.P. All authors have read and agreed to the published version of the manuscript.

Funding: This research received no external funding.

Data Availability Statement: Not applicable.

Acknowledgments: The authors would like to acknowledge the Dondi group for providing the under-row mower and technical support, the Centre for Agri-environmental Research "Enrico Avanzi" for hosting the trials and for the technical support, Andrea Sbrana for the technical support and the use of leaf wetness sensors.

Conflicts of Interest: The authors declare no conflict of interest.

References

1. Velten, S.; Leventon, J.; Jager, N.; Newig, J. What Is Sustainable Agriculture? A Systematic Review. *Sustainability* **2015**, *7*, 7833–7865. [CrossRef]
2. Sarri, D.; Lombardo, S.; Pagliai, A.; Perna, C.; Lisci, R.; De Pascale, V.; Rimediotti, M.; Cencini, G.; Vieri, M. Smart Farming Introduction in Wine Farms: A Systematic Review and a New Proposal. *Sustainability* **2020**, *12*, 7191. [CrossRef]
3. Vieri, M.; Lisci, R.; Rimediotti, M.; Sarri, D. The RHEA-Project Robot for Tree Crops Pesticide Application. *J. Agric. Eng.* **2013**, *44*, 359–362. [CrossRef]
4. USDA Agroforestry. Available online: https://www.usda.gov/topics/forestry/agroforestry (accessed on 26 April 2022).
5. FAO Agroforestry. Available online: https://www.fao.org/forestry/agroforestry/80338/en/#:~{}:text=Definition,spatial%20 arrangement%20or%20temporal%20sequence (accessed on 26 April 2022).
6. Hallema, D.W.; Rousseau, A.N.; Gumiere, S.J.; Périard, Y.; Hiemstra, P.H.; Bouttier, L.; Fossey, M.; Paquette, A.; Cogliastro, A.; Olivier, A. Framework for Studying the Hydrological Impact of Climate Change in an Alley Cropping System. *J. Hydrol.* **2014**, *517*, 547–556. [CrossRef]
7. Biswas, B.; Chakraborty, D.; Timsina, J.; Bhowmick, U.R.; Dhara, P.K.; Ghosh (Lkn), D.K.; Sarkar, A.; Mondal, M.; Adhikary, S.; Kanthal, S.; et al. Agroforestry Offers Multiple Ecosystem Services in Degraded Lateritic Soils. *J. Clean. Prod.* **2022**, *365*, 132768. [CrossRef]
8. Wezel, A.; Casagrande, M.; Celette, F.; Vian, J.-F.; Ferrer, A.; Peigné, J. Agroecological Practices for Sustainable Agriculture. A Review. *Agron. Sustain. Dev.* **2014**, *34*, 1–20. [CrossRef]
9. Schroeder, W.R. Planting and Establishment of Shelterbelts in Humid Severe-Winter Regions. *Agric. Ecosyst. Environ.* **1988**, *22/23*, 441–463. [CrossRef]
10. Gulden, R.H.; Shirtliffe, S.J. Weed Seed Banks: Biology and Management. *Prairie Soils Crops J.* **2009**, *2*, 46–52.
11. Lauer, D.K.; Glover, G.R. Stand Level Pine Response to Occupancy of Woody Shrub and Herbaceous Vegetation. *Can. J. For. Res.* **1999**, *29*, 979–984. [CrossRef]
12. Schroeder, W.R.; Naeem, H. Effect of Weed Control Methods on Growth of Five Temperate Agroforestry Tree Species in Saskatchewan. *For. Chron.* **2017**, *93*, 271–281. [CrossRef]

13. Boinot, S.; Fried, G.; Storkey, J.; Metcalfe, H.; Barkaoui, K.; Lauri, P.É.; Mézière, D. Alley Cropping Agroforestry Systems: Reservoirs for Weeds or Refugia for Plant Diversity? *Agric. Ecosyst. Environ.* **2019**, *284*, 106584. [CrossRef]

14. Mézière, D.; Boinot, S.; de Waal, L.; Cadet, E.; Fried, G. Arable Weeds in Alley Cropping Agroforestry Systems—Results of a First Year Survey. In Proceedings of the 3rd European Agroforestry Conference, Montpellier SuperAgro, France, 23–25 May 2016.

15. Burgess, P.; Incoll, L.D.; Hart, B.; Beaton, A.; Piper, R.W.; Seymour, I.; Reynolds, F.H.; Wright, C.; Pilbeam, D.J.; Graves, A.R. *The Impact of Silvoarable Agroforestry with Poplar on Farm Profitability and Biological Diversity. Final Report to DEFRA*; Cranfield University, University of Leeds, Royal Agricultural College: London, UK, 2003.

16. Vityi, A.; Schettrer Péter, A.; Kiss-Sziget, N.; Marosvölgyi, B. *I Benefici Della Pacciamatura con Biomassa Erbacea*; AGFORWARD Agroforestry for Europe; European Union: Brussels, Belgium, 2017.

17. Jones, A.; Fortier, J.; Gagnon, D.; Truax, B. Trading Tree Growth for Soil Degradation: Effects at 10 Years of Black Plastic Mulch on Fine Roots, Earthworms, Organic Matter and Nitrate in a Multi-Species Riparian Buffer. *Trees For. People* **2020**, *2*, 100032. [CrossRef]

18. Ghouse, P. Mulching: Materials, Advantages and Crop Production. In *Protected Cultivation and Smart Agriculture*, 1st ed.; Maitra, S., Gaikwad, D.J., Shankar, T., Eds.; New Delhi Publishers: New Delhi, India, 2020; pp. 55–66. ISBN 978-81-948993-2-7.

19. Hammermeister, A.M. Organic Weed Management in Perennial Fruits. *Sci. Hortic.* **2016**, *208*, 28–42. [CrossRef]

20. Curran, W.S.; Lingenfelter, D.D.; Garling, L. An Introduction to Weed Management for Conservation Tillage Systems. *Pennstate Coop. Ext. Coll. Agric. Sci.* **2009**, *2*, 1–8.

21. Sheley, R. Mowing to Manage Noxious Weeds. *Montguide Mont. State Univ. Ext.* **2017**, *0517SA*, 1–2.

22. Chen, D.; Wang, Y.; Zhang, X.; Wei, X.; Duan, X.; Muhammad, S. Understory Mowing Controls Soil Drying in a Rainfed Jujube Agroforestry System in the Loess Plateau. *Agric. Water Manag.* **2021**, *246*, 106703. [CrossRef]

23. Pergher, G.; Gubiani, R.; Mainardis, M. Field Testing of a Biomass-Fueled Flamer for In-Row Weed Control in the Vineyard. *Agriculture* **2019**, *9*, 210. [CrossRef]

24. Vityi, A.; Marosvölgyi, B.; Kiss, A.; Schetterer, P. *System Report: Alley Cropping in Hungary*; AGFORWARD Agroforestry for Europe; European Union: Brussels, Belgium, 2015.

25. Radbourne, A.; Reinsch, S.; Norris, D.; Critchley, W.; Studer, R.M. Alley Cropping—Agroforestry [France]. Available online: https://qcat.wocat.net/en/wocat/technologies/view/technologies_5645/ (accessed on 4 May 2022).

26. Nobili BE Mulcher. Available online: https://www.nobili.com/be/sb0f2deee (accessed on 7 May 2022).

27. Deleks Sideshift Flail Mower, Offset Side Mulcher for Tractors LEO-140. Available online: https://www.deleks.it/en/p/106/sideshift-flail-mower-offset-side-mulcher-for-tractors-deleks (accessed on 7 May 2022).

28. McRandal, D.M.; McNulty, P.B. Impact Cutting Behaviour of Forage Crops I. Mathematical Models and Laboratory Tests. *J. Agric. Eng. Res.* **1978**, *23*, 313–328. [CrossRef]

29. Yiljep, Y.D.; Mohammed, U.S. Effect of Knife Velocity on Cutting Energy and Efficiency during Impact Cutting of Sorghum Stalk. *Agric. Eng. Int. CIGR E J.* **2005**, *7*, 1–10.

30. Lei, X.; Qi, Y.; Zeng, J.; Yuan, Q.; Chang, Y.; Lyu, X. Development of Unilateral Obstacle-Avoiding Mower for Y-Trellis Pear Orchard. *Int. J. Agric. Biol. Eng.* **2022**, *15*, 71–78. [CrossRef]

31. Stanisavljević, R.; Vuković, A.; Petrović, D.V.; Radojević, R.L.; Barać, S.; Mileusnić, Z.; Tadić, V. Efficiency of Alfalfa Hay Mowing Machines under the Dryland Conditions. *Teh. Vjesn. Tech. Gaz.* **2021**, *28*, 1503–1510. [CrossRef]

32. Bertucci, M.; Boyd, J.; Patton, A. Mowing Your Lawn. *Div. Agric. Res. Ext. Univ. Ark. Syst.* **2018**, *FSA6023*, 1–6.

33. Fransgard Chain Mower FKR Series. Available online: https://www.fransgard.dk/en-gb/gt210/fkr-150 (accessed on 9 May 2022).

34. Wessex International Forestry Rotary Cutter SM-66. Available online: https://www.wessexintl.com/machines/compact-tractor-attachments/rotary-slasher/sm-66/ (accessed on 9 May 2022).

35. Ventura Forestry Machines DLB 410—DADU—Side Belt Mower. Available online: https://www.venturamaq.com/en/mowers-of-blades-and-chains/dlb-410-dadu-side-belt-mower/ (accessed on 9 May 2022).

36. Kellfri Chain Mulcher Three-Point Linkage. Available online: https://www.kellfri.com/agriculture/grassland-machinery/chain-mulchers/chain-mulcher-three-point-linkage (accessed on 9 May 2022).

37. Patrignani, A.; Ochsner, T.E. Canopeo: A Powerful New Tool for Measuring Fractional Green Canopy Cover. *Agron. J.* **2015**, *107*, 2312–2320. [CrossRef]

38. R Core Team. *R: A Language and Environment for Statistical Computing*; R Foundation for Statistical Computing: Vienna, Austria, 2021.

39. de Mendiburu, F. *Agricolae: Statistical Procedures for Agricultural Research*; R Package Version 1.3-5; 2021. Available online: https://cran.r-project.org/web//packages/agricolae/agricolae.pdf (accessed on 14 September 2022).

40. Limberger, F.W. Grass Trimmer. U.S. Patent 2,676,448, Filed 11 September 1950, and Issued 27 April 1954. Available online: https://www.freepatentsonline.com/2676448.pdf (accessed on 14 September 2022).

41. Guerra, B.; Steenwerth, K. Influence of Floor Management Technique on Grapevine Growth, Disease Pressure, and Juice and Wine Composition: A Review. *Am. J. Enol. Vitic.* **2012**, *63*, 149–164. [CrossRef]

42. Beard, J.B. *Turfgrass: Science and Culture*, 1st ed.; Prentice Hall. Inc.: Engelwood Cliffs, NJ, USA, 1973.

43. Steinegger, D.H.; Shearman, R.C.; Riordan, T.P.; Kinbacher, E.J. Mower Blade Sharpness Effects on Turf. *Agron. J.* **1983**, *75*, 479–480.

44. Tolosana, E.; Bados, R.; Laina, R.; Bacescu, N.M.; de la Fuente, T. Forest Biomass Collection from Systematic Mulching on Post-Fire Pine Regeneration with BioBaler WB55: Productivity, Cost and Comparison with a Conventional Treatment. *Forests* **2021**, *12*, 979. [CrossRef]

45. Jahr, A.; Bongartz, R.; Sim, H.; Pillmann, A. A Novel Test Method to Determine the Wear Resistance of Agricultural Cutting Tools with an Analyzing Method Based on an Ellipse-Fit Algorithm. In Proceedings of the European Symposium on Friction, Wear, and Wear Protection, Karlsruhe, Germany, 6–8 May 2014.

46. Kakahy, A.N.N.; Ahmad, D.; Akhir, M.D.; Sulaiman, S.; Ishak, A. Effects of Rotary Mower Blade Cutting Angles on the Pulverization of Sweet Potato Vine. In Proceedings of the 2nd International Conference on Agricultural and Food Engineering, CAFEi2014, Kuala Lumpur, Malaysia, 1–3 December 2014.

 *agronomy*

Communication

Continuous Mowing for *Erigeron canadensis* L. Control in Vineyards

Andrea Peruzzi, Lorenzo Gagliardi, Marco Fontanelli, Christian Frasconi, Michele Raffaelli and Mino Sportelli *

Department of Agriculture, Food and Environment, University of Pisa, Via del Borghetto 80, 56124 Pisa, Italy
* Correspondence: mino.sportelli@phd.unipi.it

Abstract: *Erigeron canadensis* L. directly competes with vines for nutrients, light, and water, and its management represents a challenge, especially under a vineyard trellis. Conventional weed control in the under-trellis area is achieved by cultivation or multiple herbicides applications, thus leading to relevant environmental issues. For this reason, several eco-friendly or nature-based weed control strategies such as the use of cover crops (CC) that become more relevant in last years. A two-year trial was conducted on a vineyard aimed at evaluating the effect of CC (sown both inter-rows and under-trellis) managed with an autonomous mower (AM) on *E. canadensis* under trellis control. The combination of CC and AM provided an *E. canadensis* reduction between 61 and 84% compared to conventional management. The AM work when managing a spontaneous cover provided a density reduction of 26%. Moreover, an analysis of the trampling effect of the AM on the vineyard floor and *E. canadensis* density was conducted.

Keywords: living mulch; mechanical weed control; autonomous machines; trampling analysis

Citation: Peruzzi, A.; Gagliardi, L.; Fontanelli, M.; Frasconi, C.; Raffaelli, M.; Sportelli, M. Continuous Mowing for *Erigeron canadensis* L. Control in Vineyards. *Agronomy* **2023**, *13*, 409. https://doi.org/10.3390/agronomy13020409

Academic Editor: Anestis Karkanis

Received: 12 December 2022
Revised: 23 January 2023
Accepted: 29 January 2023
Published: 30 January 2023

1. Introduction

Erigeron canadensis L. is a winter or summer annual weed belonging to the Asteraceae family [1]. It is native to North America and is widespread in several countries in Africa, Asia-Pacific, and Europe [2]. In some cases, *E. canadensis* may develop resistance to herbicides such as glyphosate [3], paraquat [4], and triazines [5]; however, it does not represent a troublesome weed in agricultural systems where tillage is provided [6]. *E. canadensis* seeds germinability significantly decreases as burial depth increases [7], thus is considered a major weed in contexts such as roadsides, berms, fallow fields, no-till and perennial cropping systems (i.e., vineyards) [8,9]. In vineyards *E. canadensis* directly compete with vines for nutrients, light, and water, causing a reduction in vines' vigor [10]. Holm et al. [11] reported a yield reduction of 28% due to *E. canadensis* infestation. In the Mediterranean basin, vineyard weed control is conventionally fulfilled by soil tillage in the inter-rows and under vines trellis [12], while herbicides applications are usually provided under vines-trellis [13]. However, environmental issues derived from soil tillage and herbicide application serve as a driving force to find more sustainable practices. A permanent soil cover by means of cover crops (CC) has shown potential as a sustainable weed control strategy that also brings a series of ecosystem services (i.e., soil and water quality improvement and biodiversity enhancement) [14]. Usually, permanent CCs are sown in vineyards inter-rows while no vegetal cover under vines-trellis is preferred to avoid competition When selecting CC species for a complete floor cover competition level, height and their management should be considered [15]. A planned under-trellis CC management has been shown to significantly improve weed control efficacy [16]. In addition, operations efficiency can be improved if small autonomous machines are employed [17]. Magni et al. [18] tested an autonomous mower (AM) for grass CC complete cover in a vineyard funding this solution advantageous in terms of power consumption and CO_2 emissions. AM's small size and intense mowing perfectly match weeds and CC management in vineyards. The aim of this trial was to evaluate how the combination between CC species and continuous mowing management could improve vineyard sustainability in terms of noxious weed control. The present study provides additional evidence to that presented by Sportelli et al. [19]. This study was based on preliminary observations of intensive mowing effects on *E. canadensis* in vineyards inter-rows and under-trellis.

2. Materials and Methods

2.1. Experimental Design

A two years trial was conducted at the Tuscany Association of Viticulture Producers (Tos.Sco.Vit) in S. Piero a Grado, Pisa, Italy (43°39′ N, 10°20′ E). The investigation was carried out on a *Vitis vinifera* L. cv. Sangiovese N. vineyard, was established in 2004 and arranged in a 0.9 × 2.5 m planting layout with rows of 40 vine plants. Vineyard floor conventional management provided multiple inter-row flail mowing of spontaneous species and post-emergence non-selective herbicide under-trellis applications. In this study, four vineyard floor management systems were compared for their capacity of *E. canadensis* control. All the management systems provided CC or resident species managed with repeated mowing treatments so as to obtain a living mulch (LM). Thus, the four-floor management systems were: living mulch 1 managed with an AM (LM1-AM), living mulch 2 managed with an AM (LM2-AM), living mulch 3 managed with an AM (LM3-AM), and living mulch 3 conventionally managed (LM3-CM). Each management system was replicated four times (16 plots in total) and each plot measured 30 m^2 (12 × 2.5 m). Experimental plots managed by AM have been settled in a 740 m^2 area of the vineyard, including the AM plots and a cushion area, to ease AM base station and boundary wire installation. The cushion area consisted of a buffer area covered by resident species and the first vines row. In the next four rows, three plots per row were defined by maintaining the vine row in the middle and measuring 1.25 m on both sides. Neighboring plots shared their edges between treatments and the three AM different treatments were randomly placed within the row. A Husqvarna AM 535 AWD (Husqvarna, Stockholm, Sweden) was used to fulfill autonomous mowing in AM plots. AM was set to work at a mowing height of 5 cm, 5 days per week and 5 h per day (charging time included). For both years AM was employed as a management strategy form May to November. The four replicates of CM experimental plots were defined in four different rows of the vineyard, one per each row (so as to obtain the four replication). Additional details on CM and AM plots operations are reported in Table 1, figures of the experimental design can be found in Sportelli et al. [19]. LM1 and LM2 sowing was carried out in November 2018 and November 2019 (Table 1). In LM3 plots, floor cover consisted of growing resident species. A total of 35 different resident species were found and major species were *Erodium cicutarium* (L.) L'Hér., *Malva sylvestris* L., *Matricaria chamomilla* L., *Schedonorus arundinaceus* (Schreb.) Dumort, *Veronica persica* Poir, *Bellis perennis* L., *Capsella bursa-pastoris* (L.) Medik., *Poa annua* L., *Stellaria media* (L.) Vill., *Geranium molle* L., *Erigeron canadensis* L., *Symphyotrichum squamatum* (Spreng.) G.L. Nesom, *Taraxacum officinale* Weber, *Euphorbia prostrata* Aiton, *Digitaria sanguinalis* (L.) Scop., *Cynodon dactylon* (L.) Pers, *Portulaca oleracea* L. and *Paspalum dilatatum* Poir. According to [20,21], only *V. persica*, *E. canadensis*, *S. squamatum* and *P. dilatatum* were considered invasive and aggressive species. However, this study focuses only on *E. canadensis* since it was considered the major threat from vineyard managers and it showed promising results from preliminary tests.

Table 1. Trial operations program for 2019 and 2020.

Operation	Number of Operations	Product/Equipment	Rate	Plots			
				LM1–AM	LM2–AM	LM3–AM	LM3–CM
2019							
Herbicide application	2	Glyphos Dakar	1.8 kg ha^{-1}	-	-	-	✓
Conventional mowing	3	Celli TCB/S Mulcher	n.a.	-	-	-	✓
Autonomous mowing	132	Husqvarna automower 535 AWD	5 h d^{-1}	✓	✓	✓	-
Living mulch hand sowing *	1	-Living mulch 1 (LM1) blend: *Lolium perenne* L. cv Tetragreen (50%) *L. perenne* L. cv Dasher 3 (50%). -Living mulch 2 (LM2) blend: *Trifolium repens* L. cv Huia (95%) *T. repens* L. cv Pertina (5%).	LM1: 50·g·m^{-2} LM2: 20 g m^{-2}	✓	✓	-	-
2020							
Herbicide application	2	Glyphos Ultra	7 kg ha^{-1}	-	-	-	✓
Conventional mowing	4	Celli TCB/S Mulcher	n.a.	-	-	-	✓
Autonomous mowing	131	Husqvarna automower 535 AWD	5 h d^{-1}	✓	✓	✓	-

* Living mulch hand sowing operations were analogus for year 2018.

2.2. Assessments

During both 2019 and 2020, from May to October, *E. canadensis* plant density was established in the inter-row (IR) and under-trellis (UT). Plants were manually counted from the four fixed quadrats of 0.25 m^2 (50 × 50 cm) positioned in the IR and four UT. Plant numbers from the four quadrants in the same position were summed to obtain the plant density values in one m^2. Two Emlid Reach RTK (Emlid Ltd., Hong Kong) devices [22] were used to record AM operative performances. Subsequently, recorded data were processed with a custom-built software "Robot mower tracking data calculator" (Qprel srl, Pistoia, Italy) version 1.8.0.0 [23] to compute the number of times the AM passed on the same position. This last parameter was used to assess the AM trampling effect on the inter-rows and under the vineyard trellis. A total of 50 measurements for each position were carried out in every repetition. At the end of the trial, 400 measurements for each position were used to estimate the autonomous mower trampling action in the studied vineyard.

2.3. Statistical Analysis

Data were analyzed using the statistical software SPSS (IBM Corp, Armonk, NY, USA). Repeated-measures analysis was performed to evaluate how treatments and positions affected *E. canadensis* plant density. Data were fitted into a generalized linear mixed model (GENLINMIXED) with a Poisson distribution and a log link function. Months and years were considered as repeated factors. Treatments, position, and the interaction between treatments and positions were considered fixed effects, while the experimental plot repetition was included as a random effect. The comparisons between pairs of estimated values were computed by estimating the 95% confidence interval of the difference between the values based on Bonferroni's test (Equation (1)):

$$CI\ (difference) = (x_1 - x_2) \pm 1.96 \sqrt{(SE_{x1})^2 + (SE_{x2})^2} \tag{1}$$

where (x_1) is the mean of the first value, (x_2) is the mean of the second value, (SE_{x1}) is the standard error of (x_1), and (SE_{x2}) is the standard error of (x_2). If the resulting 95% confidence interval (CI) of the difference between values did not cross the value 0, the null hypothesis that the compared values were not different was rejected.

3. Results

Repeated-measures analysis revealed treatment ($p < 0.001$), position ($p < 0.001$), and their interaction ($p < 0.01$) significantly affected *E. canadensis* plant density. Pairwise comparisons and mean separation results have been reported in Table 2.

Over the two years, the highest *E. canadensis* plant density was recorded under a vineyard trellis of the LM3-CM plots with an average of 7.14 (±0.37) plants m^{-2}. The lowest plant density, instead, was obtained on the inter-row of the LM2-AM plots with an average of 0.1 (±0.04) plants m^{-2} (Figure 1). In general, *E. canadensis* plant density was higher under a vineyard trellis with 4.2 (±0.15) plants m^{-2} compared to the inter-rows 0.5 (±0.07) plants m^{-2}.

Within inter-rows areas, no differences emerged between conventional management and AM areas provided with a spontaneous cover. In contrast, the presence of LM significantly reduced *E. canadensis* plant density in the inter-rows. Among the two LM adopted in this trial, *T. repens* resulted in a lower plant density compared to *L. perenne* ($p < 0.001$). Indeed, *E. canadensis* density reduction resulted in approximately 61% for *L. perenne* LM and 84% for *T. repens*. LM significantly boosted the weed control effect also under vineyard trellis compared to spontaneous cover plots ($p < 0.001$), while no differences were detected between the two LM typologies.

The trampling effect in terms of the number of times the AM passed in the same position was analyzed both in the inter-rows and under the vineyard trellis (Figure 2).

Table 2. Results of pairwise comparisons for *Erigeron canadensis* plant density in function of treatments (LM1–AM, LM2–AM, LM3–AM, LM3–CM) and positions (IR and UT). LM1–AM: living mulch 1 autonomously mowed; LM2–AM: living mulch 2 autonomously mowed; LM3–AM: living mulch 3 autonomously mowed; LM3–CM: living mulch 3 conventionally managed. IR: inter-rows, UT: Under vineyards trellis. Mean separations have been carried out separately for positions.

Position	Treatment	Plant Density Mean Values (Plant m^{-2})	Standard Error	95% CI	
				Lower Bond	Upper Bond
IR	LM1–AM	0.48 b [1]	0.09	0.32	0.70
	LM2–AM	0.09 c	0.04	0.03	0.22
	LM3–AM	1.02 a	0.14	0.78	1.33
	LM3–CM	1.38 a	0.16	1.09	1.74
UT	LM1–AM	3.07 c	0.24	2.63	3.58
	LM2–AM	2.70 c	0.23	2.29	3.18
	LM3–AM	5.29 b	0.32	4.70	5.95
	LM3–CM	7.14 a	0.37	6.45	7.90

[1] Different letters within the same position group (IR or UT) represent mean values significantly different based con 95% Confidence Interval (CI).

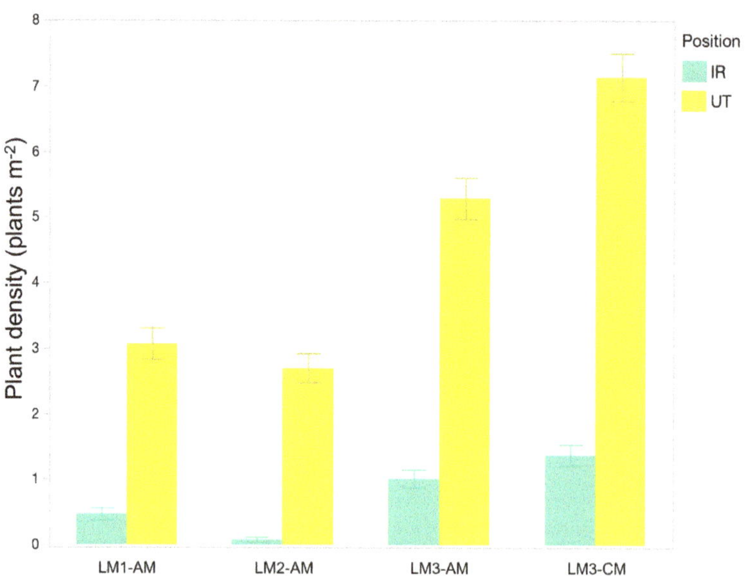

Figure 1. *Erigeron canadensis* plant density mean values during the two years trial in function of treatments (LM1–AM, LM2–AM, LM3–AM, LM3–CM) and positions (IR and UT). LM1–AM: living mulch 1 autonomously mowed; LM2–AM: living mulch 2 autonomously mowed; LM3–AM: living mulch 3 autonomously mowed; LM3–CM: living mulch 3 conventionally managed. IR: inter-rows, UT: Under vineyards trellis. Error bar represents the upper and lower limits of the Standard error.

Trampling effect of the AM managed plot resulted higher under vineyards trellis with an average of 24 passages (ranging from 7 to 53) compared to the inter-row area with an average of 4 passages (ranging from 0 to 14). The number of passages were counted after 5 h of work.

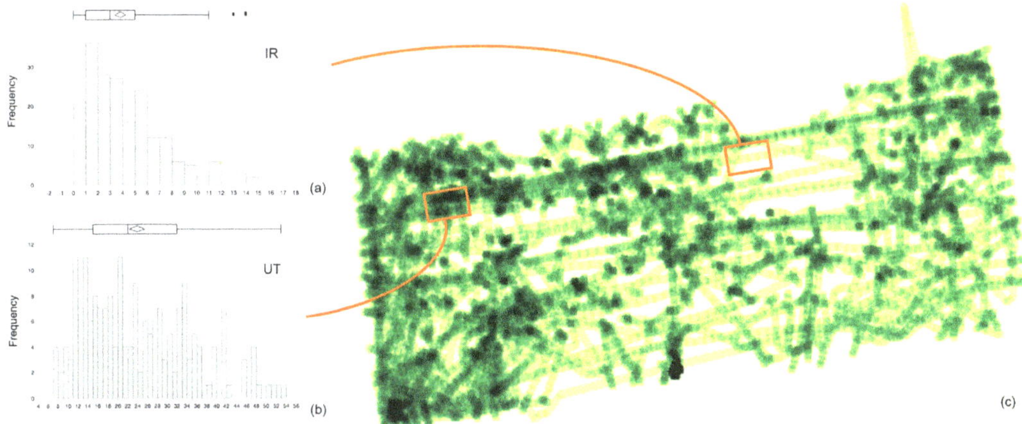

Figure 2. Number of passages distributions on vineyards positions. (**a**) Number of passages distribution on the inter-rows area. (**b**) Number of passages distribution under vineyard trellis. (c) An example of trampling distribution on the whole experimental area managed by autonomous mowers.

4. Discussion

The aim of this trial was to evaluate the effect of different LM managed by an autonomous mower on *E. canadensis*. Plant density was recorded over the 2 years trial and in general, a higher *E. canadensis* plant density was detected in plots conventionally managed compared to the AM-managed plots and under vineyards trellis (4.2 plants m^{-2}) compared to the inter-rows (0.5 plants m^{-2}). Moreover, *E. canadensis* can grow inside the vine canopy and up to two meters high [17], thus causing operations and product quality issues. According to these findings, the presence of an LM showed to be particularly effective for *E. canadensis* suppression. Sanguankeo et al. [24] found that CC can suppress several major weeds in vineyards, including *E. canadensis*, by up to 48% compared with herbicides application and cultivation weed control methods. Wallace et al. [25] also studied the effect of different CC on *E. canadensis* density and obtained a reduction due to CC ranging from 52% to 86%. A reduction of 26% was obtained in plots with spontaneous cover and managed with AM, thus resulting in a significantly lower number of plants growing in LM3-CM plots ($p < 0.001$). When compared to LM3-CM plots, AM-managed plots maintained a 5 cm high floor canopy and a lower plant dry biomass (results from Sportelli et al. [19]). Providing this CC or LM management may be beneficial in maintaining an optimal balance between vines, weeds, and CC [26] with low energy requirements [27,28]. AM management was characterized by a high overlapping; however, differences were detected between positions (Figure 2). Indeed, the AM achieved an average of 24 passages under a vineyard trellis and compared to the inter-row area with an average of 4 passages. The higher trampling under the vineyard trellis occurred because the AM became stuck in this area and because of its random working pattern, multiple maneuvers were required to reach inter-rows. Despite AM trampling resulting in up to six times higher under vineyards trellis compared to inter-rows, this did not affect *E. canadensis* plant density. However, a few things need to be considered: (i) AM moved following random trajectories, thus reducing its working efficiency when employed in contexts with a large number of obstacles [29]. (ii) Due to safety regulations, AM working width (24 cm) is significantly smaller than its cowling width (55 cm) and weeds can develop close to the trunks of the vines because of the blades and cowling distance gap. (iii) AM hit the vines' trunks several times before exiting the under-trellis area. Whenever the AM hit an obstacle, it stops moving, and it stops the cutting disk as well; thus, the majority of weed control under the trellis was exploited by AM trampling. (iv) In general, plants that can stand trampling cope with this stress with radial growth. Dumitrașcu et al. [30] studied the effect of different levels of trampling in various herbaceous species and found an increase in the number of lateral branches and the number of shoots that reached fruition in *E. canadensis*. The combination of continuous mowing and intense trampling obtained in AM plots together with the effect of LM provided a satisfactory *E. canadensis* control. These findings confirm that small and light autonomous machines have the potential to enhance weed control and overall sustainability [31]. Eventually, a good balance between tractor-implements combination [32] together with small and light autonomous machines will improve the top-soil compaction distribution and the overall vineyard sustainability.

Author Contributions: Conceptualization, M.S. and A.P.; methodology, M.S.; software, M.S.; validation, C.F., M.R. and L.G.; formal analysis, M.S., investigation, M.S.; resources, M.F.; data curation, M.S.; writing—original draft preparation, L.G.; writing—review and editing, M.S.; visualization, M.S.; supervision, M.R.; project administration, M.F.; funding acquisition, A.P. All authors have read and agreed to the published version of the manuscript.

Funding: This research received no external funding.

Institutional Review Board Statement: Not applicable.

Informed Consent Statement: Not applicable.

Data Availability Statement: Not applicable.

Acknowledgments: The authors would like to thank the Husqvarna group for providing the autonomous mower and technical support; the Tuscany Association of Viticulture Producers (Tos.Sco.Vit) for their technical support.

Conflicts of Interest: The authors declare no conflict of interest.

References

1. Buhler, D.D.; Owen, M.D.K. Emergence and Survival of Horseweed (*Conyza canadensis*). *Weed Sci.* **1997**, *45*, 98–101. [CrossRef]
2. Sansom, M.; Saborido, A.A.; Dubois, M. Control of *Conyza* Spp. with Glyphosate—A Review of the Situation in Europe. *Plant Prot. Sci.* **2013**, *49*, 44–53. [CrossRef]
3. Schramski, J.A.; Sprague, C.L.; Patterson, E.L. Environmental Cues Affecting Horseweed (*Conyza canadensis*) Growth Types and Their Sensitivity to Glyphosate. *Weed Sci.* **2021**, *69*, 412–421. [CrossRef]
4. Smisek, A.; Doucet, C.; Jones, M.; Weaver, S. Paraquat Resistance in Horseweed (*Conyza canadensis*) and Virginia Pepperweed (*Lepidium virginicum*) from Essex County, Ontario. *Weed Sci.* **1998**, *46*, 200–204. [CrossRef]
5. Lehoczki, E.; Laskay, G.; Pölös, E.; Mikulás, J. Resistance to Triazine Herbicides in Horseweed (*Conyza canadensis*). *Weed Sci.* **1984**, *32*, 669–674. [CrossRef]
6. Brown, S.M.; Whitwell, T. Influence of Tillage on Horseweed, *Conyza canadensis*. *Weed Technol.* **1988**, *2*, 269–270. [CrossRef]
7. Nandula, V.K.; Eubank, T.W.; Poston, D.H.; Koger, C.H.; Reddy, K.N. Factors Affecting Germination of Horseweed (*Conyza canadensis*). *Weed Sci.* **2006**, *54*, 898–902. [CrossRef]
8. Alcorta, M.; Fidelibus, M.W.; Steenwerth, K.L.; Shrestha, A. Competitive Effects of Glyphosate-Resistant and Glyphosate-Susceptible Horseweed (*Conyza canadensis*) on Young Grapevines (*Vitis vinifera*). *Weed Sci.* **2011**, *59*, 489–494. [CrossRef]
9. Alcorta, M.; Fidelibus, M.W.; Steenwerth, K.L.; Shrestha, A. Effect of Vineyard Row Orientation on Growth and Phenology of Glyphosate-Resistant and Glyphosate-Susceptible Horseweed (*Conyza canadensis*). *Weed Sci.* **2011**, *59*, 55–60. [CrossRef]
10. Shrestha, A.; Hembree, K.; Wright, S. *Biology and Management of Horseweed and Hairy Fleabane in California*; University of California. Oakland, CA, USA, 2008; pp. 1–9. [CrossRef]
11. Holm, L.G.; Plucknett, D.L.; Pancho, J.V.; Herberger, J.P. *The World's Worst Weeds: Distribution and Biology*; University Press of Hawaii: Honolulu, HI, USA, 1977.
12. Travlos, I.S.; Bilalis, D.J.; Katsenios, N.; De Prado, R. Sustainable Weed Control in Vineyards. In *Weed Control*; CRC Press: Boca Raton, FL, USA, 2018; pp. 526–542.
13. Guerra, J.G.; Cabello, F.; Fernández-Quintanilla, C.; Peña, J.M.; Dorado, J. How Weed Management Influence Plant Community Composition, Taxonomic Diversity and Crop Yield: A Long-Term Study in a Mediterranean Vineyard. *Agric. Ecosyst. Environ.* **2022**, *326*, 107816. [CrossRef]
14. Antichi, D.; Sbrana, M.; Martelloni, L.; Chehade, L.A.; Fontanelli, M.; Raffaelli, M.; Mazzoncini, M.; Peruzzi, A.; Frasconi, C. Agronomic Performances of Organic Field Vegetables Managed with Conservation Agriculture Techniques: A Study from Central Italy. *Agronomy* **2019**, *9*, 810. [CrossRef]
15. Vanden Heuvel, J.; Centinari, M. Under-Vine Vegetation Mitigates the Impacts of Excessive Precipitation in Vineyards. *Front. Plant Sci.* **2021**, *12*, 1542. [CrossRef] [PubMed]

16. Pergher, G.; Gubiani, R.; Mainardis, M. Field Testing of a Biomass-Fueled Flamer for in-Row Weed Control in the Vineyard. *Agriculture* **2019**, *9*, 210. [CrossRef]
17. Bechar, A.; Vigneault, C. Agricultural Robots for Field Operations. Part 2: Operations and Systems. *Biosyst. Eng.* **2017**, *153*, 110–128. [CrossRef]
18. Magni, S.; Sportelli, M.; Grossi, N.; Volterrani, M.; Minelli, A.; Pirchio, M.; Fontanelli, M.; Frasconi, C.; Gaetani, M.; Martelloni, L.; et al. Autonomous Mowing and Turf-Type Bermudagrass as Innovations for an Environment-Friendly Floor Management of a Vineyard in Coastal Tuscany. *Agriculture* **2020**, *10*, 189. [CrossRef]
19. Sportelli, M.; Frasconi, C.; Fontanelli, M.; Pirchio, M.; Raffaelli, M.; Magni, S.; Caturegli, L.; Volterrani, M.; Mainardi, M.; Peruzzi, A. Autonomous Mowing and Complete Floor Cover for Weed Control in Vineyards. *Agronomy* **2021**, *11*, 538. [CrossRef]
20. Catizone, P.; Zanin, G. *Malerbologia*, 1st ed.; Patròn Editore: Bologna, Italy, 2001.
21. Portal to the Flora of Italy. Available online: https://dryades.units.it/floritaly/index.php (accessed on 23 January 2023).
22. Emlid. A Reach RTK Docs. Specification. Available online: https://docs.emlid.com/reach/specs/ (accessed on 28 January 2023).
23. Sportelli, M.; Luglio, S.M.; Caturegli, L.; Pirchio, M.; Magni, S.; Volterrani, M.; Frasconi, C.; Raffaelli, M.; Peruzzi, A.; Gagliardi, L.; et al. Trampling Analysis of Autonomous Mowers: Implications on Garden Designs. *AgriEngineering* **2022**, *4*, 592–605. [CrossRef]
24. Sanguankeo, P.P.; Leon, R.G.; Malone, J. Impact of Weed Management Practices on Grapevine Growth and Yield Components. *Weed Sci.* **2009**, *57*, 103–107. [CrossRef]
25. Wallace, J.M.; Curran, W.S.; Mortensen, D.A. Cover Crop Effects on Horseweed (*Erigeron canadensis*) Density and Size Inequality at the Time of Herbicide Exposure. *Weed Sci.* **2019**, *67*, 327–338. [CrossRef]
26. MacLaren, C.; Bennett, J.; Dehnen-Schmutz, K. Management Practices Influence the Competitive Potential of Weed Communities and Their Value to Biodiversity in South African Vineyards. *Weed Res.* **2019**, *59*, 93–106. [CrossRef]
27. Sportelli, M.; Frasconi, C.; Fontanelli, M.; Pirchio, M.; Gagliardi, L.; Raffaelli, M.; Peruzzi, A.; Antichi, D. Innovative Living Mulch Management Strategies for Organic Conservation Field Vegetables: Evaluation of Continuous Mowing, Flaming, and Tillage Performances. *Agronomy* **2022**, *12*, 622. [CrossRef]
28. Pirchio, M.; Fontanelli, M.; Labanca, F.; Sportelli, M.; Frasconi, C.; Martelloni, L.; Raffaelli, M.; Peruzzi, A.; Gaetani, M.; Magni, S.; et al. Energetic Aspects of Turfgrass Mowing: Comparison of Different Rotary Mowing Systems. *Agriculture* **2019**, *9*, 178. [CrossRef]
29. Sportelli, M.; Pirchio, M.; Fontanelli, M.; Volterrani, M.; Frasconi, C.; Martelloni, L.; Caturegli, L.; Gaetani, M.; Grossi, N.; Magni, S.; et al. Autonomous Mowers Working in Narrow Spaces: A Possible Future Application in Agriculture? *Agronomy* **2020**, *10*, 553. [CrossRef]
30. Dumitraşcu, M.; Marin, A.; Preda, E.; Ţîbîrnac, M.; Vădineanu, A. Trampling Effects on Plant Species Morphology. *Plant Biol.* **2010**, *55*, 89–96.
31. Slaughter, D.C.; Giles, D.K.; Downey, D. Autonomous Robotic Weed Control Systems: A Review. *Comput. Electron. Agric.* **2008**, *61*, 63–78. [CrossRef]
32. Pessina, D.; Galli, L.E.; Santoro, S.; Facchinetti, D. Sustainability of Machinery Traffic in Vineyard. *Sustainability* **2021**, *13*, 2475. [CrossRef]

Article

Development of a "0-Pesticide Residue" Grape and Wine Production System for Standard Disease-Susceptible Varieties

Mihaela Roškarič [1,*], Andrej Paušič [1], Janez Valdhuber [2], Mario Lešnik [1] and Borut Pulko [2]

[1] Department of Phytomedicine, Faculty of Agriculture and Life Sciences, University of Maribor, Pivola 10, 2311 Hoče, Slovenia

[2] Department of Viticulture and Enology, Faculty of Agriculture and Life Sciences, University of Maribor, Pivola 10, 2311 Hoče, Slovenia

* Correspondence: mihaela.roskaric1@um.si

Abstract: In order to realize the goals of the EU Farm to Fork strategy, grape growers are introducing new grape-growing technologies. Among the new trends, "0-pesticide residue" protection is quite a promising one. Field trials were carried out in vineyards located in the Mediterranean part of Slovenia in 2021 and 2022 to test the "0-pesticide residue" (ZPR) grape protection system with the goal of producing wine without pesticide residues above the limit of 0.001 mg kg^{-1}. The standard integrated grape protection program (IP) was compared to the ZPR program. The level of infection of leaves and grapes by fungal pathogens did not significantly increase due to the implementation of the ZPR spray program. The amount of yield and quality of yield were not decreased significantly, but a small financial loss of EUR 70–400 ha^{-1} appeared at ZPR grape production when compared to the IP production system. The ZPR system enabled a significant decrease in pesticide residue concentration in wine at a rate of 27 applied pesticide active substances in a rage from 20% to 99%. The goal of producing wine without pesticide residues above the limit concentration of 0.001 mg kg^{-1} was not completely achieved in these experiments, but we came very close to it with the tested spraying programs. Further finetuning of pesticide positioning and alternative plant protection products in 0-pesticide residue systems is needed.

Keywords: grape; wine; pesticide; reduction; economics; IPM; alternative plant protection products

Citation: Roškarič, M.; Paušič, A.; Valdhuber, J.; Lešnik, M.; Pulko, B. Development of a "0-Pesticide Residue" Grape and Wine Production System for Standard Disease-Susceptible Varieties. *Agronomy* **2023**, *13*, 586. https://doi.org/10.3390/agronomy13020586

Academic Editors: Christian Frasconi, Marco Fontanelli and Daniele Antichi

Received: 23 January 2023
Revised: 16 February 2023
Accepted: 16 February 2023
Published: 18 February 2023

1. Introduction

Grape growers are confronted with many challenges related to climate change and global economic and health crises. One of the additional important challenges is a request of the European community to significantly reduce the amount of applied synthetic chemical pesticides in order to address the issues of environmental and human health protection. Overall, the amount of pesticides used in the EU is largely related to grape growing, which is why changes in grape-growing systems significantly impact the overall EU pesticide usage statistics. The request to reduce chemical (target 1) and hazardous (target 2) pesticide use by 50% was presented in the Farm to Fork strategy. To reach these ambitious goals, growers will have to use a broad toolbox of integrated pest management solutions (IPM). Commonly proposed solutions are: cultivation of disease-resistant varieties (PIWI varieties; in German Pilzwiderstandsfähige Rebsorten), switch to organic production, the introduction of all possible digitalization tools, better pesticide application techniques and further upgrading of IPM concepts such as the introduction of better decision support systems [1–5].

One of the possible solutions to reduce the amount of applied synthetic chemical pesticides is introducing a "0-pesticide residue" grape protection system (ZPR). The "0-pesticide residue" protection concept was developed decades ago in apple fruit production in England to reduce the residues in fruits and is currently part of a well-established integrated and low-input fruit production system with developed marketing brands [6].

The first developments in marketing "0-pesticide residue" wines were documented in France during the last three years. Several publications about the endorsement of "0-pesticide residue" wines are available on wine producers' Internet sites. French wine producers started to sell wines under the market brand "Zero pesticide residue within the limits of quantification".

The basic concept of 0-pesticide residue grape protection is relatively simple. It is considered a kind of upgraded integrated production system where we apply all possible measures to reduce synthetic chemical pesticide application. Among the applied pesticides, we prefer to use active substances (a.s.) that degrade fast and have a low environmental and human toxicological impact. The backbone of the ZPR protection concept of disease-susceptible grape varieties that still prevails in our vineyards is dividing the growing season into two parts. In the first part, which usually lasts until two weeks after the finished flowering, chemical pesticides are applied with the highest possible efficacy. During the next two to three weeks, only preparations with active substances that decompose very fast (DT_{50} less than two weeks) within the prolonged pre-harvest interval (PHI), biological pesticides, or low-risk substances registered for organic grape production are applied. Afterward, until harvest, preparations used in the organic production system (biological products and low-risk substances) are applied. This concept significantly reduces the amount of applied synthetic chemical pesticides and pesticide residues in grapes and wine.

In integrated grape production, we usually expect grapes to contain 3 to 12 pesticide residues at a concentration from 0.005 to 2 mg kg^{-1} and wines containing 2–5 residues at levels 0.001–0.25 mg kg^{-1} [7–13]. Regarding pesticide residue concentration, the ZPR system aims to produce grapes with residues lower than 0.01 mg kg^{-1} and wines with residues lower than 0.001 mg kg^{-1}. The already endorsed marketing approach developed by the Collective Noveaux Champs in France follows this mentioned limit. They put the statement "without residues above limits of detection" on the wine brand label; in French, "Sans Résidu de Pesticide Détecté". Food analysis laboratories usually declare the limit of 0.001 mg kg^{-1} as the standard pesticide detection limit. The goal in 0-residue production is to produce wine with residues lower than the mentioned 0.001 mg kg^{-1} limit.

Like any grapevine protection system, the ZPR system has advantages and disadvantages, and there is a lack of information about them in available literature sources. Some information on the advantages and disadvantages of ZPR production is available for fruit production systems, but those are not entirely comparable to grape production systems [14]. The variety of ZPR concepts and features is well presented in the "Zero residue agriculture: the "third way" presentation, which is gaining ground in value-added production. It is described on the Spanish company SEIPASA [15]'s website and in many other online project presentations such as [16–18]. When examining these mentioned sources, it could be concluded that the reduction in applied chemical pesticides and, thus, a lower toxicological burden for the environment and humans is one positive side of ZPR. If farmers practiced this form of grapevine protection, they would have greater chances of obtaining environmental subsidies and better opportunities for wine marketing. Some weaknesses of ZPR are the increase in the cost of protecting the grapevine from harmful organisms and the slightly increased occurrence of some diseases and pests.

Research reports dealing with the economic analysis of ZPR grape protection and production are not available at the moment. The ZPR system is under investigation and is being further developed. The majority of materials on ZPR seem to be nonscientific opinions found on websites produced by the wine industry. The differences between the costs of chemical pesticide-based and alternative preparation-based spray programs and between different countries are very large, and the range of possible economic results due to the introduction of ZPR is very variable. Wine industry online postings rate that ZPR is paying off. According to wine industry opinions, the feasibility of ZPR is primarily a matter of successful marketing and not so much related to the costs of protecting the vines with alternative preparations. If marketing is successful, the higher cost of spray programs can easily be covered.

There is a lack of information on the performance of ZPR, and our research aimed to test the ZPR concept's executability in practice and to demonstrate its effects on; (1) yield, (2) diseases and pest control efficacy, (3) amount of pesticide residues in grape and wine, and (4) partially on financial results. Our research contributes to the development of a more sustainable grape and wine production system for disease-sensitive varieties, following the EU goals related to the Farm to Fork strategy, the sub-goal of a significant reduction of pesticide use (target 1) [19].

2. Materials and Methods

2.1. Trial Design and Statistics

Field trials were carried out in three vineyards in 2021 and in one vineyard in 2022. A standard randomized block design was used with randomly arranged plots consisting of 400–450 grapevines (approx. 1000 m^2 area each). One plot consisted of seven 70–85 m long rows. All evaluations were performed only in the middle row. For data analysis, standard analysis of variance (one-way ANOVA general linear model; F-test) was used. The difference significance among treatment means was tested via the Tukey HSD significance post hoc test ($p < 0.05$) at the assessments of disease severity. Student's *t*-test for independent means ($p < 0.05$) was used to test the difference significance between treatments at yield and pesticide concentration assessments. The SPSS Statistics software (IBM SPSS Statistics V20, Chicago, IL, USA) was used to perform the analysis.

The following assessments were carried out: disease infestation and pest attack rate, amount of grape yield, and analysis of pesticide residues in grapes and wine. We analyzed the infestation rate of three fungi and the attack rate of three pests. The studied fungi were downy mildew (DM) (*Plasmopara viticola* Berk. and M.A. Curtis), powdery mildew (PM) (*Erysiphe necator* Schwein.), and gray mold (GM) (*Botrytis cinerea* Pers. = *Botryotinia fuckeliana* (de Bary)). The analyzed pests were spotted wing drosophila *Drosophila suzukii* Matsumura, European grapevine moth *Lobesia botrana* Denis and Schiffermüller, and mite *Panonychus ulmi* Koch. Diseases and pest attack rate analysis were performed according to EPPO (European Plant Protection Organisation) standards by direct visual scouting of infestation/attack rate on 200 randomly chosen leaves or grape clusters per plot four times per season. Not all obtained data during seasonal assessments are presented in this manuscript. The followed EPPO standards were: PP1/4(4) (*E. necator*), PP1/11(3) (*Eupoecilia ambiguella* and *L. botrana*), PP1/17(3) (*B. fuckeliana* on grapevine); PP1/31(3) (*P. viticola*), PP1/133(2) (Tetranychid mites in vineyards), and PP1/281(1) (*D. suzukii*) [20]. Only data from the last disease infestation evaluations directly prior to harvest are presented in this manuscript.

Two spray programs were tested, which were composed to control all major diseases and pests that usually appear in the area of the experimental vineyards. The first, based on the applications of synthetic chemical pesticides throughout the whole season, was called the IP program, and the second was based on chemical synthetic pesticides applied during the first part of the season and on biological and low-risk preparations during the second part of the season. Some biostimulants were added to relieve plant stress. It was named the 0-pesticide residue program (ZPR). Spray programs are presented in Tables 1–4. We also had smaller, randomly scattered plots that were not treated with plant protection products (control plots). The main aim of the trial was to test the possibility of reducing the amount of applied conventional pesticides by 20–25% without compromising grape yield and financial results. Secondly, we wanted to test as many different active substances as possible, regardless of their toxicological properties (classified as less or more hazardous for the environment or human health), to see the final transfer rate of residues to wine. We wanted to obtain as much information as possible about where we can or can not reach the target of all residues in wine being below 0.001 mg kg^{-1}. We tested the IP spray programs winegrowers in Slovenia practically implement. We wanted to reach target 1 from the "From Field to Fork" strategy, i.e., a reduction of applied chemical pesticide amount per ha per year. The seasonal amount of pesticide a.s. at IP Pinot Gris spray program was 31.34 kg ha^{-1} and 26.77 kg ha^{-1} at ZPR (−14.58%). Amounts of a.s. (kg ha^{-1}) for

Sauvignon, Rebula, and Merlot spray programs were (IP 33.71 vs. 24.03 ZPR (-28.72%)), (IP 26.61 vs. 21.00 ZPR (-21.01%)) and (IP 35.37 vs. 28.24 ZPR (-20.15%)) respectively. Yield determination was performed by hand harvesting 50 vines, randomly chosen in the middle row of each plot. At harvest, parts of the grapes attacked by diseases or pests (e.g., gray mold) were removed from bunches, so we weighed only completely healthy bunches. For simple financial result estimation, we calculated the value of grapes and the cost of the spray program, considering only the cost of preparations and not the cost of application since the number of applications was practically the same in both spraying programs. We assumed that the price for a kilogram of grapes is the same for IP and ZPR as long as the quality parameters of the grapes are comparable. Information about prices of preparations (VAT included) was obtained from several local cooperatives that sell plant protection products. We took the average price for big packages (5–10 kg, L). Information on grape prices was obtained from local wineries.

Table 1. Spray program for the Pinot Gris vineyard in the 2021 season at integrated (IP) and 0-pesticide residue (ZPR) grape protection systems. Preparation was applied if marked with X.

Date	Commercial Name	Active Ingredient	Dosage (kg, L/ha)	IP	ZPR
16 May	Kumulus [1]	Sulfur (80%)	8	X	X
9 May	Kumulus [1]	Sulfur (80%)	6	X	X
	Wetcit [2]	Alcohol ethoxylate (8.15% *w/w*)	0.2	X	X
21 May	Delan pro [1]	Dithianon (12.5%) + Al-fosetyl (56.12%)	3	X	X
	Sercadis plus [1]	Difenconazole (5%) + Fluxapyroxad (7.5%)	0.15	X	X
	Vitanica SI [3]	Algae extract with Si and micronutrients	2		X
	Plantonic [18]	*Salix* + *Urtica* plant extract	4		X
28 May	Delan pro [1]	Dithianon (12.5%) + Al-fosetyl (56.12%)	3	X	X
	Vivando [1]	Metrafenone (50%)	0.2	X	X
	Vitanica SI [3]	Algae extract with Si and microelements	2	X	X
3 June	Cabrio Top [1]	Metiram (55%) + Pyraclostrobin (5%)	2	X	X
	Kumulus [1]	Sulfur (80%)	2	X	X
	Basfoliar Force [3]	Algae extract + Mn + Zn	2	X	
	Basfoliar Active [3]	Algae extract + NPK	2		X
13 June	Cabrio Top [1]	Metiram (55%) + Pyraclostrobin (5%)	2	X	X
	Kumulus [1]	Sulfur (80%)	2	X	X
	Collis [1]	Boscalid (20%) + Kresoxim-methyl (10%)	0.4	X	X
25 June	Cabrio Top [1]	Metiram (55%) + Pyraclostrobin (5%)	2	X	X
	Kumulus [1]	Sulfur (80%)	3	X	X
	Basfoliar Active [3]	Algae extract + NPK	3	X	
	Orvego [1]	Ametoctradin (30%) + Dimetomorph (22.5%)	0.8		X
	Sivanto prime [4]	Flupyradifurone (20%)	0.5		X
8 July	Orvego [1]	Ametoctradin (30%) + Dimetomorph (22.5%)	0.8	X	
	Collis [1]	Boscalid (20%) + Kresoxim-m. (10%)	0.4	X	
	Basfoliar Force [3]	Algae extract + Mn + Zn	2.5	X	X
	Sivanto prime [4]	Flupyradifurone (20%)	0.5	X	
	Kumulus [1]	Sulfur (80%)	3		X
24 July	Cabrio Top [1]	Metiram (55%) + Pyraclostrobin (5%)	2	X	
	Kumulus [1]	Sulfur (80%)	3	X	
	Basfoliar force [3]	Algae extract + Mn + Zn	2		X
	Wetcit [2]	Alcohol ethoxylate (8.15% *w/w*) + botanical oil	2		X

Table 1. *Cont.*

Date	Commercial Name	Active Ingredient	Dosage (kg, L/ha)	IP	ZPR
31 July	Vivando [1]	Metrafenone (50%)	0.2	X	
	Vitisan [9]	Potassium hydrogen carbonate (99.49%)	8		X
	Wetcit [2]	Alcohol ethoxylate (8.15% w/w)	1		X
22 August	Kumulus [1]	Sulfur (80%)	3	X	
	Vitisan [9]	Potassium hydrogen carbonate (99.49%)	8		X
	Wetcit [2]	Alcohol ethoxylate (8.15% w/w) + botanical oil	2		X
	Vitanica SI [3]	Algae extract with Si and microelements	3		X

Product producer: [1]—BASF Germany, [2]—Oro Agri USA, [3]—Compo Germany, [4]—BAYER Germany, [5]—Syngenta Switzerland, [6]—Idai Nature Spain, [7]—Corteva Bulgaria, [8]—Adama Belgium, [9]—Biofa Switzerland, [10]—Belchim Crop Protection Belgium, [11]—Tradecorp Spain, [12]—Novozymes Denmark, [13]—Intermag Poland, [14]—ISK Bioscience Belgium, [15]—K + S Germany, [16]—Sumito Chemical Agro Europe, [17]—Aspanger GmbH Austria, [18]—OGET GmbH Austria.

Table 2. Spray program for the Sauvignon vineyard in the 2021 season at integrated (IP) and 0-pesticide residue (ZPR) grape protection systems. Preparation was applied if marked with X.

Date	Commercial Name	Active Ingredient	Dosage (kg, L/ha)	IP	ZPR
16 April	Kumulus [1]	Sulfur (80%)	8	X	X
10 May	Thiovit jet [5]	Sulfur (80%)	6	X	X
	Folpan [8]	Folpet (80%)	0.7	X	X
18 May	Luna exper. [4]	Fluopyram (20%) + Tebuconazole (20%)	0.4	X	X
	Thiovit jet [5]	Sulfur (80%)	4	X	X
	Universalis [5]	Azoxystrobin (9.35%) + Folpet (50%)	2	X	X
28 May	Mikal pr. F [4]	Folpet (25%) + Iprovalicarb (4%) + Al-fosetyl (50%)	3	X	X
	Nativo [4]	Tebuconazole (50%) + trifloyxstrobin (25%)	0.16	X	X
5 June	Mikal pr. F [4]	Folpet (25%) + Iprovalicarb (4%) + Al-Fosetl (50%)	3	X	X
	Karathane gold [7]	Meptyldinocap (35%)	0.3	X	X
13 June	Mikal flash [4]	Folpet (25%) + Al-fosetyl (50%)	3	X	X
	Vivando [1]	Metrafenone (50%)	0.2	X	X
25 June	Orvego [1]	Ametoctradine (30%) + Dimetomorph (22.5%)	0.8	X	X
	Sercadis plus [1]	Difenconazole (5%) + Fluxapyroxad (7.5%)	0.15	X	X
	Sivanto prime [4]	Flupyradifurone (20%)	0.5		X
8 July	Thiovit jet [5]	Sulfur (80%)	3	X	
	Nativo [4]	Tebuconazole (50%) + trifloyxstrobin (25%)	0.16	X	
	Profiler [4]	Flupicolide (4.44%) + Al-fosetyl (66.67%)	3	X	
	Sivanto prime [4]	Flupyradifurone (20%)	0.5	X	
	Quitobasic [6]	Chitosan hydrochloride (5%)	3		X
	Vitisan [9]	Potassium hydrogen carbonate (99.49%)	6		X
	Vegex beta [6]	Plant extract + essential oils	1.5		X
24 July	Thiovit jet [5]	Sulfur (80%)	2.5	X	
	Basfoliar force [3]	Algae extract + Mn + Zn	3		X
	Pergado F [5]	Mandipropamid (5%) + Folpet (40%)	2.5	X	
	Equibasic [6]	*Equisetum* extract	2		X

Table 2. *Cont.*

Date	Commercial Name	Active Ingredient	Dosage (kg, L/ha)	IP	ZPR
31 July	Thiovit jet [5]	Sulfur (80%)	2.5	X	
	Basfoliar force [3]	Algae extract + Mn + Zn	3	X	
	Quitobasic [6]	Chitosan hydrochloride (5%)	3		X
	Vitisan [9]	Potassium hydrogen carbonate (99.49%)	6		X
	Vegex beta [6]	Plant extract + essential oils	1.5		X
7 August	Vitisan [9]	Potassium hydrogen carbonate (99.49%)	4		X
	Wetcit [2]	Alcohol ethoxylate (8.15% w/w) + botanical oil	1		X
	Quitobasic [6]	Chitosan hydrochloride (5%)	2		X

Product producer: [1]—BASF Germany, [2]—Oro Agri USA, [3]—Compo Germany, [4]—BAYER Germany, [5]—Syngenta Switzerland, [6]—Idai Nature Spain, [7]—Corteva Bulgaria, [8]—Adama Belgium, [9]—Biofa Switzerland, [10]—Belchim Crop Protection Belgium, [11]—Tradecorp Spain, [12]—Novozymes Denmark, [13]—Intermag Poland, [14]—ISK Bioscience Belgium, [15]—K + S Germany, [16]—Sumito Chemical Agro Europe, [17]—Aspanger GmbH Austria, [18]—OGET GmbH Austria.

Table 3. Spray program for the Rebula vineyard in the 2021 season at integrated (IP) and 0-pesticide residue (ZPR) grape protection system. Preparation was applied if marked with X.

Date	Commercial Name	Active Ingredient	Dosage (kg, L/ha)	IP	ZPR
16 April	Kumulus [1]	Sulfur (80%)	8	X	X
11 May	Thiovit jet [5]	Sulfur (80%)	6	X	X
	Delan pro [1]	Dithianon (12.5%) + Al-fosetyl (56.12%)	3	X	X
21 May	Karathane gold [7]	Meptyldinocap (35%)	0.5	X	X
	Universalis [5]	Azoyxstrobin (9.35%) + Folpet (50%)	1.5	X	X
27 May	Ampexio [5]	Mandipropamid (25%) + Zoxamide (2%)	0.5	X	X
	Dynali [5]	Ciflufenamid (3%) + Difenconazole (6%)	2.5	X	X
3 June	Karathane gold [7]	Meptyldinocap (35%)	0.5	X	X
	Ridomil gold [5]	Metalaxyl-m (3.88%) + Mancozeb (64%)	2.5	X	X
13 June	Ampexio [5]	Mandipropamid (25%) + Zoxamide (24%)	0.5	X	X
	Dynali [5]	Ciflufenamid (3%) + Difenconazole (6%)	0.65	X	X
	Delfan plus [11]	Amino acid-based foliar fertilizer	1		X
25 June	Ridomil gold [5]	Metalaxyl-m (3.88%) + Mancozeb (64%)	2.5	X	X
	Topas 100 EC [5]	Penconazole (10%)	0.3	X	X
	Trafos K [11]	P_2O_5 (30% w/w) + K_2O (20% w/w)	2		X
	Sivanto prime [4]	Flupyradifurone (20%)	0.5		X
8 July	Trafos K [11]	P_2O_5 (30% w/w) + K_2O (20% w/w)	2.5	X	
	Thiovit jet [5]	Sulfur (80%)	3	X	
	Ampexio [5]	Mandipropamid (25%) + Zoxamide (24%)	0.5	X	
	Quitobasic [6]	Chitosan hydrochloride (5%)	3		X
9 July	Sivanto prime [4]	Flupyradifurone (20%)	0.5	X	
24 July	Pergado F [5]	Mandipropamid (5%) + Folpet (40%)	2.5	X	
	Topas 100 EC [5]	Penconazole (10%)	0.3	X	
	Quitobasic [6]	Chitosan hydrochloride (5%)	3		X
	Thiovit jet [5]	Sulfur (80%)	3		X
	Affirm [5]	Emamectin (0.95%)	1,5	X	X
31 July	Pergado F [5]	Mandipropamid (5%) + Folpet (40%)	2.5	X	
	Thiovit jet [5]	Sulfur (80%)	2.5	X	

Table 3. *Cont.*

Date	Commercial Name	Active Ingredient	Dosage (kg, L/ha)	IP	ZPR
31 August	Thiovit jet [5]	Sulfur (80%)	3	X	
	Equibasic [6]	*Equisetum* extract	3		X
7 September	Taegro [12]	*Bacillus amyloliquefaciens* strain FZB24 (13%)	0.375		X
	Quitobasic [6]	*Chitosan hydrochloride (5%)*	3		X

Product producer: [1]—BASF Germany, [2]—Oro Agri USA, [3]—Compo Germany, [4]—BAYER Germany, [5]—Syngenta Switzerland, [6]—Idai Nature Spain, [7]—Corteva Bulgaria, [8]—Adama Belgium, [9]—Biofa Switzerland, [10]—Belchim Crop Protection Belgium, [11]—Tradecorp Spain, [12]—Novozymes Denmark, [13]—Intermag Poland, [14]—ISK Bioscience Belgium, [15]—K + S Germany, [16]—Sumito Chemical Agro Europe, [17]—Aspanger GmbH Austria, [18]—OGET GmbH Austria.

Table 4. Spray program for the Merlot vineyard in the 2022 season at integrated (IP) and 0-pesticide residue (ZPR) grape protection system. Preparation was applied if marked with X.

Date	Commercial Name	Active Ingredient	Dosage (kg, L/ha)	IP	ZPR
18 April	Kumulus [1]	Sulfur (80%)	10	X	X
	Ovitex [10]	Paraffin oil (81.7%)	10	X	X
3 May	Thiovit jet [5]	Sulfur (80%)	6	X	X
10 May	Delan pro [1]	Dithianon (12.5%) + Al-fosetyl (56.12%)	3	X	X
	Talendo extra [7]	Proquinazide (16%) + Tetraconazole (8%)	0.25	X	X
20 May	Cymbal [10]	Cymoxanil (45%)	0.2	X	X
	Karathane gold [7]	Meptyldinocap (35%)	0.4	X	X
24 May	Mikrovit B [13]	Boron (11%)	1	X	X
	Phylgreen [11]	Algae extract (15% w/w)	1	X	X
31 May	Delan pro [1]	Dithianon (12.5%) + Al-fosetyl (56.12%)	3	X	X
	Vivando [1]	Metrafenone (50%)	0.2	X	X
11 June	Kusabi [14]	Pyriofenone (30%)	0.25	X	X
	Profiler [4]	Fluopicolide (4.44%) + Al-fosetyl (66.67%)	3	X	X
16 June	Delfan plus [11]	Amino acid-based fertilizer	1	X	
	Epso top [15]	MgO (16%) + SO_3 (32.5%)	3	X	
21 June	Ampexio [5]	Mandipropamid (25%) + Zoxamide (24%)	0.5		X
	Dynali [5]	Ciflufenamid (3%) + Difenconazole (6%)	2.5		X
25 June	Ampexio [5]	Mandipropamid (25%) + Zoxamide (24%)	0.5	X	
	Dynali [5]	Ciflufenamid (3%) + Difenconazole (6%)	2.5	X	
30 June	Collis [1]	Boscalid (20%) + Kresoxim-m. (10%)	0.4	X	
	Orvego [1]	Ametoctradin (30%) + Dimetomorph (22.5%)	0.8	X	
	Plantonic [18]	*Salix* + *Urtica* extract	4		X
	Quitobasic [6]	Chitosan hydrochloride (5%)	4		X
14 July	Prolectus [16]	Fenpyrazamine (30%)	1.2		
	Decis 2.5 EC [4]	Deltamethrin (2.5%)	0.5	X	X
	Aspanger [17]	Muscovite clay	6	X	X
	S-system [6]	SO_3 (32%) + Mn (1%) + Zn (1%) + polycarboxylic acid	2		X
31 July	Cabrio top [1]	Metiram (55%) + Pyraclostrobin (5%)	2	X	
	Kumulus [1]	Sulfur (80%)	3	X	
	Vitisan [9]	Potassium hydrogen carbonate (99.49%)	9		X
	Wetcit [2]	Alcohol ethoxylate (8.15% w/w) + botanical oil	0.8		X

Table 4. *Cont.*

Date	Commercial Name	Active Ingredient	Dosage (kg, L/ha)	IP	ZPR
3 August	Delfan plus [11]	Amino acid-based fertilizer	1		X
	Final K [11]	K_2O (46.5% *w/w*)	3		X
5 August	Vivando [1]	Metrafenone (50%)	0.2	X	
	Quitobasic [6]	Chitosan hydrochloride (5%)	4		X
14 August	Kumulus [1]	Sulfur (80%)	3	X	
	Laser 240 SC [7]	Spinosad (24%)	0.22		X
	Serenade ASO [4]	*Bacillus amyloliquefaciens* (1.396%)	4		X

Product producer: [1]—BASF Germany, [2]—Oro Agri USA, [3]—Compo Germany, [4]—BAYER Germany, [5]—Syngenta Switzerland, [6]—Idai Nature Spain, [7]—Corteva Bulgaria, [8]—Adama Belgium, [9]—Biofa Switzerland, [10]—Belchim Crop Protection Belgium, [11]—Tradecorp Spain, [12]—Novozymes Denmark, [13]—Intermag Poland, [14]—ISK Bioscience Belgium, [15]—K + S Germany, [16]—Sumito Chemical Agro Europe, [17]—Aspanger GmbH Austria, [18]—OGET GmbH Austria.

2.2. Trial Locations, Vineyard Characteristics, and Pesticide Application

The vineyards were located in the western part of Slovenia, called the Vipava valley district, with a mixed semi-Mediterranean and sub-alpine climate. The vineyard with the Sauvignon variety was located in Zavino (45°51′17.43″ N, 13°50′13.38″ E, WGS 84), the vineyards with the varieties Rebula and Pinot Gris were located in the village Draga (45°52′49.03″ N, 13°43′20.20″ E, WGS 84), and the vineyard with the Merlot vines was situated in Merljaki (45°53′12.07″ N, 13°38′50.74″ E, WGS 84). Some characteristics of vineyards are presented in Table 5. We chose older vineyards managed extensively with considerable disease pressure in past seasons and less frequent pesticide application. Pesticides were applied with standard vineyard axial fan cross-flow sprayer Zupan DT (Zupan sprayers, Slovenia) equipped with electrostatic support, which delivered 300 L of spray per ha using an Albuz ATR 80 nozzle operated at 12.2 bars. The spray droplet VMD_{50} value was 80 μm. The timing of pesticide application was adopted according to the advice of the local plant protection advisory and disease prognostic service.

Table 5. Characteristics of vineyards where the trials were performed.

Variety	Age (Years)	Planting Density	No. of Vines per ha	Training System	Max. GRV m^3/ha
Pinot Gris	15	0.9 m × 2.7 m	4100	Single Guyot	6500
Rebula	22	0.8 m × 2.6 m	4800	Short cordon	8300
Merlot	18	0.8 m × 2.4 m	5200	Short cordon	9000
Sauvignon	19	0.8 m × 2.7 m	4600	Single Guyot	7000

GRV—maximum grapevine row volume during the summer period.

2.3. Pesticide Residue Analysis

The pesticide residue determinations were performed in the internationally validated food analytics laboratory Wagner Lebensmittel analytic GmbH located in Lebring (Austria) using the QuEChERS (Quick, Easy, Cheap, Effective, Rugged, and Safe) extraction method, followed by ultra-high-performance liquid chromatography coupled with tandem mass spectrometry (LC-MS/MS) and gas chromatography with mass spectrometry (GC-MS/MS). The sample preparation was carried out according to the European standard EN 15662, "Foods of plant origin Multimethod for determining pesticide residues using GC- and LC-based analysis following acetonitrile extraction/partitioning and clean-up by dispersive SPE—Modular QuEChERS-method" [21]. The pesticides were extracted with acetonitrile, followed by a dispersive cleaning step. We delivered samples of grapes (5 kg) for laboratory analysis within 24 h after harvest. Before the analysis, samples were kept in cooling boxes at 2 °C. The same procedures were applied to the wine samples, amounting to 0.5 L each. The limit of quantification of pesticides was 0.003 mg kg^{-1}, and the limit of detection was

0.001 mg kg^{-1} for grapes and wine. The analysis was performed in four repetitions for each treatment.

In the case of the active substances where the average residue concentration was at a level of 0.001 mg kg^{-1} or higher, we also calculated the degree of reduction in the concentration of pesticide residues. In cases where the average residue concentration at ZPR and IP was equal to or higher than 0.001 mg kg^{-1}, the following formula was used: pesticide concentration reduction rate (%) = 100 − ((concentration ZPR/concentration IP) × 100). In cases where the average concentration at IP was above 0.001 mg kg^{-1} and at ZPR below 0.001 mg kg^{-1}, we assessed the theoretical reduction of the concentration of residues according to the following formula: approximate minimum pesticide concentration reduction rate (%) = 100 − ((0.0009/concentration IP) × 100). A theoretical estimate of the degree of reduction was thus obtained, and the calculated value was marked with the $\simeq$ symbol, which means that the estimated value asymptotically approaches some approximation of the minimal residue concentration reduction degree.

2.4. Grape Juice Production, Analysis, and Winemaking Procedure

From each plot, 300–450 grape clusters were chosen randomly from 100 grapevines in the middle row of plots to obtain a random sample weighing 50 kg. Must was prepared in four repetitions. Prior to pressing, the grapes were not destemmed. They were crushed, and the grape juice was sulfited immediately after the grape pressing (5 g hL^{-1}). For the grape pressing, a small stainless-steel basket press was used (Obst/Berren Spindel-Korb-Presse V20, Fischer Germany). Before starting the fermentation, the fresh juice was analyzed for sugar content (total soluble solids TSS; °Oe) using a digital refractometer (ATAGO 4487 PAL-87S; Atago Inc., Bellewue, WA, USA) with an automatic temperature compensation (ATC). Total titratable acids content (total acidity TA; g L^{-1}) was determined using a titration method by application of the standard base and color indicator bromothymol blue [22]. Fermentation and vinification were performed in miniature tanks. For all wine samples, the vinification process was the same. After the grape pressing, the grape juice was clarified for 24 h and was later decanted and then fermented for two weeks at a temperature of around 18 °C by adding dry yeast. For the fermentation, the dry yeast (*Saccharomyces cerevisiae* Meyen ex E.C. Hansen) strain ZYMAFLORE® X5 Laffort (France) was used (20 g hL^{-1}). After the finished fermentation, fresh wine was decanted from lees and transferred to new tanks, where it was aged and clarified for 3 months. Finning during the aging was not performed. In January 2022 and 2023 (Merlot), the clarified wine was filtered, and samples were taken for pesticide residue analysis. Before the pesticide analysis, the filtration of the wine samples was performed via a Hettich® ROTOFIX 320A centrifuge (Hettich GmbH, Tuttlingen, Germany) at 4000 RPM for 8 min.

3. Results

3.1. Weather Conditions

According to Koppen's climate classification, the climate features at all research locations belong to the climate type Cfa (a humid subtropical climate characterized by hot and humid summers and cool to mild winters). These climates normally lie between latitudes of 25° and 40° and are located poleward from adjacent tropical climates. It is also characterized as a warm temperate climate in some climate classifications.

Figures 1 and 2 show the climatic charts obtained from the meteorological station Vipolže (for the year 2021) and from the meteorological station Miren (for the year 2022). The stations are located close to the experimental vineyards.

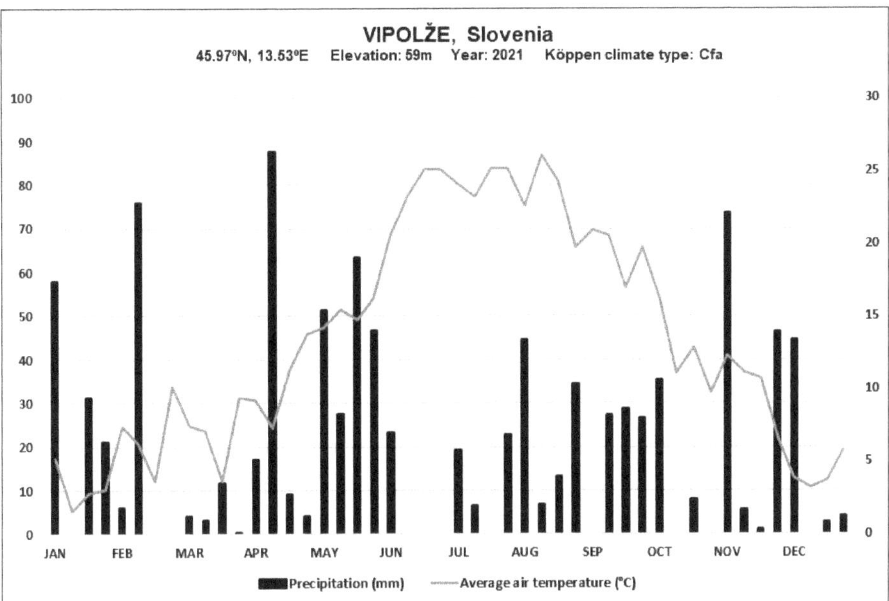

Figure 1. The climatic diagram shows the annual rainfall and air temperature regime. The bars represent the average weekly precipitation (mm), and the line is the average weekly air temperature (°C) at the meteorological station Vipolže (Slovenia) in 2021.

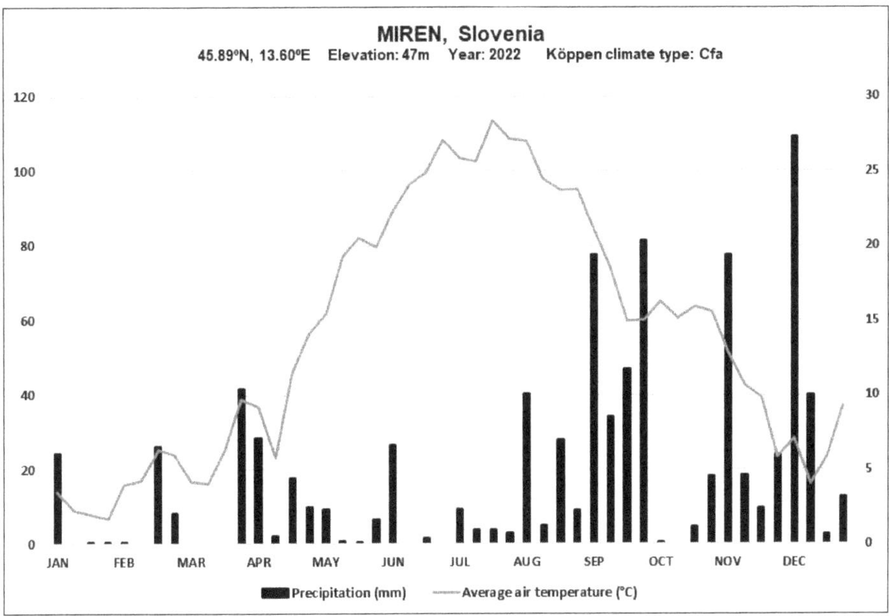

Figure 2. The climatic diagram shows the annual rainfall and air temperature regime. The bars represent average weekly precipitation (mm), and the line is the average weekly air temperature (°C) at the meteorological station Miren (Slovenia) in 2022.

The spring growing period in 2021 was unusually wet, with a low-temperature period from the end of February to the beginning of March (Figure 1). The second lower-temperature period occurred at the end of April and the beginning of May. The first part of May was moderately dry. These conditions caused a delayed flowering period and also delayed primary infections by downy and powdery mildew. The drought in 2021 started in the third week of June and lasted almost until the end of August. Rains in mid-August and September created optimal conditions for the development of gray mold and powdery mildew (Figure 1).

At the beginning of the growing period in 2022, the weather conditions were less suitable for grape and disease development. There were periods of lower temperatures during the end of April and the beginning of May. The first part of May was cold and moderately dry (Figure 2). The grapevine developed slowly, and flowering started with a slight delay. The temperatures in the first two-thirds of June were suitable for grape development. The rainfall enabled a proper supply of water until the last week of June, after which a long drought period started. During the first part of June, conditions for infections with downy mildew were very suitable. During the summer, the plants were exposed to drought until the harvest. During the second part of summer, there was some rain but not enough to completely stabilize the physiological conditions in the grape plants. The rain occurring at the end of August enabled the development of powdery mildew and gray mold (Figure 2).

3.2. Disease Infestation Rate and Pest Attack Rate at Harvest

The pest attack rate was very low, so the pest attack rate data are not presented. Small populations of the following pests were detected: *D. suzukii*, *L. botrana*, and *P. ulmi*. We think that pests had no significant influence on the amount of yield and disease development. The data on disease infestation rates assessed by visual scouting (% infested area) a day before harvest are presented in Table 6. The infestation rate of downy mildew (DM) (*P. viticola*), powdery mildew (PM) (*E. necator*), and gray mold (GM) (*B. cinerea*) at the control nontreated plots on Rebula grapes was moderate (DM 13.85%, PM 12.05%, GM 16.18%). The differences in infestation rates between ZPR and IP programs were not statistically significant in any of the studied diseases. The ZPR program caused a minor increase in infestation rate on grapes only at PM (IP 2.11% vs. 2.93% ZPR) and GM (IP 6.30% vs. 6.76% ZPR). In the Pinot Gris vineyard, the control plots infestation rate was a little higher, especially at DM (21.83% on grapes). PM attack rate in the control grapes was 14.22%, and the GM infestation rate was 11.56%. Differences in attack rates on grapes were noticed (at DM IP 1.89% vs. 4.22% ZPR; at PM IP 5.10% vs. 5.23% ZPR, at GM IP 4.88% vs. ZPR 6.42%) but were not statistically significant. The DM and PM infestation rates were equally severe on leaves and grapes in the Sauvignon vineyard (see Table 6). The GM attack rate at the Sauvignon control plot was 11.18%. The differences between ZPR and IP were small and statistically non-significant. The highest difference was noticed at GM (IP 3.30% vs. ZPR 6.76%). In the Merlot vineyard in 2022, the weather conditions were suitable for disease development. The DM control plots infestation rate on the grapes was over 60%. PM and GM infections were at a lower level but were present in a significant amount of bunches. The differences between IP and ZPR plots were not statistically different for any of the studied diseases (see Table 6). Results for all four varieties demonstrate a particular increase in disease infestation rate due to the ZPR implementation, but in general, the rate of increase is not statistically significant. The population of pests was small and did not contribute significantly to disease development or yield loss.

Table 6. Disease infestation rates (% infested area ± SE) at harvest for two grape protection systems (ZPR—0-pesticide residue protection; IP—integrated protection) in four grape varieties.

Grape Variety	Variant	*Plasmopara viticola* DM Leaf	*Plasmopara viticola* DM Grapes	*Uncinula necator* PM Leaf	*Uncinula necator* PM Grapes	*Botrytis cinerea* GM Grapes
Rebula	ZPR	0.25 ± 0.09 b	1.89 ± 0.17 b	2.45 ± 0.67 b	2.93 ± 0.36 b	6.76 ± 1.04 b
	IP	0.46 ± 0.08 b	2.10 ± 0.20 b	2.89 ± 0.49 b	2.11 ± 0.57 b	6.30 ± 0.95 b
	Control	20.85 ± 1.76 a	13.85 ± 1.93 a	32.60 ± 3.99 a	12.05 ± 2.31 a	16.18 ± 2.71 a
Pinot Gris	ZPR	7.51 ± 1.06 b	4.22 ± 0.88 b	5.78 ± 0.77 b	5.23 ± 0.84 b	6.42 ± 0.84 a
	IP	3.12 ± 0.88 b	1.89 ± 0.07 b	5.64 ± 0.71 b	5.10 ± 0.44 b	4.88 ± 1.02 a
	Control	33.68 ± 3.23 a	21.83 ± 3.15 a	18.25 ± 2.73 a	14.22 ± 1.19 a	11.56 ± 2.95 a
Sauvignon	ZPR	3.56 ± 0.88 b	1.90 ± 0.46 b	4.83 ± 0.81 b	4.89 ± 0.89 b	6.76 ± 1.42 ab
	IP	2.45 ± 0.29 b	2.10 ± 0.40 b	4.00 ± 1.32 b	3.90 ± 0.42 b	3.30 ± 0.81 b
	Control	23.6 ± 4.58 a	20.0 ± 1.88 a	23.55 ± 1.73 a	17.55 ± 4.92 a	11.18 ± 1.27 a
Merlot	ZPR	9.03 ± 0.80 b	6.32 ± 0.47 b	1.26 ± 0.22 b	1.99 ± 0.11 b	1.05 ± 0.16 b
	IP	6.62 ± 1.02 b	5.1 ± 0.86 b	0.92 ± 0.44 b	1.58 ± 0.24 b	0.76 ± 0.16 b
	Control	78.97 ± 3.09 a	61.0 ± 3.16 a	8.74 ± 0.48 a	7.13 ± 0.24 a	3.63 ± 0.52 a

Means marked with the same letters within the same parameters and grape variety are not significantly different according to the Tukey HSD test ($p < 0.05$).

3.3. Results of Yield and Financial Loss Analysis

The results of the yield assessments are presented in Table 7. In the Rebula variety, the differences in the average amount of yield (9636 kg ha^{-1} IP vs. 9269 kg ha^{-1} ZPR), TSS (88.50 °Oe IP vs. 87.33 °Oe ZPR), and in total acidity TA (7.45 gL^{-1} IP vs. 8.00 g L^{-1} ZPR) were not statistically significant. The average yield loss caused by the implementation of ZPR amounted to 367 kg ha^{-1}, and the financial loss was EUR 326 ha^{-1}. Similar results were obtained in Pinot Gris, in which differences in the average amount of yield (6295 kg ha^{-1} IP vs. 6236 kg ha^{-1} ZRP), TSS (84.60 °Oe IP vs. 84.80 °Oe ZRP), and total acidity TA (10.08 g L^{-1} IP vs. 10.05 g L^{-1} ZRP) were also not statistically significant. The implementation of the ZPR concept caused a non-significant loss of yield, amounting to 59 kg ha^{-1}, and a financial loss of EUR 78 ha^{-1}. In the Sauvignon variety, the ZPR concept did not reduce the average yield significantly (10181 kg ha^{-1} IP vs. 9772 kg ha^{-1} ZPR; −410 kg ha^{-1} loss). The reduction in average TSS content (104.6 °Oe IP vs. 91.8 °Oe ZRP) and the increase in TA (8.00 g L^{-1} IP vs. 10.22 g L^{-1} ZPR) were significant. The implementation of the ZPR practice decreased the grape quality. The financial loss in Sauvignon, due to the performance of ZPR, amounted to EUR 267 ha^{-1}. The average yield of Merlot grapes in 2022 was high because the vine development conditions were suitable despite the drought. The ZPR implementation did not significantly reduce the amount of yield (22,754 kg ha^{-1} IP vs. 22,548 kg ha^{-1} ZRP) or grape quality. The average TSS content (96.25 °Oe IP vs. 96.63 °Oe ZPR) and TA of grapes (5.19 g L^{-1} IP vs. 5.53 g L^{-1} ZPR) were similar. Because of the higher yield and higher costs of the spray program, the nominal financial loss in Merlot EUR 437 ha^{-1} was higher than in the other three varieties in season 2021.

3.4. Pesticide Residues in Grapes

Results about the concentration of pesticide residues in grapes at harvest are presented in Tables 8–11. 13 active substances were applied during the 2021 season in the Rebula vineyard (see Table 8). On average, 8.0 a.s. with a concentration above the 0.001 mg kg^{-1} limit were found in IP and 7.0 in the ZPR grapes. In 3 of the detected a.s. (cyflufenamid, folpet and mandipropamid), the concentration was significantly lower in ZPR grapes than in IP grapes. Concentration reduction rates ranged from 24% to 90%. The concentrations of azoxystrobin, meptyldinocap, difenconazole, and penconazol fell under the 0.001 mg kg^{-1} level (marked with <0.001 ND). It is interesting that no fungicide penconazol at ZPR was detected, despite the fact that it was applied quite late in the season. With the ZPR system, only two-thirds of the a.s. were at a level below 0.01 mg kg^{-1}, and with the IP system, only three a.s. residues were below that level. The basic concept of 0.0-residue has not been fully achieved, i.e., all a.s. residues in grape shells being below a concentration level of 0.01 mg kg^{-1}. Al-fosetyl, flupyradifuron, and cyflufenamid were over the limit of 0.01 mg kg^{-1}.

Table 7. Yield and must parameters (mean $\pm$ SE) (TSS—total soluble solids, TA—total acidity) for two grape protection systems (ZPR—0-pesticide residue protection; IP—integrated protection) in four grape varieties.

Variant	Healthy Grapes kg/ha	TSS Oe°	TA g/L	Value of Grapes (*VG*) EUR/ha	Cost of Spray Program EUR/ha (*CS*)	Financial Loss EUR/ha (*VG* **IP**–*VG* **ZPR**) + (*CS* **ZPR**–*CS* **IP**)
Rebula; harvest 28 September 2021; price of 1 kg of grape EUR 0.65						
ZPR	9269 $\pm$ 235 a	87.3 $\pm$ 1.37 a	8.00 $\pm$ 0.26 a	6025 $\pm$ 153 a	808	239 + 87 = 326
IP	9636 $\pm$ 295 a	88.5 $\pm$ 0.56 a	7.45 $\pm$ 0.23 a	6264 $\pm$ 192 a	721	/
Sauvignon; harvest 21September 2021; price of 1 kg of grape EUR 0.70						
ZPR	9772 $\pm$ 189 a	91.8 $\pm$ 2.30 b	10.22 $\pm$ 0.22 a	6840 $\pm$ 132 a	743	287–20 = 267
IP	10,181 $\pm$ 272 a	104.6 $\pm$ 1.28 a	8.00 $\pm$ 0.2 b	7127 $\pm$ 190 a	763	/
Pinot Gris; harvest 9 September 2021; price of 1 kg of grape EUR 0.70						
ZPR	6236 $\pm$ 299 a	84.8 $\pm$ 2.10 a	10.05 $\pm$ 0.25 a	4366 $\pm$ 210 a	771	41 + 37 = 78
IP	6295 $\pm$ 231 a	84.6 $\pm$ 2.23 a	10.08 $\pm$ 0.18 a	4407 $\pm$ 162 a	734	/
Merlot; harvest 3 October 2022; price of 1 kg of grape EUR 0.65						
ZPR	22,548 $\pm$ 1173 a	96.63 $\pm$ 1.22 a	5.53 $\pm$ 0.14 a	14656 $\pm$ 763 a	2024	134 + 303 = 437
IP	22,754 $\pm$ 1596 a	96.25 $\pm$ 0.75 a	5.19 $\pm$ 0.09 a	14790 $\pm$ 1038 a	1721	/

Means marked with the same letters within the same parameters and grape variety are not significantly different according to the independent-sample t-test ($p < 0.05$).

Table 8. Average concentration (mean $\pm$ SE) of pesticide residues in Rebula grapes. A result of <0.001 ND means that a.s. was not detected at a level higher than 0.001 mg kg^{-1}. /—concentration reduction rate and significance of difference between IP (integrated) and ZPR (0-pesticide residue) means cannot be calculated. Pesticide concentration reduction rate (%) = 100 $-$ ((ZPR/IP) $\times$ 100) (see explanation in Section 2.4).

Active Substance	ZPR mg/kg	IP mg/kg	% Reduction Rate
$\sum$ Dithiocarbamates	<0.001 ND	<0.001 ND	/
Al-fosetyl	<0.001 ND	<0.001 ND	/
Al-fosetyl + metabolites	0.7663 $\pm$ 0.067 a	1.327 $\pm$ 0.289 a	42.25
Phosphonic acid	0.5667 $\pm$ 0.050 a	0.9833 $\pm$ 0.210 a	42.36
Ciflufenamid	0.028 $\pm$ 0.001 b	0.039 $\pm$ 0.003 a	28.20
Difenconazole	<0.001 ND	<0.001 ND	/
Emamectin	<0.003	<0.003	/
Fludioxonil	0.0085 $\pm$ 0.00454 a	0.0157 $\pm$ 0.00067 a	45.85
Azoxystrobin	<0.001 ND	<0.001 ND	/
Flupyradifurone	0.0222 $\pm$ 0.016 a	0.0393 $\pm$ 0.009 a	43.51
Folpet	0.009 $\pm$ 0.002 b	0.095 $\pm$ 0.015 a	90.52
Mandipropamid	0.003 $\pm$ 0.0004 b	0.0122 $\pm$ 0.002 a	75.41
Meptyldinocap	<0.001 ND	<0.001 ND	/
Metalaxyl-m	0.0033 $\pm$ 0.001 a	0.0049 $\pm$ 0.0002 a	32.65
Penconazole	<0.001 ND	0.008 $\pm$ 0.001	$\simeq$88.75
Zoxamide	0.0044 $\pm$ 0.0003 a	0.0058 $\pm$ 0.0008 a	24.13
AN of found a.s. > 0.01 mg/kg	3.00 $\pm$ 0.58 b	4.67 $\pm$ 0.33 a	35.76
AN of found a.s. > 0.001 mg/kg	7.00 $\pm$ 0.00 a	8.00 $\pm$ 0.00 a	12.50

Means marked with the same letters at a specific a.s. do not differ significantly according to the independent-samples t-test ($p < 0.05$). AN is the average number of found active substances.

Table 9. Average concentration (mean ± SE) of pesticide residues in Pinot Gris grapes. A result of <0.001 ND means that a.s. was not detected at a level higher than 0.001 mg kg^{-1}. /—concentration reduction rate and significance of difference between IP (integrated) and ZPR (0-pesticide residue) means cannot be calculated. Pesticide concentration reduction rate (%) = 100 − ((ZPR/IP) × 100) (see explanation in Section 2.4).

Active Substance	ZPR mg/kg	IP mg/kg	% Reduction Rate
$\sum$ Dithiocarbamates	<0.001 ND	0.01127 ± 0.0024	≃92.01
Al-fosetyl	<0.001 ND	<0.001 ND	/
Al-fosetyl + metabolites	0.009 ± 0.003 a	0.011 ± 0.009 a	18.18
Phosphonic acid	0.003 ± 0.001 a	0.007 ± 0.002 a	57.14
Ametoctradin	0.00833 ± 0.0019 b	0.0179 ± 0.0087 a	53.46
Boscalid	0.0034 ± 0.0007 b	0.0607 ± 0.01 a	94.39
Difenconazole	<0.001 ND	<0.001 ND	/
Dimetomorph	0.0066 ± 0.001 b	0.01273 ± 0.003 a	48.15
Dithianon	<0.001 ND	<0.001 ND	/
Flupyradifurone	0.00333 ± 0.0006 a	0.00503 ± 0.001 a	33.79
Fluxapyroxad	<0.003	<0.003	/
Kresoxim-methyl	<0.001 ND	<0.001 ND	/
Metrafenone	<0.001 ND	0.061 ± 0.007	≃98.52
Pyraclostrobin	0.0058 ± 0.0004 b	0.01967 ± 0.004 a	70.51
AN of found a.s. > 0.01 mg/kg	0.33 ± 0.33 b	5.67 ± 0.33 a	94.17
AN of found a.s. > 0.001 mg/kg	7.00 ± 0.00 b	9.00 ± 0.00 a	22.22

Means marked with the same letters at a specific a.s. do not differ significantly according to the independent-samples *t*-test ($p < 0.05$). AN is the average number of found active substances.

Table 10. Average concentration (mean ± SE) of pesticide residues in Sauvignon grapes. A result of <0.001 ND means that a.s. was not detected at a level higher than 0.001 mg kg^{-1}. /—concentration reduction rate and significance of difference between IP (integrated) and ZPR (0-pesticide residue) means cannot be calculated. Pesticide concentration reduction rate (%) = 100 − ((ZPR/IP) × 100) (see explanation in Section 2.4).

Active Substance	ZPR (mg/kg)	IP (mg/kg)	% Reduction Rate
Al-fosetyl	0.019 ± 0.011 a	0.038 ± 0.017 a	50.00
Al-fosetyl + metabolites	2.100 ± 0.176 a	2.900 ± 0.223 a	27.58
Phosphonic acid	0.750 ± 0.033 a	0.900 ± 0.080 a	16.60
Ametoctradin	<0.003	0.0086 ± 0.002	/
Boscalid	<0.003	0.003 ± 0.003	/
Azoxystrobin	<0.001 ND	<0.001 ND	/
Difenconazole	<0.001 ND	<0.001 ND	/
Dimetomorph	<0.003	<0.003	/
Fluopyram	<0.003	<0.003	/
Fluopicolide	<0.003	<0.003	/
Flupyradifurone	<0.003	<0.003	/
Fluxapyroxad	<0.003	<0.003	/
Folpet	0.006 ± 0.002 b	0.018 ± 0.004 a	66.66
Iprovalicarb	<0.001 ND	<0.003	/
Mandipropamid	<0.003	0.029 ± 0.004	/
Metrafenone	<0.001 ND	<0.001 ND	/
Tebuconazole	<0.001 ND	0.011 ± 0.002	≃91.81
Zoxamide	0.0054 ± 0.0016 a	0.0084 ± 0.0043 a	35.71
Trifloxystrobin	<0.001 ND	0.009 ± 0.0006 a	≃90.00
AN of found a.s. > 0.01 mg/kg	1.33 ± 0.38 b	4.33 ± 0.38 a	69.28
AN of found a.s. > 0.001 mg/kg	7.33 ± 0.88 b	10.33 ± 0.88 a	29.04

Means marked with the same letters at a specific a.s. do not differ significantly according to the independent-samples *t*-test ($p < 0.05$). AN is the average number of found active substances.

Table 11. Average concentration (mean ± SE) of pesticide residues in Merlot grapes. A result of <0.001 ND means that a.s. was not detected at level higher than 0.001 mg kg^{-1}. /—concentration reduction rate and significance of difference between IP (integrated) and ZPR (0-pesticide residue) means cannot be calculated. Pesticide concentration reduction rate (%) = 100 − ((ZPR/IP) × 100) (see explanation in Section 2.4).

Active Substance	ZPR (mg/kg)	IP (mg/kg)	% Reduction Rate
$\sum$ Dithiocarbamates	<0.001 ND	0.0057 ± 0.0009	≃84.21
Al-fosetyl	<0.001 ND	<0.001 ND	/
Al-fosetyl + metabolites	4.400 ± 2.570 b	13.730 ± 2.160 a	67.95
Phosphonic acid	5.810 ± 2.610 a	10.230 + 1.610 a	43.20
Ametoctradin	<0.001 ND	0.180 ± 0.036	≃99.50
Boscalid	<0.001 ND	0.011 ± 0.003	≃91.81
Cyflufenamid	0.010 ± 0.004 a	0.030 ± 0.014 a	66.66
Cymoxanil	<0.001 ND	<0.001 ND	/
Delatamethrin	<0.001 ND	<0.001 ND	/
Difenconazole	0.015 ± 0.006 a	0.041 ± 0.019 a	63.41
Dimetomorph	<0.001 ND	0.050 ± 0.015 a	≃98.20
Dithianon	0.002 ± 0.001 b	0.006 ± 0.001 a	66.66
Fenpyrazamine	<0.001 ND	0.011 ± 0.004 a	≃91.81
Fluopicolide	0.015 ± 0.002 a	0.025 ± 0.007 a	40.00
Kresoxim-methyl	<0.001 ND	<0.001 ND	/
Mandipropamid	<0.003	0.024 ± 0.0078	/
Meptyldinocap	<0.001 ND	<0.001 ND	/
Metrafenone	<0.001 ND	0.012 ± 0.002	≃92.50
Pyraclostrobin	<0.001 ND	0.032 ± 0.009	≃97.18
Pyriofenone	<0.003	0.004 ± 0.0005	/
Proquinazide	<0.001 ND	<0.001 ND	/
Zoxamide	0.027 ± 0.0100 a	0.040 ± 0.0130 a	32.50
Spinosad	<0.001 ND	<0.001 ND	/
AN of found a.s. > 0.01 mg/kg	4.25 ± 1.03 b	9.75 ± 0.48 a	56.41
AN of found a.s. > 0.001 mg/kg	6.75 ± 0.48 b	15.00 ± 0.00 a	55.00

Means marked with the same letters at a specific a.s. do not differ significantly according to the independent-samples *t*-test ($p < 0.05$). AN is the average number of found active substances.

The analysis results of the Pinot Gris grapes are presented in Tables 9 and 12; active substances were applied during the 2021 season. On average, 7.00 a.s. at a level higher than 0.001 mg/kg were found in ZPR grapes and 9.00 a.s. in IP grapes. The concentration of difenconazole, metrafenon, ditianon, metiram, and kresoxim-methyl fell to such a low level that there was no detection (below 0.001 mg kg^{-1}; marked with <0.001 ND). At 5 a.s., the ZPR system resulted in a decrease in residue concentration (metiram, ametoctradin, boscalid, dimethomorph, metrafenone, and pyraclostrobin). In grapes from the ZPR plots, practically all residues were below the level of 0.01 mg kg^{-1}, and in the IP system, only five a.s. residues were below this level. The basic concept of ZPR has been achieved in the ZPR grapes, i.e., all residues were below 0.01 mg kg^{-1}. Different levels of degradation speed among different a.s. were noticed. Flupyradifuron degraded faster than we expected. Metrophenone disintegrated quickly. Despite its use at the end of July, the concentration was low (0.0061 mg kg^{-1} in IP). Amethoctradine and dimethomorph degraded slowly. Metiram was degraded at a moderate speed. The same applies to fluxapiroxad.

The analysis results of Sauvignon grapes are presented in Table 10. The spraying program for the Sauvignon variety was very intensive, and we found many residues of a.s. We applied 16 different a.s. In the IP grapes, 10.33 residues above the limit of 0.001 mg/kg were found on average, of which two residues were at a level above 0.01 mg kg^{-1}. In the ZPR grapes, 7.33 residues above 0.001 mg kg^{-1} were found on average, of which 1 residue (Al-fosetyl) was above 0.01 mg kg^{-1}. We did not reach the 0.0-residue goal completely because there were some residues found above the 0.01 mg kg^{-1} level. Residues were generally at a very low level. An extensive reduction of concentration was achieved, especially for the substances fluopyram, flupyradifuron, folpet, and fluopicolid. Five theoretically fairly persistent substances were not found at a level higher than 0.001 mg kg^{-1} (metrafenone, azoxystrobin, difenconazole, tebuconazole, and trifloxystrobin).

Table 12. Average concentration (mean $\pm$ SE) of pesticide residues in Rebula wine. A result of <0.001 ND means that a.s. was not detected at a level higher than 0.001 mg kg^{-1}. /—concentration reduction rate and significance of difference between IP (integrated) and ZPR (0-pesticide residue) means cannot be calculated. Pesticide concentration reduction rate (%) = 100 − ((ZPR/IP) × 100) (see explanation in Section 2.4).

Active Substance	ZPR (mg/kg)	IP (mg/kg)	% Reduction Rate
$\sum$ Dithiocarbamates	<0.001 ND	<0.001 ND	/
Al-fosetyl	0.008 $\pm$ 0.003 b	0.150 $\pm$ 0.130 a	94.66
Al-fosetyl + metabolites	3.300 $\pm$ 1.270 a	5.200 $\pm$ 0.400 a	36.53
Phosphonic acid	2.450 $\pm$ 0.950 a	4.250 $\pm$ 0.530 a	42.35
Azoxystrobin	<0.001 ND	0.011 $\pm$ 0.010	$\simeq$91.81
Cyflufenamid	<0.001 ND	<0.003	/
Difenconazole	<0.001 ND	<0.001 ND	/
Emamectin	<0.001 ND	<0.001 ND	/
Flupyradifurone	<0.003	<0.003	18.18
Folpet	<0.001 ND	0.004 $\pm$ 0.001	$\simeq$77.50
Mandipropamid	<0.001 ND	0.005 $\pm$ 0.002	$\simeq$82.00
Meptyldinocap	<0.001 ND	<0.001 ND	/
Metalaxyl-M	<0.003	<0.003	/
Penconazole	<0.001 ND	<0.001 ND	/
Zoxamide	<0.001 ND	<0.001 ND	/
AN of found a.s. > 0.001 mg/kg	1.50 $\pm$ 0.58 b	4.25 $\pm$ 0.96 a	64.70

Means marked with the same letters at a specific a.s. do not differ significantly according to the independent-samples *t*-test ($p < 0.05$). AN is the average number of found active substances.

The results of the Merlot grape analysis from season 2022 are presented in Table 11. The beginning of the 2022 season was rainy, and fungicides were therefore applied frequently. We applied many different a.s. to obtain much data on a.s. dissipation rates. A total of 21 different a.s. were applied. In the IP program, 15 were detected, on average, at levels higher than 0.001 mg kg^{-1}, and in the ZPR, 6.75 a.s. were detected. The ZPR program reduced the number of detected substances significantly. Implementing the ZPR program reduced the concentration of the following a.s.: metiram, Al-fosetyl, ametoctradin, boscalid, dithianon, difenconazole, dimethomorph, metrafenone, piraclostrobin, and pyriofenone. Reduction rates were in range between 30% and 90% (see Table 11). The goal to have all residues at a level lower than 0.01 mg kg^{-1} in ZPR was not achieved, but a decrease from 9.75 a.s. found on average at IP to 4.25 a.s. found in ZPR grapes was significant.

3.5. Pesticide Residues in Wine

The results of the residue analysis in wine are presented in Tables 12–15. In the Rebula wine from the ZPR plots, only Al-fosetyl metabolites, flupyradifurone, and metalaxyl-M were found at a level higher than 0.001 mg kg^{-1} (see Table 12). In the IP wine, residues of azokxystrobin, Al-fosetly, cyflufenamid, flupyradifurone, folpet, mandipropamid, and metalxyl-M were found. In the IP wine, 4.25 a.s. with a concentration over 0.001 mg kg^{-1} and 1.5 a.s. at ZPR were found on average. The average number of found a.s. over the level 0.001 mg kg^{-1} was 64.7% reduced at ZPR. With the ZPR system in the Rebula plots, the goal of all residues being below 0.001 mg kg^{-1} was almost achieved but was not reached completely. The substances flupyradifuron and Al-fosetyl were used a little too late in the season. In one of the four repetitions in ZPR, the goal of all a.s. being at a level lower than 0.001 mg kg^{-1} was reached.

Table 13. Average concentration (mean $\pm$ SE) of pesticide residues in Sauvignon wine. Result < 0.001 ND means a.s. was not detected at a level higher than 0.001 mg kg^{-1}. /—concentration reduction rate and significance of difference between IP (integrated) and ZPR (0-pesticide residue) means cannot be calculated. Pesticide concentration reduction rate (%) = 100 − ((ZPR/IP) × 100) (see explanation in Section 2.4).

Active Substance	ZPR (mg/kg)	IP (mg/kg)	% Reduction Rate
$\sum$ Dithiocarbamates	<0.001 ND	<0.001 ND	/
Al-fosetyl	0.033 $\pm$ 0.016 b	0.910 $\pm$ 0.438 a	96.33
Al-fosetyl + metabolites	2.420 $\pm$ 0.810 b	4.930 $\pm$ 0.357 a	50.81
Phosphonic acid	1.730 $\pm$ 0.598 b	4.000 $\pm$ 0.280 a	56.69
Ametoctradin	<0.001 ND	<0.003	$\simeq$77.50
Azoxystrobin	<0.001 ND	<0.001 ND	/
Difenconazole	<0.001 ND	<0.001 ND	/
Dimetomorph	<0.001 ND	0.004 $\pm$ 0.001	$\simeq$77.50
Fluopyram	<0.001 ND	<0.003	$\simeq$55.00
Fluopicolide	<0.001 ND	<0.003	$\simeq$59.09
Flupyradinefurone	<0.001 ND	<0.001 ND	/
Fluxapyroxad	<0.001 ND	<0.003	/
Folpet	<0.001 ND	0.008 $\pm$ 0.0006	$\simeq$88.75
Iprovalicarb	<0.001 ND	<0.003	/
Mandipropamid	<0.001 ND	<0.003	/
Metrafenone	<0.001 ND	<0.001 ND	/
Tebuconazole	<0.001 ND	0.011 $\pm$ 0.001	$\simeq$91.81
Trifloxystrobin	<0.001 ND	<0.001 ND	/
AN of found a.s. > 0.001 mg/kg	1.00 $\pm$ 0.00 b	7.50 $\pm$ 1.23 a	86.67

Means marked with the same letters at a specific a.s. do not differ significantly according to the independent-samples *t*-test ($p < 0.05$). AN—average number of found active substances.

Table 14. Average concentration (mean $\pm$ SE) of pesticide residues in Pinot Gris wine. Result < 0.001 ND means a.s. was not detected at a level higher than 0.001 mg kg^{-1}. /—concentration reduction rate and significance of difference between IP (integrated) and ZPR (0-pesticide residue) means cannot be calculated. Pesticide concentration reduction rate (%) = 100 − ((ZPR/IP) × 100) (see explanation in Section 2.4).

Active Substance	ZPR (mg/kg)	IP (mg/kg)	% Reduction Rate
$\sum$ Dithiocarbamates	<0.001 ND	<0.001 ND	/
Al-fosetyl	0.079 $\pm$ 0.030 a	0.134 $\pm$ 0.016 a	41.04
Al-fosetyl + metabolites	10.600 $\pm$ 2.220 b	62.500 $\pm$ 20.362 a	83.04
Phosphonic acid	7.720 $\pm$ 1.712 b	38.650 $\pm$ 13.818 a	80.02
Ametoctradin	<0.003	0.005 $\pm$ 0.001	/
Boscalid	<0.001 ND	0.004 $\pm$ 0.002	$\simeq$77.50
Difenconazole	<0.001 ND	<0.001 ND	/
Dimetomorph	<0.003	0.003 $\pm$ 0.0003	/
Dithianon	<0.001 ND	<0.001 ND	/
Flupyradifurone	<0.003	<0.003	/
Fluxapyroxad	<0.001 ND	<0.001 ND	/
Kresoxim-methyl	<0.001 ND	<0.001 ND	/
Metrafenon	<0.001 ND	<0.001 ND	/
Pyraclostrobin	<0.001 ND	<0.001 ND	/
AN of found a.s. > 0.001 mg/kg	3.00 $\pm$ 0.71 a	4.50 $\pm$ 0.50 a	33.33

Means marked with the same letters at a specific a.s. do not differ significantly according to the independent-samples *t*-test ($p < 0.05$). AN is the average number of found active substances.

Table 15. Average concentration (mean $\pm$ SE) of pesticide residues in Merlot wine. A result of <0.001 ND means that a.s. was not detected at a level higher than 0.001 mg kg^{-1}. /—concentration reduction rate and significance of difference between IP (integrated) and ZPR (0-pesticide residue) means cannot be calculated. Pesticide concentration reduction rate (%) = 100 − ((ZPR/IP) $\times$ 100) (see explanation in Section 2.4).

Active Substance	ZPR (mg/kg)	IP (mg/kg)	% Reduction Rate
$\sum$ Dithiocarbamates	<0.001 ND	<0.001 ND	/
Al-fosetyl	0.00325 $\pm$ 0.003 b	0.0210 $\pm$ 0,0025 a	84.52
Al-fosetyl + metabolites	1.475 $\pm$ 0.871 b	8.625 $\pm$ 0,427 a	82.85
Phosphonic acid	1.100 $\pm$ 0.647 b	6.500 $\pm$ 0.303 a	83.07
Ametoctradin	<0.001 ND	0.0582 $\pm$ 0.024	$\simeq$98.45
Boscalid	<0.001 ND	<0.003	/
Cyflufenamid	<0.001 ND	<0.001 ND	/
Cymoxanil	<0.001 ND	<0.001 ND	/
Delatamethrin	<0.001 ND	<0.001 ND	/
Difenconazole	<0.001 ND	<0.001 ND	/
Dimetomorph	<0.001 ND	<0.003	/
Dithianon	<0.001 ND	<0.001 ND	
Fenpyrazamine	<0.001 ND	<0.003	/
Fluopicolide	<0.003	<0.003	/
Kresoxim-methyl	<0.001 ND	<0.001 ND	/
Mandipropamid	<0.003	<0.003	/
Meptyldinocap	<0.001 ND	<0.001 ND	/
Metrafenone	<0.001 ND	<0.003	/
Pyraclostrobin	<0.001 ND	<0.003	/
Pyriofenone	<0.001 ND	<0.001 ND	/
Proquinazide	<0.001 ND	<0.001 ND	/
Zoxamide	<0.001 ND	<0.003	/
Spinosad	<0.001 ND	<0.001 ND	/
AN of found a.s. > 0.001 mg/kg	2.25 $\pm$ 0.25 b	9.75 $\pm$ 0.25 a	76.92

Means marked with the same letters at a specific a.s. do not differ significantly according to the independent-samples t-test ($p < 0.05$). AN is the average number of found active substances.

The results for the Sauvignon wine are presented in Table 13. In the Sauvignon wine from the ZPR managed plots, only Al-fosetyl metabolites were found. If we did not use Al-fosetyl, we would produce wine practically without pesticide residues and reach the 0-pesticide residue requirements. In the IP Sauvignon wine, 7.5 residues were found at levels higher than 0.001 mg kg^{-1} and 1 a.s. in ZPR on average. The average number of found a.s. over the level 0.001 mg kg^{-1} was reduced by 86.67% at ZPR. Here, we can see a clear difference between the results in the IP and ZPR concepts. The only a.s. found in ZPR was Al-fosetyl.

The results for Pinot Gris are presented in Table 14. In the Pinot wine from the ZPR plots, four residues were found (Al-fosetyl, ametoctradin, dimethomorph, and flupyradifurone). The goal of 0-pesticide residue requirements was not reached. In the IP Pinot wine, 4.5 a.s. were found with concentration over 0.001 mg kg^{-1}, and in ZPR wine, 3.0 a.s. on average. In Pinot we had the worst results of all trials. The average number of found a.s. over the level 0.001 mg kg^{-1} was reduced by only by 33.33% in ZPR. Maybe the reason for that was the grapes of the Pinot being harvested earlier than the grapes of the other tested varieties. The available residue deterioration time was shorter than in other varieties.

The results for the Merlot wine are presented in Table 15. In the Merlot plots, we applied many a.s., and therein lies the reason why many residues were found at a moderate concentration level. In the IP wine, we found 9.75 residues, and in ZPR, 2.25 residues were above the level of 0.001 mg kg^{-1} on average. We did not reach the ZPR goal, but the difference between IP and ZPR was statistically significant. In ZPR wine, we found only residues of Al-fosetyl, fluopicolide, and mandipropamid at a level between 0.003 and 0.001 mg kg^{-1}. We detected only 3 a.s. out of 21 applied. This is a suitable result; we almost reached the goal of ZPR. It looks as though the mentioned three substances have quite a high grape-to-wine transfer rate.

4. Discussion

4.1. Efficacy of Disease and Pest Control

The downy mildew disease pressure in the 2021 season was moderate, and in 2022, high. Despite the reduction in the number of conventional pesticide applications and of the cumulative dose of pesticides, of a.s. per ha per season for approx. 20–25% and the supplementation of chemical pesticides with alternative products, we provided a high level of protection for the grapevine. Only a statistically non-significant increase in disease infestation rate in all three studied diseases was noticed. The differences in infestation rates at IP and ZPR plots were mostly not significant. Some studies demonstrate the consequences of different levels of pesticide reduction on the efficacy of grapevine pest and disease control [23–31]. We cannot compare mentioned studies dealing with the development of new production methods with our research directly because we focused only on the alteration of the spray program as the only measure to reach a higher level of sustainability of grape production. In other studies, they studied the broader spectrum of vineyard management strategies, plus pesticide reduction. We think the 0-pesticide residue approach fits in with the new developments of IPM, enabling a more sustainable integrated production. The efficacy we achieved with ZPR in our trials is close to the level of standard integrated spray programs with incorporated alternative plant protection products, as reported by [14,24,26–28].

The comparison results between the IP and ZPR systems also depend on how well and intensive the IP spray program is. If we have a highly efficient IP spray program, the difference to ZPR will be bigger than if we compare less efficient IP spray programs to ZPR. Disease pressure has a very significant influence. When the disease pressure is moderate, more favorable results for ZPR are obtained. Considering short-term results, it seems that the ZPR protection system is equally efficient as standard IP. Some grape growers in the region where the trials were performed executed IP protection trials with a significant reduction in the number of pesticide applications per season for a couple of years. Their approach was to start spraying as late as possible and to cancel the last 2–3 sprays before harvest. By doing this, they reduced the annual dose of applied pesticide a.s. per ha for approximately 30%. The described practice resulted in a significant increase in infestation rates of *Phomopsis viticola* (Sacc.) Sacc., *Elsinoë ampelina* Shear, *Guignardia bidwellii* (Ellis) Viala and Ravaz, and sour bunch rot (combined yeast and bacterial infection) (oral information provided by growers). The same problems arose in growing PIWI-resistant grape varieties in systems with any application of pesticides (oral information provided by growers). Considering this information, the important question accompanying the introduction of ZPR is, what are the long-term consequences of a significant chemical pesticide use reduction rate for the health status and fitness of the vineyards, also considering climate change and the introduction of new diseases and pests? In the short term, it looks like ZPR holds a population of harmful organisms at an acceptable level, but we have to be aware of the serious threats in grape growing, such as phytoplasma diseases (Flavescence dorée) [32], rickettsia caused by Pierce's decline disease (*Xylella fastidiosa* Wells et al.) [33], and trunk diseases from the ESCA complex [34]. These diseases, if not held at a low population level, can completely destroy vineyards. In the literature, we can mostly find results of pesticide reduction studies lasting 2–3 years in integrated production [4,35] or studies about conversions to organic production [24,25]. Studies that last long and analyze long time scale dynamics of grapevine pests and diseases are rare, and it is therefore not possible to predict the consequences of long-term ZPR system usage. Long-term studies are needed. Pragmatically speaking, we could say that if organic vineyards sustain the pressure of diseases, then the 0-pesticide residue vineyards with reduced pesticide usage will also sustain. The long-term consequences are also related to the availability of efficient alternative plant protection products. If these are available and we start to grow more resistant grape varieties, then there is no fear of a large-scale increase in disease and pest populations.

4.2. Yield

The trials demonstrated that the implementation of the ZPR concept causes a small yield reduction between 2% and 5% due to the reduced disease and pest control efficiency. In one of the four trials, the quality of must was slightly reduced, but in three of four, it was not. We expect that the ZPR wine has a comparable sensorics quality to the IP wine. In the literature, many sources are available on the comparisons between integrated and organic grape production. Reports are also available on the newly developed low-input integrated grape production with significantly reduced intensity of pesticide use. Our ZPR-tested protection system can be compared with such low-input production trials. One of the latest publications on low-input grape growing was published by Perria et al. in 2022 [35]. They tested the "Green Grapes" concept. The significant reduction of the pesticide concept was feasible only in low or moderate disease pressure conditions. The overall results of the three-year study indicated that disease management protocols based solely or predominantly on the use of alternative and resistance-inducing substances do not appear to ensure effective protection against diseases in disease-susceptible varieties. The trials have confirmed that we need highly efficient control at the beginning of the season and that the intensity of pesticide use can possibly be significantly reduced in the second part of the season. The conclusions in these other studies resemble the results of our research. The other two publications that are comparable with our results were published by Pertot et al. in 2017 [1] and Fouillet et al. in 2022 [4]. They studied new strategies for a grapevine cropping system redesign and found out that cropping system redesign entails drastic changes in the vineyards; new, less-susceptible grape varieties and new non-chemical plant protection products with higher levels of efficacy than the existing ones would have to be used since most of the existing alternative solutions are only partially effective. This means that pest and disease pressure is reduced, but pests are not eradicated to a level where we could guarantee avoiding economically relevant yield losses. IPM strategies were mentioned similarly to ZPR but were not considered as 0-pesticide residue production.

4.3. Financial Loss

We were not able to find any scientific reports on the feasibility of the ZPR system. The ZPR system is in its early stages of development. Some wine producer websites, for example, vitisphere.com or decanter.com, have indicated the possibility that 0-pesticide residue wines could be sold for a 10–15% higher price than standard wines from integrated production. They also posted that the level of pesticide residues in ZPR wines is comparable to wines from organic production and think that the quality of ZPR wines is comparable to or of higher quality than wines from organic production. In the case of wines selling at a 10% higher price than standard conventional IP wines, grape growers could very likely compensate for a smaller yield reduction (2–3%) and a 5% increase in pest and disease control costs. Some studies show that growers are very reluctant to accept the risks involved in introducing new cropping systems [36]. We did the simple compensation calculation for our four trials considering the obtained yields, pressing efficacy of 75% (grape/must ratio), and a price of EUR 4.5 per 1 L of bottled IP wine. The results indicate that an equal financial result at IP and ZPR production is reached if the ZPR Rebula wine is sold at a 3.95% higher price, the Pinot wine at a 0.94% higher price, the Sauvignon wine at a 4.19% higher price and the Merlot at a 0.90% higher price. We believe that such an increase in price can be realized by innovative marketing. The professionally performed ZPR concept seems to be feasible. It could be especially feasible if supported with subsidies issued for reducing pesticide use. Research shows that by using advanced prognostic models to predict the need for spraying and by using improved application techniques, it is possible to reduce pesticide use by as much as

40% and production costs by up to 20% [36,37]. If we combine the ZPR concept and the mentioned technical support systems, then we can improve the economics of ZPR cultivation and probably further reduce pesticide residues in grapes.

4.4. Level of Residue Reduction in Grapes and Wine

Many factors influence pesticide dissipation dynamics from application to harvest (DR) and the transfer rate from grapes to wine (TR) during the winemaking processes. In several scientific publications, much data on DR and TR are available and can serve as tools for decisions on the latest time periods of specific preparation applications in ZPR. Factors that significantly impact DR are: weather, method and timing of pesticide application, pesticide formulation, and chemical properties of pesticide [7–9,12]. Factors that significantly impact the TR are: chemical and physical properties of pesticide (p.e. water solubility and pK_{ow} value (octanol-water partition coefficient), grape pressing technique, methods of wine clearing and filtration, and many others [10,12,13,38–45]. It was proven several times that TR of specific a.s. are lower in red wines than in white wines. The maceration step in the preparation of red wines has a significant impact on residue dissipation/transfer. In this manuscript, we do not present data on the DR and TR we obtained. We tested more than 25 active substances. In half of them, the results are similar to the data obtained by researchers in the abovementioned publications, and in the other half, the differences were quite big (more than ±30%). This means that the data on DR and TR are quite variable, and in order to provide suitable advice to grape growers, local trials need to be carried out for several seasons. The growers should systematically collect data on weather, the timing of pesticide application, and vinification processes, so they can gain the needed experience for planning ZPR spray program usage. It is not an easy task since studies show that growers are very reluctant to implement novelties [37,38]. Any long-term behavioral change in plant protection requires efforts from many stakeholders [37–39].

The goal of our trials was to produce grapes with residues lower than 0.01 mg kg^{-1} and wine with residues lower than 0.001 mg kg^{-1}. The lower concentration of residues in ZPR grapes is a consequence of several factors: on the one hand, prolonged PHI, and on the other hand, changes in residue dissipation kinetics. The range of available alternative plant protection preparations is very wide, and they have different chemical and physical effects on the deposit of pesticides on the surface of vine organs and also on the physiological processes inside the plant tissues [46]. That means that the choice of alternative products for spraying in the second half of the growing season can cause very different impacts on pesticide dissipation kinetics. Many studies are needed to understand the effects of chitosan, laminarins, carbonates, algae and plant extracts, clay minerals, silicon polymers, and many others [47,48]. Some alternative products significantly alter the pH value of plant surfaces, and others act as natural detergents. Humectants increase the penetration of pesticides, plant resins, and silicon polymers and prolong residue deposit stability. The described processes cause a great variability of results and make giving suitable advice very difficult. This means that we need to provide much education for the growers and advisors.

Growing PIWI (in German Pilzwiderstandsfähige Rebsorten) varieties would fit perfectly into the ZPR concept and would make it easier to implement. Grape growers need suitable support in at least two directions. They need information about the dissipation kinetics and grape-to-wine transfer rates of all pesticide substances we use in the vineyard. A decision support system of adopted prolonged pre-harvest waiting intervals (PHI) needs to be developed, so the growers can plan the last possible period of application of a specific pesticide a.s. to reach residue concentrations lower than 0.001 mg kg^{-1}. With many a.s., the PHIs need to be prolonged for at least 2–3 weeks.

The first step of ZPR implementation should be to provide simple, concise information on a.s. that have a fast DR and low TR. As a result of our research, we estimate that a.s. with a TR lower than 10% are usually not found in wine at concentrations higher than

0.001 mg kg^{-1} if we respect the PHI on the pesticide label. Such a.s. are: deltamethrin, emmamectin, azoxystrobin, cyflufenamid, difenconazole, penconazol, krezoxim-methyl, folpet, fluxapyroxad, zooxamid, spinosad, and metalaxyl-M. We have to pay attention to the following a.s. with a TR higher than 20%, which will appear in wine in concentrations higher than 0.001 mg kg^{-1} if the PHI stated on the product label is followed: fluopyram, Al-fosetyl, boscalid, cyprodinil, fludixonil, pyrimethanil, dimethomorph, ametoctradin, tebuconazol, fluopicolid, mandipropamid, and flupyradifurone. As mentioned a.s., we need a significant prolongation of the PHI stated on the pesticide label. Special attention needs to be paid to Al-fosetyl. Our advice is not to use Al-fosetyl in a ZPR system. Residues are easily found in wine despite it being used just early in the season.

Some applications are already available on websites that offer growers calculations of pesticide residue concentrations in wine according to the date of pesticide application and vinification method type. When growers enter the data about the pesticide application dates, they receive information about the expected concentration of pesticide residues in grapes and wine. With such applications, growers can quite precisely determine which preparations they can or cannot use in order to achieve the goal of having all pesticide residues below 0.001 mg kg^{-1}. An example of a freely available smartphone application is available on the website [49].

The second condition for implementing ZPR is a suitable market availability of alternative plant protection preparations from the low-risk substance category and biological pesticides at a sufficiently low competitive price. If the prices of alternative preparations are high, the cost of alternative spray programs could be too high compared to the cost of a standard IP spray program. In such cases, the feasibility of a 0-residue approach decreases. Help in the form of subsidies for the purchase of alternative products is always welcome.

5. Conclusions

Our research aimed to test the ZPR concept's executability and feasibility. The results obtained in the four trials demonstrate that reaching the goal of all residues in wine being at a concentration level lower than 0.001 mg kg^{-1} is not easily achieved when growing disease-susceptible varieties that need to be sprayed frequently. We would have had to restrict the application of chemical pesticides more than we did. ZPR provided suitable diseases control, and yield reduction was not statistically significant. We conducted only four experiments, and from the obtained results, it is not possible to conclude that ZPR offers completely comparable financial results to standard integrated IPM-steered cultivation.

It is necessary to perform a significant number of experiments in order to create a comprehensive opinion on the applicability of the ZPR approach and to provide growers with technical guidance, i.e., provide them rules for the preparation of spray programs based on a.s. dissipation rate data and grape-to-wine residue transfer rate data.

Author Contributions: Conceptualization M.L. and M.R.; methodology M.L. and A.P.; software statistics A.P. and M.R.; validation A.P. and M.R.; investigation M.R., M.L., A.P., B.P. and J.V.; writing—original draft preparation M.R., M.L., A.P., J.V. and B.P.; writing—review and editing, M.R., M.L. and A.P.; supervision, M.L.; funding acquisition M.L. All authors have read and agreed to the published version of the manuscript.

Funding: This research was funded by the Slovenian national research agency ARRS and the Ministry of Agriculture, Forestry and Food of the Republic of Slovenia (grant no. CRP V4-2226) via the project "Measures to reduce the use of chemical plant protection products in viticulture".

Institutional Review Board Statement: Not applicable.

Informed Consent Statement: Not applicable.

Data Availability Statement: Not applicable.

Acknowledgments: The authors are thankful to the company Merum (Branik, Slovenia) for providing vineyards, technical equipment, and stuff for pesticide application and to the technician Marjan Sirk, who organized the grape harvest and pressing.

Conflicts of Interest: The authors declare no conflict of interest. The funders had no role in any stage of research or preparation of the manuscript.

References

1. Pertot, I.; Caffi, T.; Rossi, V.; Mugnai, L.; Hoffmann, C.; Grando, M.S.; Gary, C.; Lafond, D.; Duso, C.; Thiery, D.; et al. A critical review of plant protection tools for reducing pesticide use on grapevine and new perspectives for the implementation of IPM in viticulture. *J. Crop Prot.* **2017**, *97*, 70–84. [CrossRef]
2. Merot, A.; Fermaud, M.; Gosme, M.; Smits, N. Effect of Conversion to Organic Farming on Pest and Disease Control in French Vineyards. *Agronomy* **2020**, *10*, 1047. [CrossRef]
3. Román, C.; Perisb, M.; Esteve, J.; Tejerina, M.; Cambray, J.; Vilardell, P.; Planas, S. Pesticide dose adjustment in fruit and grapevine orchards by DOSA3D: Fundamentals of the system and on-farm validation. *Sci. Total Environ.* **2022**, *808*, 152158. [CrossRef] [PubMed]
4. Fouillet, E.; Deliere, L.; Chartier, N.; Munier-Jolain, N.; Cortel, S.; Rapidel, B.; Merot, A. Reducing pesticide use in vineyards. Evidence from the analysis of the French DEPHY network. *Eur. J. Agron.* **2022**, *136*, 8. [CrossRef]
5. Jacquet, F.; Jeuffroy, M.H.; Jouan, J.; Le Cadre, E.; Litrico, I.; Malausa, T.; Reboud, X.; Huyghe, C. Pesticide-free agriculture as a new paradigm for research. *Agron. Sustain. Dev.* **2022**, *42*, 8. [CrossRef]
6. Cross, J.V.; Berrie, A.M. Eliminating Reportable Pesticide Residues from Apples. *Agric. Eng. Int.* **2008**, *10*, 1–9. Available online: https://cigrjournal.org/index.php/Ejounral/article/view/1242/1099 (accessed on 10 December 2022).
7. Cabras, P.; Angioni, A. Pesticide Residues in Grapes, Wine, and Their Processing Products. *J. Agric. Food Chem.* **2000**, *48*, 967–973. [CrossRef]
8. Schusterova, D.; Hajslova, J.; Kocourek, V.; Pulkrabova, J. Pesticide Residues and Their Metabolites in Grapes and Wines from Conventional and Organic Farming System. *Foods* **2021**, *2*, 307. [CrossRef]
9. Gabur, G.D.; Gabur, I.; Cucolea, E.I.; Costache, T.; Rambu, D.; Cotea, V.V.; Teodosiu, C. Investigating Six Common Pesticides Residues and Dietary Risk Assessment of Romanian Wine Varieties. *Foods* **2022**, *11*, 2225. [CrossRef]
10. Santana-Mayor, A.; Rodríguez-Ramos, R.; Socas-Rodríguez, B.; Díaz-Romero, C.; Rodríguez-Delgado, M.A. Comparison of Pesticide Residue Levels in Red Wines from Canary Islands, Iberian Peninsula, and Cape Verde. *Foods* **2020**, *9*, 1555. [CrossRef]
11. Čuš, F.; Baša Česnik, H.; Velikonja Bolta, Š. Pesticide residues, copper and biogenic amines in conventional and organic wines. *Food Control* **2022**, *132*, 108534. [CrossRef]
12. Gonzalez, P.A.; Parga Dans, E.; Acosta Dacal, A.C.; Zumbado Pena, M.; Perez Luzardo, O. Differences in the levels of sulphites and pesticide residues in soils and wines under organic and conventional production methods. *J. Food Compos. Anal.* **2022**, *112*, 104714. [CrossRef]
13. Kittelmann, A.; Müller, C.; Rohn, S.; Michalski, B. Transfer of Pesticide Residues from Grapes (Vitis vinifera) into Wine—Correlation with Selected Physicochemical Properties of the Active Substances. *Toxics* **2022**, *10*, 248. [CrossRef] [PubMed]
14. Rozman, Č.; Unuk, T.; Pažek, K.; Lešnik, M.; Prišenk, J.; Vogrin, A.; Tojnko, S. Multi Criteria Assessment of Zero Residue Apple Production. *Erwerbs Obstbau* **2013**, *55*, 51–62. [CrossRef]
15. Seipasa. Available online: https://www.seipasa.com/en/blog/zero-residue-agriculture-the-third-way-gaining-ground/ (accessed on 11 February 2023).
16. Mafa Vegetal Ecobiology. Available online: https://www.mafa.es/en/zero-residue-agriculture-sustainable-and-healthy-nutrition/ (accessed on 11 February 2023).
17. AgriculturePost.com. Available online: https://agriculturepost.com/opinion/should-you-go-organic-or-residue-free-farming/ (accessed on 11 February 2023).
18. IFNH- Institute for Food, Nutrition & Health. CleanFruit Project. Available online: https://research.reading.ac.uk/ifnh/cases/cleanfruit-standardization-of-innovative-pest-control-strategies-to-produce-zero-residue-fruit-for-baby-food-and-other-fruit-produce/ (accessed on 11 February 2023).
19. Farm to Fork Strategy. A Farm to Fork Strategy for a Fair, Healthy and Environmentally-Friendly Food System. Communication from the Commission to the European Parliament, the Council, the European Economic and Social Committee and the Committee of the Regions; COM/2020/381. Available online: https://ec.europa.eu/food/sites/food/files/safety/docs/f2f_action-plan_20 20_strategy-info_en.pdf (accessed on 20 December 2022).
20. EPPO Standards–PP1 Efficacy Evaluation of Plant Protection Products—Summary List of Approved Standards (2022-09). Available online: https://www.eppo.int/RESOURCES/eppo_standards/pp1_list (accessed on 20 December 2020).
21. British Standard BS EN 15662:2008—Foods of Plant Origin- Determination of Pesticide Residues Using GC MS and/or LC-MS/MS Following Cetonitrile Extraction Partitioning and Cleanup by Dispersive SPE-QuEChERS-Method. Available online: http://www.chromnet.net/Taiwan/QuEChERS_Dispersive_SPE/QuEChERS_%E6%AD%90%E7%9B%9F%E6%96%B9 %E6%B3%95_EN156622008_E.pdf (accessed on 20 December 2020).

22. Tyl, C.; Sadler, G.D. pH and Titratable Acidity. In *Food Analysis*; Nielsen, S.S., Ed.; Food Science Text Series; Springer: Cham, Switzerland, 2017; pp. 389–406. [CrossRef]

23. Mailly, F.; Hossard, L.; Barbier, J.M.; Thiollet-Scholtus, M.; Gary, C. Quantifying the impact of crop protection practices on pesticide use in wine-growing systems. *Eur. J. Agron.* **2017**, *84*, 23–34. [CrossRef]

24. Merot, A.; Ugaglia, A.; Barbier, J.M.; Del'homme, B. Diversity of conversion strategies for organic vineyards. *Agron. Sustain. Dev.* **2019**, *39*, 16. [CrossRef]

25. Merot, A.; Smits, N. Does conversion to organic farming impact vineyards yield? A diachronic study in southeastern France. *Agronomy* **2020**, *10*, 1626. [CrossRef]

26. Calzarano, F.; Seghetti, L.; Pagnani, G.; Metruccio, E.G.; Di Marco, S. Control of Grapevine Downy Mildew by an Italian Copper Chabasite-Rich Zeolitite. *Agronomy* **2022**, *12*, 1528. [CrossRef]

27. Rotolo, C.; De Miccolis, A.R.M.; Dongiovanni, C.; Pollastro, S.; Fumarola, G.; Di Carolo, M.; Perrelli, D.; Natale, P.; Faretra, F. Use of biocontrol agents and botanicals in integrated management of Botrytis cinerea in table grape vineyards. *Pest Manag. Sci.* **2018**, *74*, 715–725. [CrossRef]

28. Delaunois, B.; Farace, G.; Jeandet, P.; Clément, C.; Baillieul, F.; Dorey, S.; Cordelier, S. Elicitors as alternative strategy to pesticides in grapevine? Current knowledge on their mode of action from controlled conditions to vineyard. Environ. *Sci. Pollut. Res.* **2013**, *21*, 4837–4846. [CrossRef]

29. Bleyer, G.; Lösch, F.; Schumacher, S.; Fuchs, R. Together for the Better: Improvement of a Model Based Strategy for Grapevine Downy Mildew Control by Addition of Potassium Phosphonates. *Plants* **2020**, *2*, 710. [CrossRef]

30. Deliere, L.; Miclot, A.S.; Sauris, P.; Rey, P.; Calonnec, A. Efficacy of fungicides with various modes of action in controlling the early stages of an Erysiphe necator-induced epidemic. *Pest Manag. Sci* **2010**, *66*, 1367–1373. [CrossRef]

31. Valdés-Gómez, H.; Araya-Alman, M.; De La Fuente, C.P.; Verdugo-Vásquez, N.; Lolas, M.; Acevedo-Opazo, C.; Gary, C.; Calonnec, A. Evaluation of a decision support strategy for the control of powdery mildew (Erysiphe necator [Schw.] Burr.) in grapevine in the central region of Chile. *Pest Manag. Sci.* **2017**, *73*, 1813–1821. [CrossRef]

32. Jeger, M.; Bragard, C.; Caffier, D.; Candresse, T.; Chatzivassiliou, E.; Dehnen-Schmutz, K.; Gilioli, G.; Jaques, J.; Macleod, A.; Navajas, M.; et al. Risk to plant health of Flavescence dorée for the EU territory. *EFSA J.* **2016**, *14*, 4603. [CrossRef]

33. Trkulja, V.; Tomić, A.; Iličić, R.; Nožinić, M.; Popović Milovanović, T. *Xylella fastidiosa* in Europe: From the Introduction to the Current Status. *Plant Pathol. J.* **2022**, *38*, 551–571. [CrossRef] [PubMed]

34. Mondello, V.; Songy, A.; Battiston, E.; Pinto, C.; Coppin, C.; Trotel-Aziz, P.; Clement, C.; Mugnai, L.; Fontaine, F. Grapevine trunk diseases: A review of fifteen years of trials for their control with chemicals and biocontrol agents. *Plant Dis.* **2018**, *102*, 1189–1217. [CrossRef]

35. Perria, R.; Ciofini, A.; Petrucci, W.A.; D'Arcangelo, M.E.M.; Valentini, P.; Storchi, P.; Carella, G.; Pacetti, A.; Mugnai, L. A Study on the Efficiency of Sustainable Wine Grape Vineyard Management Strategies. *Agronomy* **2022**, *12*, 392. [CrossRef]

36. Maddalena, G.; Marone Fassolo, E.; Bianco, P.A.; Toffolatti, S.L. Disease Forecasting for the Rational Management of Grapevine Mildews in the Chianti Bio-District (Tuscany). *Plants* **2023**, *12*, 285. [CrossRef] [PubMed]

37. Aka, J.; Ugaglia, A.A.; Lescot, J.M. Pesticide Use and Risk Aversion in the French Wine Sector. *J. Wine Econ.* **2018**, *13*, 451–460. [CrossRef]

38. Alister, C.; Araya, M.; Morandé, J.; Volosky, C.; Kogan, M. Disipación de plaguicidas utilizados en uva vinífera y traspaso de sus residuos al vino. *Fitosanidad* **2020**, *22*, 16–18. Available online: https://sidal.cl/assets/pdf-13.pdf (accessed on 5 December 2022).

39. Alister, C.; Araya, M.; Morandé, J.E.; Volosky, C.; Saavedra, J.; Cordova, A.; Kogan, M. Effects of wine grape cultivar, application conditions and the winemaking process on the dissipation of six pesticides. *Cienc. Investig. Agrar.* **2014**, *41*, 375–386. [CrossRef]

40. Chen, X.; He, S.; Gao, Y.; Ma, Y.; Hu, J.; Liu, X. Dissipation behavior, residue distribution and dietary risk assessment of field-incurred boscalid and pyraclostrobin in grape and grape field soil via MWCNTs-based QuEChERS using an RRLC-QqQ-MS/MS technique. *Food Chem.* **2019**, *274*, 291–297. [CrossRef]

41. Xu, T.; Feng, X.; Pan, L.; Jing, J.; Zhang, H. Residue and risk assessment of fluopicolide and cyazofamid in grapes and soil using LC-MS/MS and modified QuEChERS. *RSC Adv.* **2018**, *62*, 8. [CrossRef]

42. Edder, P.; Ortelli, D.; Viret, O.; Cognard, E.; De Montmollin, A.; Zali, O. Control strategies against grey mould (*Botrytis cinerea* Pers.) and corresponding fungicide residues in grapes and wines. *Food Addit. Contam.* **2009**, *26*, 719–725. [CrossRef] [PubMed]

43. Pazzirota, T.; Martin, L.; Mezcua, M.; Ferrer, C.; Fernandez-Alba, A.R. Processing factor for a selected group of pesticides in a winemaking process: Distribution of pesticides during grape processing. *Food Addit. Contam.* **2013**, *30*, 1752–1760. [CrossRef] [PubMed]

44. Navarro, S.; Barba, A.; Oliva, J.; Navarro, G.; Pardo, F. Evolution of residual levels of six pesticides during elaboration of red wines. Effect of winemaking procedures in their dissappearance. *J. Agric. Food Chem.* **1999**, *47*, 264–270. [CrossRef]

45. Cabras, P.; Conte, E. Pesticide residues in grapes and wine in Italy. *Food Addit. Contam.* **2001**, *18*, 880–885. [CrossRef]

46. Jindo, K.; Goron, T.L.; Pizarro-Tobias, P.; Sanchez-Monedero, M.A.; Audette, Y.; Deolu-Ajayi, A.O.; Van der Werf, A.; Goitom Teklu, M.; Shenker, M.; Pombo Sudre, C.; et al. Application of biostimulant products and biological control agents in sustainable viticulture: A review. *Front. Plant. Sci.* **2022**, *18*, 13. [CrossRef]

47. Dumitriu Gabur, G.D.; Teodosiu, C.; Cotea, V.V. *Management of Pesticides from Vineyard to Wines: Focus on Wine Safety and Pesticides Removal by Emerging Technologies. Grapes and Wine*; IntechOpen: London, UK, 2021; pp. 3–25. [CrossRef]

48. Chen, Y.; Alcala Herrera, R.; Benitez, E.; Hoffmann, C.; Möth, S.; Paredes, D.; Plaas, E.; Popescu, D.; Rascher, S.; Rusch, A.; et al. Winegrowers' decision-making: A pan-European perspective on pesticide use and inter-row management. *J. Rural Stud.* **2022**, *94*, 37–53. [CrossRef]
49. Aplicación Móvil Control Plaguicidas. Available online: https://twgroup.cl/portfolio/laboratorio/ (accessed on 5 January 2023).

Article

Legume Cover Crop Alleviates the Negative Impact of No-Till on Tomato Productivity in a Mediterranean Organic Cropping System

Lara Abou Chehade [1,*], Daniele Antichi [1,2], Christian Frasconi [1], Massimo Sbrana [1,2], Lorenzo Gabriele Tramacere [1,2], Marco Mazzoncini [1,2] and Andrea Peruzzi [1]

[1] Department of Agriculture, Food and Environment, University of Pisa, Via Del Borghetto 80, 56124 Pisa, Italy
[2] Centre for Agri-Environmental Research "Enrico Avanzi", University of Pisa, Via Vecchia di Marina 6, San Piero a Grado, 56122 Pisa, Italy
* Correspondence: lara.abouchehade@agr.unipi.it

Abstract: The ecosystem services a cover crop (CC) provides depend enormously on species choice and tillage system. Here, we evaluated the impact of (a) three winter CCs—rye (*Secale cereale* L.) and squarrose clover (*Trifolium squarrosum* L.) monocultures and their mixture, and (b) two tillage systems—roller-crimping of CC residue as dead mulch for no-till (NT) systems and incorporating CC residue into the soil as green manure for conventional tillage (CT) systems—on the performance of organic processing tomato, i.e., plant growth, nutrient accumulation, fruit yield, and weed biomass. The assessments took place over two years in field experiments conducted under Mediterranean conditions. At the termination time, rye and mixture were the most productive and the best weed-suppressive CCs. During tomato growing season, squarrose clover regardless of tillage system stimulated tomato growth, Nitrogen content and uptake, and the yield relative to the other cover crops. Nevertheless, NT generally impaired the tomato nutritional status and increased weed biomass compared to CT despite some potential weed control by cover crops. These two aspects caused a significant drop in tomato yield in all NT systems. The results suggested that, despite the multiple benefits the compared CCs can offer in Mediterranean agroecosystems, legume CCs could be the key to developing feasible organic vegetable no-till systems.

Keywords: residue management; cereal-legume mixture; agrobiodiversity; conservation agriculture; integrated weed management

Citation: Abou Chehade, L.; Antichi, D.; Frasconi, C.; Sbrana, M.; Tramacere, L.G.; Mazzoncini, M.; Peruzzi, A. Legume Cover Crop Alleviates the Negative Impact of No-Till on Tomato Productivity in a Mediterranean Organic Cropping System. *Agronomy* **2023**, *13*, 2027. https://doi.org/10.3390/agronomy13082027

Academic Editors: Hailin Zhang and Xiaobing Liu

Received: 31 May 2023
Revised: 25 July 2023
Accepted: 27 July 2023
Published: 30 July 2023

1. Introduction

Cropping practices that enhance and preserve long-term soil fertility are central to organic farming. Realizing the harmful consequences of frequent soil inverting and stirring on soil fertility has raised questions regarding the use of conventional tillage in organic agroecosystems. In fact, decades of research and application have documented the positive impacts of conservation agriculture techniques, i.e., no-till or reduced tillage, residue retention on the soil surface, and cropping system diversification, on soil quality including soil stability, chemical, and biological fertility [1]. In the European Mediterranean region, organic farmers primarily focus on growing cover crops in winter, whose residue eventually turns into green manure, to promote soil health, biodiversity, and weed management [2–4].

Previous research has emphasized the difficulties in implementing no-till systems in organic farming, pointing particularly to ineffective weed control and nutrient management [5,6]. Thus, finding cover crop species capable of targeting these two aspects to sustain crop yield is needed [2]. Cover crops that generate dense, ideally long-lasting biomass [7] and/or release allelochemicals [8] are favored for the first target.

It is known that the magnitude of cover crop services changes as cover crop traits differ among cover crop species and their families. Legumes can symbiotically fix Nitrogen

(N) and add it significantly to agroecosystems though they often yield low biomass [7]. Cereals are more efficient in resource use, have a rapid establishment, and can produce more residue with a slow decay rate [8,9]. One of the most effective cover crops is cereal rye (*Secale cereale* L.), which has excelled in no-till systems owing to these characteristics and its remarkable capacity to control weeds [10]. Ascribed to soil N immobilization [11,12] and the allelochemicals released during residue decomposition [13], rye has occasionally been reported to injure crops and reduce yields if not appropriately managed [11–13].

One of the means to achieve balanced ecosystem services while maintaining reduced tillage is using cover crop mixtures. Besides combining the advantages of the single species, multispecies cover crops, based mainly on functional complementarity, favor distinct niche occupations and resource use [14,15]. As numerous studies have documented [16–19], among the primary ecosystem services consequently provided is increased biomass production, i.e., overyielding. Therefore, mixtures are more likely to improve weed suppression [18,20], N retention, and N supply compared to pure cover crop stands [9,17,21]. Cereal and legume bicultures have been popular for these purposes and since they are simpler to manage in fields than multispecies cover crops [9,22,23]. These mixtures can be combined to include species with distinct traits, both above- and below-ground, such as those related to growth habits and nutrient-foraging strategies. Legumes such as vetches (*Vicia* sp.) with a tall climbing habit and spreading roots can be coupled with grass species with extensive shallower fibrous root systems and an upright canopy, such as the well-known rye. By exploiting these traits among others, mixing rye with hairy vetch can be a valid substitution for monocultures, able to outyield them by more than 60% and build up as much N as the vetch [9]. Other legumes such as clovers (*Trifolium* sp.), which have a deep and branched taproot and grow relatively quickly to cover the soil and control weeds [18], can also be used as part of other cereal–legume combinations. Mixtures with the right ratio of legume to cereal can guarantee a low residue C: N ratio, which is fundamental for N to be available to the cash crop, especially in situations with slow mineralization rates such as in no-till systems [21,24–26].

Here, we seek to assess the agronomic impacts of reducing tillage in Mediterranean organic vegetable production. We focused on processing tomato, a cash crop that accounts for most of the organic vegetable production in Italy [27]. Over two growing seasons, we evaluated (i.) the performance of three cover crops before termination—monoculture of squarrose clover (*Trifolium squarrosum* L.), monoculture of rye, and a combination of both—in terms of biomass production, nutrient uptake, and weed suppression, and (ii.) the effects of these cover crops under two tillage systems—conventional tillage and no-till—on organic processing tomato performance, i.e., crop growth, nutrient status, nutrient uptake, yields, and residual weed biomass.

2. Materials and Methods

2.1. Site Description

The study was established for two seasons (2019–2020 and 2020–2021) on certified organic fields at the Center for Agri-environmental Research "Enrico Avanzi" of the University of Pisa, situated in the lower valley of the Arno River in San Piero a Grado in Pisa, Central Italy (43°40′ N Lat; 10°19′ E Long; 1 m above mean sea level and 0% slope). The site is characterized by a Mediterranean climate with an average annual precipitation of 895 mm, mainly in spring and fall, and a mean annual temperature of 14.9 °C (1993–2020). Monthly average temperatures and total precipitations registered during the study and long-term average values (1993–2020) are represented in Figure 1.

Warmer temperatures occurred during the autumn–winter of 2019–2020 and summer 2021 compared with the multiannual trend. The hottest temperatures in both years occurred in August, reaching almost 31 °C, and the coolest ones were registered in January with 3.6 °C in 2020 and 3.0 °C in 2021. Different distributions in precipitation compared to the multiannual average occurred in both years. In particular, a wetter November and December and a drier spring than the multiannual trend were recorded in 2019–2020. In

2020–2021, the season was mainly characterized by unusual rainfall in December and January and a particularly dry March. In 2020, precipitations were roughly double that of the multiannual average in early summer. A drier summer (June to September) compared to the multiannual trend and 2020 was encountered in 2021.

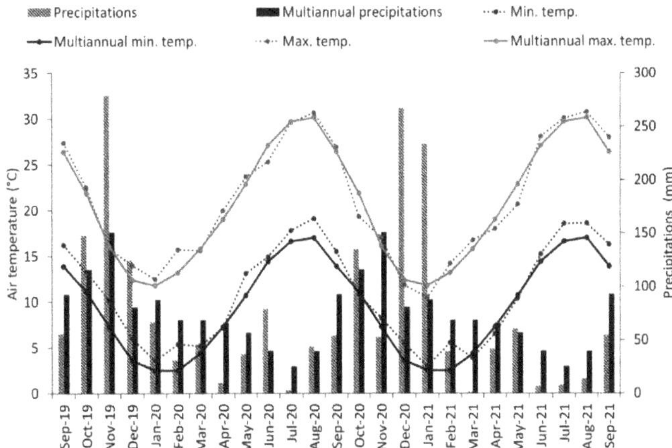

Figure 1. Monthly average minimum and maximum temperatures (°C) and monthly total precipitations (mm) during the trials at San Piero a Grado (2019–2021) compared with the long-term average values (1993–2020).

The experiment was set up in different fields each year. Both trials followed a fallow year. The soil, a Typic Xerofluvent, had a loamy sand texture in the first site (2019–2020) and a sandy loam texture in the second (2020–2021). Soil chemical characteristics (0–30 cm depth) at the beginning of the experiment are represented in Table 1.

Table 1. Initial soil characteristics (0–30 cm depth) at trial sites.

Characteristic	2019–2020	2020–2021
pH	6.72	8.27
Organic matter (g 100 g^{-1})	2.19	1.45
Total N (g 100 g^{-1})	0.12	0.09
Available P (mg 100 g^{-1})	0.49	0.91
Cation exchange capacity (meq 100 g^{-1})	10.81	3.56
Electrical conductivity (µS cm^{-1})	113.54	54.45

Organic matter was determined with the Walkley–Black method, total Nitrogen with the Kjeldahl method, and available Phosphorus with the Olsen method.

2.2. Experimental Set-Up

The experiment included eight treatments within two factors, cover crop species and tillage system. Treatments were arranged over plots following a randomized complete block design (RCBD) with three replicates. Plots measured 6 m × 10 m. Cover crop treatments consisted of monocultures of rye (*S. cereale* cv. Dukato), squarrose clover (*T. squarrosum* cv. OK), their mixture, and control kept with natural vegetation during winter. Cover crops and control plots were compared under two tillage systems (i) flail mowing and soil incorporation to almost 15 cm depth with a rotary tiller/cultivator in two passes as a standard conventional tillage system (CT) and (ii) rolling with two passes of a roller-crimper (Eco-Roll®, Clemens Technologies, Wittlich, Germany) followed by two flaming interventions (MAITO Srl., Arezzo, Italy, based on a prototype designed and realized at the University of Pisa) as a no till (NT) system (Figure 2).

Figure 2. Cover crop termination with a roller-crimper (**left**) and a flame-weeding machine (**right**) in the no-till systems (NT).

Before cover crop establishment, the soil was tilled by a combined harrow at 15 cm depth, followed by one pass of a rotary harrow to complete seedbed preparation. Cover crop seeds were hand-broadcast on the plots on 25 October 2019 and 23 November 2020, respectively. Cover crops were seeded at a rate of 50 kg ha^{-1} for squarrose clover and 180 kg ha^{-1} for rye. The corresponding mixture consisted of half the seeding rate of the single species (25:90 kg ha^{-1}). Cover crops grew rainfed without fertilization and weed control interventions and were terminated on 28 May 2020 and 7 June 2021, respectively, for the first and second years. At termination time, squarrose clover was almost at the full flowering stage (BBCH 65), and rye was at the early milk stage (BBCH 73) [28].

Tomato seedlings (*Solanum lycopersicum* L. cv. Elba F1) were transplanted with a vegetable transplanter (Fast model, Fedele Costruzioni, Lanciano, Chieti, Italy) adapted to each system [29]. Transplantation was conducted at a density of 2.2 plants m^{-2} (1.5 m between rows, 0.3 m in-row) on 3 and 9 June 2020 and 2021, respectively. Tomatoes were harvested manually twice in each season, starting at the end of August/early September. Fertilization, irrigation, and phytosanitary interventions were the same for all treatments compared and followed the European organic agriculture standards (ex Reg. EU 2018/848 and 2018/1584). Fertilization consisted of a total of 38.7 and 36.3 kg ha^{-1} of N given in the first and the second year, respectively, and 21.8 kg ha^{-1} K$_2$O only in the second year (VIT-ORG, Green Has Italia, Canale, Italy, 3-0-6; NUTRIGREEN, Green Has Italia, 8-0-0). Fertilizers were distributed by drip fertigation along the tomato cropping cycle. In addition, calcium (NEWCAL, Green Has Italia, 16.8%) was given in the second year at 8 kg ha^{-1} of CaO via foliar applications. Weed management during tomato growth was carried out with one manual weeding intervention each year on the intra-row space in all the treatments. Weeds in the inter-row area at early tomato growth stages were controlled by inter-row cultivation (twice in the first year and once in the second year) only in the plots where cover crops and natural vegetation were incorporated. In contrast, inter-row shredding was performed twice each year in no-till plots.

2.3. Samplings and Measurements

Cover crop biomass samplings were taken at the termination time by clipping the cover crop 2–3 cm above the soil surface over two areas of 0.5 m^2 each per plot. Weeds were hand-separated from the samples, and the biomass yield on a dry weight (DW) basis of each component (weeds and cover crops) was determined after oven-drying of aliquots at 60 °C until a constant weight. Biomass data were used to calculate the land equivalent ratio (LER) for the mixture as a metric of "outyielding" or biodiversity effects, as reported by Smith et al. [30]. A value of LER > 1 indicates that the mixture is overyielding relative to the constituent monocultures or that more land would be required for monocultures to produce equivalent biomass to the mixture. An LER < 1 instead implies that the mixture is under-yielding or that monocultures need less land than the mixture to reach the same biomass.

Dried samples of the cover crops were then ground to pass through a 1 mm sieve. Total N and P were evaluated in the dry ground samples by Kjeldhal and colorimetric methods, respectively.

Soil-Plant Analysis Development (SPAD) readings were taken on three plants in two central regions of each plot by means of a portable chlorophyll meter (SPAD-502 Plus, Konica Minolta, Tokyo, Japan) to monitor the N status of the tomato crop. The measurements were performed on three dates (SPAD1, SPAD2, and SPAD3), on the same plants every time, during each cropping season (10/07/2020, 30/07/2020, and 27/08/2020 in the first year; 19/06/2021, 22/07/2021, and 06/08/2021 in the second year). The three dates corresponded to the vegetative growth (SPAD1), the flowering (SPAD2), and the fruit development (SPAD3) stages. One fully developed leaf was chosen for each plant, and an average of three measurements per leaf was recorded. Tomato plant growth was evaluated as plant height variations on the same plants and two dates (Height1: flowering stage; Height2: fruit development stage).

Tomato fruits were harvested from each plot in two interventions at fruit maturity, over two areas of three plants each, to determine crop yield. Tomato fruits were separated into unmarketable and potentially marketable fruits (the sum of commercial and green fruits). Fruits in each category were counted, and their fresh weight was measured. The same plants were harvested during each intervention, and the data were cumulated. A sample from marketable fruits at each harvest was put in the oven at 60 °C to determine the dry matter content. Ground samples were then analyzed for total N and P in the same way as the cover crops. The same plants harvested were also sampled each year at the final harvest time to determine their total dry biomass and N and P contents. N and P accumulations were calculated by multiplying their concentration by the dry yield. Nitrogen utilization efficiency (NUtE) was calculated as follows:

$$\text{NUtE} = \frac{Y}{Nacc_t} \tag{1}$$

where Y is the fresh weight of potential tomato yield (Mg ha^{-1}), and $Nacc_t$ is the N accumulated (kg N ha^{-1}) in total aboveground biomass of the crop.

Weeds from the same 1.5 m^2 areas where tomato plants were harvested were also collected for total aboveground dry biomass determination.

2.4. Statistical Analysis

Statistical analyses and data visualization were conducted using Rstudio statistical software (R version 4.1.2). All data were first subjected to Shapiro–Wilk and Levene's tests to check for data normality and heteroscedasticity. Data that verified these assumptions were fitted to general linear models. Data that failed to meet these assumptions were fitted to generalized linear models selecting the best fit (distribution and link function) via the Akaike information criterion (AIC). Initially, mixed models were fitted including the block and the year as random factors. As the variance of the random effect was insignificant and the models were overfitted, we used linear models including only fixed factors. For cover crop variables, we considered the cover crop, the year, their interaction, and the block as model fixed terms. For tomato crop variables, the terms in the models were the cover crop, the tillage system, the year, their interaction, and the block. The same terms were adopted for SPAD and plant height besides timing and its interaction with the other terms. Mixed models were first fitted to these two variables trying to account for the correlation structure of data between the individual plants assessed. Analysis of deviance was then generated using the Likelihood Ratio test or F-test according to the distribution. Estimated marginal means (emmeans) comparison for statistically significant terms was carried out with Bonferroni-adjusted post hoc test at a significance level of 5% ($p \leq 0.05$). Data were presented as emmeans and corresponding standard errors. The significance of the main terms and their interactions are presented in Tables S1–S3 in the Supplementary Material.

3. Results

3.1. Cover Crops Performance at Termination Time: Aboveground Dry Biomass, Nutrient Accumulation, and Weed Control

The statistical significance of the effects of the main factors and their interactions on the evaluated cover crop variables are presented in Table S1 in the Supplementary Material. The cover crop biomass was not affected by the year but only by cover crop species (Figure 3). The mixture and rye produced, on average, almost 36% more biomass (8.14 and 8.51 t DW ha^{-1} for mixture and rye, respectively) than squarrose clover (6.09 t DW ha^{-1}). The mixture, which was equivalent to pure rye in biomass, had consistently in both years around 64% rye and 36% squarrose clover in dry matter composition at the termination time. The year factor did not affect the LER of the mixture. Pooled across the two years, the LER had a mean of 1.10 ± 0.04 ($p < 2 \times 10^{-16}$).

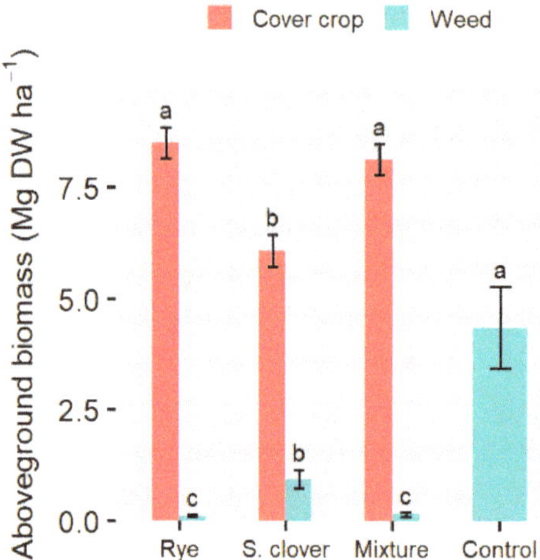

Figure 3. Cover crop and weed aboveground dry biomass at the time of cover crop termination. For each variable, bars with different letters are significantly different ($p \leq 0.05$). S. clover: Squarrose clover. Error bars represent the standard errors of the estimated marginal means.

The year and cover crop species drove changes in weed biomass in cover crop stands with no interaction between both (Table S1, Figure 3). Greater weed biomass characterized the second year (+35.3%). However, the three cover crops successfully controlled weeds in both years. The positive weed suppression service provided by the cover crops was about 97% for rye and the mixture, and 78% for squarrose clover, averaged over the two years, compared to the control kept with natural vegetation.

Different tissue N concentrations were found across cover crop species and years (Table S1). In 2021, tissue N concentration in rye was slightly greater than in 2020. In both years, squarrose clover had the highest N concentration, around 2.5 times on average more than N in rye at the termination time (Figure 4A). N concentration for the mixture (1.23 and 1.33 g 100 g^{-1}, respectively for 2020 and 2021) was intermediate between pure rye (0.72 and 0.96 g 100 g^{-1}, respectively, for 2020 and 2021) and pure squarrose clover (1.93 and 2.22 g 100 g^{-1}, respectively, for 2020 and 2021).

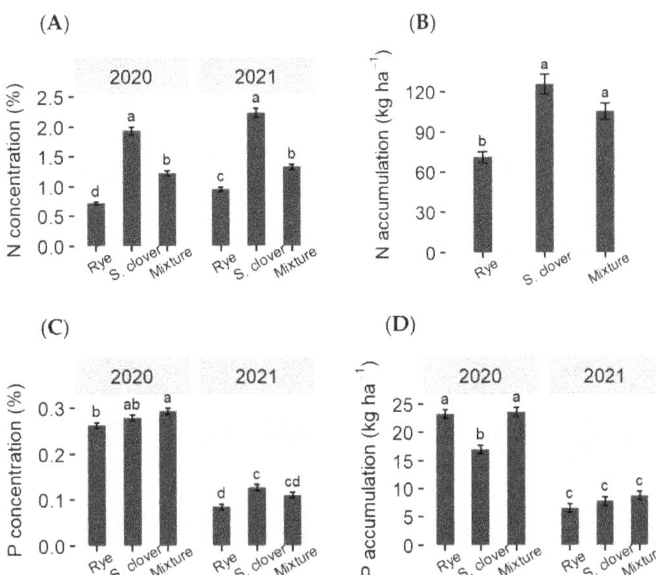

Figure 4. Interaction effects of cover crop and year (**A,C,D**) and main effects of cover crop (**B**) on tissue N and P content and accumulation at the cover crop termination time. Different letters indicate significant differences ($p \leq 0.05$) in each interaction/main effect. S. clover: Squarrose clover. Error bars represent the standard errors of the estimated marginal means.

Similarly, cover crop N accumulation differed across cover crop species and years (with no significant interaction, as shown in Table S1) and was mainly influenced by cover crop tissue N concentration (Figure 4B). Across both years, squarrose clover and the mixture accumulated the highest amount of N, while rye had the least amount), with N accumulation varying between 71.21 and 125.60 kg ha^{-1}. P concentration was drastically lower in 2021 in all the cover crops (on average, 0.28 vs. 0.11 in 2020 and 2021, respectively). The mixture in 2020 and the clover in 2021 had higher P concentration than rye, with differences accounting for up to +40% in 2021, as represented in Figure 4C. P accumulation in cover crops ranged from 16.99 to 23.63 kg ha^{-1} in 2020, with clover accumulating the least (but significantly compared to rye and the mixture only in 2020). In 2021, on average the cover crops accumulated around half the P accumulated in 2020, ranging from 7.78 to 8.75 kg ha^{-1}, with no differences between the cover crops (Figure 4D).

3.2. SPAD and Plant Height

Differences arose for SPAD values and tomato plant height between years, tillage systems, cover crop species, timing and their interactions (Table S2 in the Supplementary Material, Figures 5 and 6). Averaged over cover crop species and years (tillage × crop stage), we observed a positive effect of CT on the SPAD value compared to NT during the cropping season until fruit development (Figure 5A). Furthermore, the tillage system interacted with cover crops differently each year (Table S2). As Figure 5B shows, differences between cover crops in each tillage system (averaged over crop stages) were observed mainly in 2020, with squarrose clover and the mixture (this latter only in NT) outperforming the other cover crop treatments. None of the cover crop effects varied significantly between tillage systems despite a general increase in CT compared to NT treatments, nor between the years (Figure 5B).

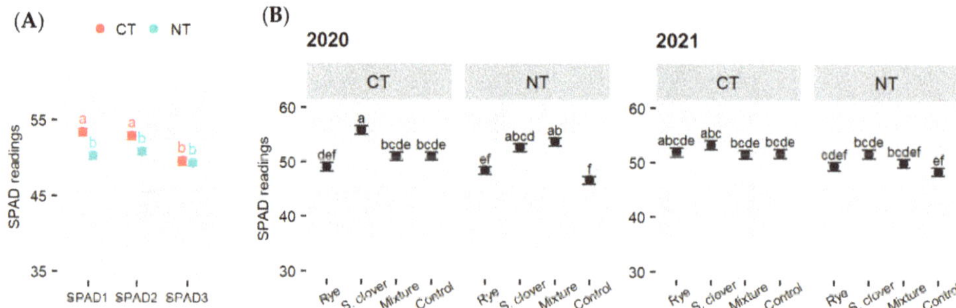

Figure 5. Variation in SPAD values during the growing season in response to the interactions of the tillage system and crop stages (**A**) and cover crop, tillage, and year (**B**). Points represent the emmeans, and bars represent their standard errors. Different letters indicate significant differences ($p \leq 0.05$) in each interaction. S. clover: squarrose clover; CT: conventional tillage; NT: no-till; SPAD1: vegetative growth stage; SPAD2: flowering stage; SPAD3: fruit development stage.

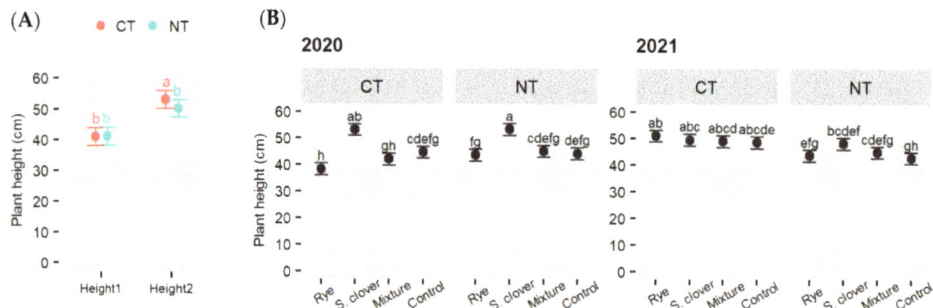

Figure 6. Variation in tomato plant height during the growing season in response to the interactions of tillage system and crop stage (**A**) and cover crop, tillage system and year (**B**). Points represent the emmeans, and bars represent their standard errors. Different letters indicate significant differences ($p \leq 0.05$) in each interaction. S. clover: squarrose clover; CT: conventional tillage; NT: no-till; Height1: flowering stage; Height2: fruit development stage.

With regards to tomato plant height, we found a significant interaction between tillage and crop stage (Figure 6A). Averaged over years and cover crops, tomato plants grown under CT were taller than NT late in the season (Figure 6A). Averaged over years and tillage (cover crop × crop stage), higher plants under squarrose clover compared to all other treatments (except for mixture during fruit development) were found during the season. Tillage interacted also with cover crops and years (Table S2, Figure 6B). Similarly to SPAD, squarrose clover plots had the tallest tomato plants among the cover crops in 2020 in both tillage systems. Differences between cover crops in each tillage system in 2021 were less evident, even under CT where a significant increase in plant height in rye and mixture treatments relative to 2020 was noticed (Figure 6B).

3.3. Tomato Fresh Yield and Aboveground Dry Biomass

Significant interactions between tillage and year, cover crop and year, and cover crop and tillage were found for tomato unmarketable, potential, and total fresh yields (Table S3 in the Supplementary Material, Figure 7A–C).

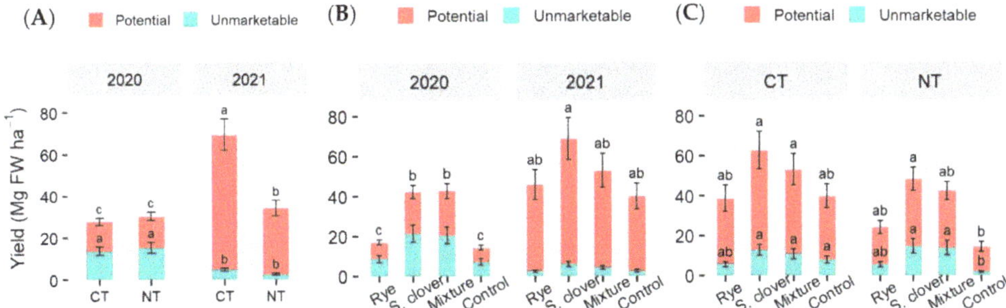

Figure 7. Interaction effects of tillage system and year (**A**), cover crop and year (**B**) and cover crop and tillage system (**C**) on tomato potential and unmarketable yields. Different letters indicate significant differences ($p \leq 0.05$) in each interaction. Interaction is not significant for the unmarketable yield in Figure (**B**). S. clover: Squarrose clover; CT: conventional tillage; NT: no-till. Error bars represent the standard errors of the estimated marginal means.

Concerning the interaction between tillage system and year (Figure 7A), a significant increase in the potential yield and a decrease in the unmarketable yield were observed for both CT and NT in 2021. However, differences between tillage systems for potential fresh yield were only observed in 2021 (Figure 7A). In 2021, the total fresh yield was 2.5 times greater than that in 2020. Compared to NT, double the total fresh yield was found in CT (69.73 vs. 34.46 t ha^{-1}). Averaged over tillage systems, the interaction between cover crop and year was significant only for the potential yield and the total yield, showing clover and mixture producing around 2.7 times more of potential and total yield than rye and control in 2020 while being statistically like them in 2021 (Figure 7B). A significant interaction between cover crop and tillage system (Figure 7C) did not reveal statistically supported differences between cover crops despite the higher values in squarrose clover and mixture treatments in both tillage systems. In contrast, clover and mixture significantly incremented NT's potential and unmarketable yields by almost 68% and 150%, averaged over both cover crops, compared to rye and control. respectively (Figure 7C). Differences between treatments were not statistically significant for the total fresh yield as well, despite a tendency for squarrose clover and mixture in CT and NT to have higher yields relative to all other treatments.

Differences in fresh yields were consistent with the trend demonstrated by the number of fruits produced per area (Tables S3 and S4 and Figure S1A,B in the Supplementary Material). For dry matter content in tomato fruits, we found it was 30% higher in the first year (7.91 vs. 6.09 g 100 g^{-1}) (Table S4). Averaged over tillage systems and years, fruit dry matter content was lower in clover and mixture treatments than the control (Table S4). Considering the interaction between years, tillage systems and cover crops, the dry matter content of fruits did not reveal differences among the three cover crops in either tillage system but only with control (Tables S3 and S4 in the Supplementary Material) and tended to be slightly higher in NT (confirmed for squarrose clover in 2021 with 6.16 vs. 5.49 g 100 g^{-1}).

Tomato total dry yield at harvest followed the same trend as fresh tomato yields each year (Figure 8A–C). Tomato shoot plant biomass was influenced by the interaction of cover crop, tillage, and year and confirmed the trend observed in yield results. As represented in Figure 8D, squarrose clover monoculture and the mixture (though the latter was statistically like rye) stimulated tomato shoot biomass in CT and NT in 2020. Despite the same tendency, no differences between cover crops were found in 2021 in both tillage systems. Although there was a tendency for higher shoot biomass in all cover crops in CT relative to NT in 2021, no statistically significant differences were reported (Figure 8 D). As with the yield, shoot biomass was greater in CT in 2021 compared to that in 2020.

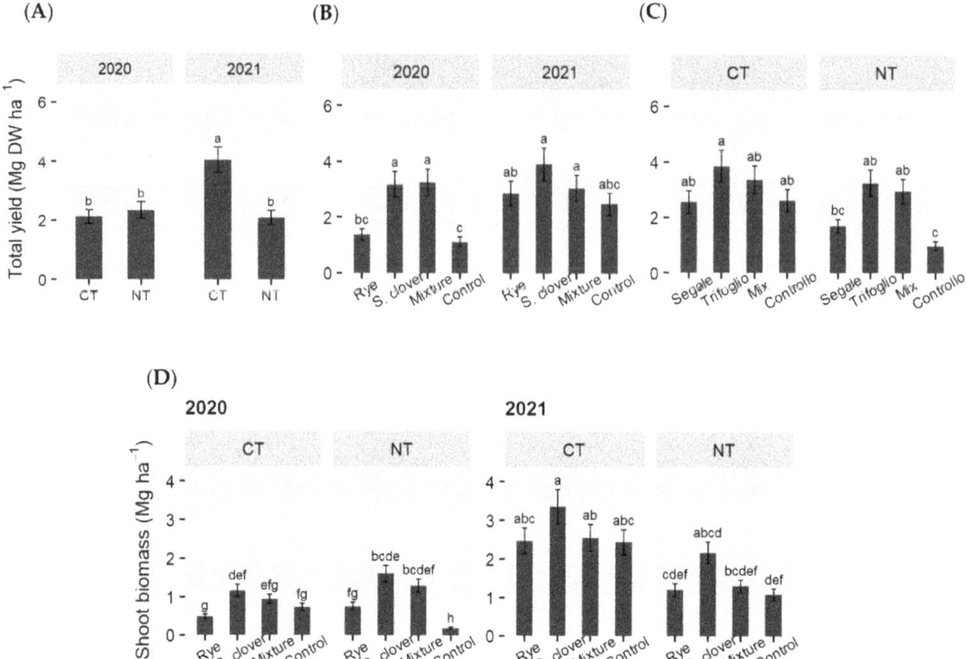

Figure 8. Interaction effects of tillage system and year (**A**), cover crop and year (**B**) and cover crop and tillage system (**C**) on dry tomato yield, and interaction effects of tillage system, cover crop and year (**D**) on tomato shoot biomass at harvest. Different letters indicate significant differences ($p \leq 0.05$) in each interaction. S. clover: Squarrose clover; CT: conventional tillage; NT: no-till. Error bars represent the standard errors of the estimated marginal means.

3.4. N and P Concentration and Accumulation in Tomato Shoots and Fruits

The significance of factors and their interaction effects on N and P concentrations and accumulations in tomato biomass components are presented in Table S3 in the Supplementary Material. Nconc in fruits was generally greater than that in shoots (Figure 9A–C). Tillage had an insignificant impact on Nconc in plant tissues but a significant effect on Nconc in fruits (Figure 9A). Under NT, tomato fruits had almost 9.2% lower N contents. Cover crop species changed Nconc in shoots, with those under squarrose clover monoculture having around 15.5 and 26.3% more compared to those under rye monoculture and mixture respectively, as represented in Figure 9B. At the same time, the interaction between cover crops and years on Nconc in fruits revealed the effects of cover crops only in 2021, with clover also increasing fruit N content (2.23 g N 100 g^{-1}) relative to all other cover crop treatments which appeared to behave similarly (1.84 g N 100 g^{-1} in rye, 1.73 g N 100 g^{-1} in control and 1.58 g N 100 g^{-1} in mixture). Nconc in fruits was generally higher in 2021, as Nconc in shoots (Figure 9C), but it was most evident only in clover (+37.6%).

Like N content and tomato yields, N accumulations (Nacc) in shoots and fruits were higher in the second year due mainly to a considerable increase in those of CT treatments. The interaction of cover crop, tillage system and year influenced Nacc in tomato shoots while that of tillage system and year, cover crop and year, and cover crop and tillage influenced that in fruits (Table S3 in the Supplementary Material). Changes in tomato yields mainly drove changes in N accumulation in the plant's organs. Squarrose clover stimulated shoot Nacc under both tillage systems compared to control (a significant difference was noticed within NT in 2020) and the other cover crops, mainly rye (significantly only under CT in 2020) (Figure 10A). Within the years, differences between tillage systems were not statistically detected for any of the cover crop treatments essayed, but for the control in 2020. Nevertheless, Nacc in shoot biomass seemed to be higher in CT than NT in 2021, and in 2021 than 2020 in CT as with yields. The amount of N accumulated in tomato fruits tended to be the highest following the monoculture of squarrose clover with 70.30 kg ha^{-1} averaged over both years and tillage systems, and the lowest in the rye and control with a total of 39.17 and 30.54 kg ha^{-1}, respectively (Figure 10B). Nacc in fruits was remarkably reduced in NT (-50%) only in 2021 (37.46 vs. 37.11 kg ha^{-1} for CT and NT, respectively, in 2020 and 77.69 vs. 39.01 kg ha^{-1} for CT and NT, respectively, in 2021). Differences were due to the increase in Nacc in shoots and fruits in CT in general in 2021 compared to 2020. On average, tomato plants accumulated 114.6 kg ha^{-1} under CT and only 55.1 kg ha^{-1} under NT in 2021, while 40.4 kg ha^{-1} of N was absorbed on average in 2020 for both NT and CT.

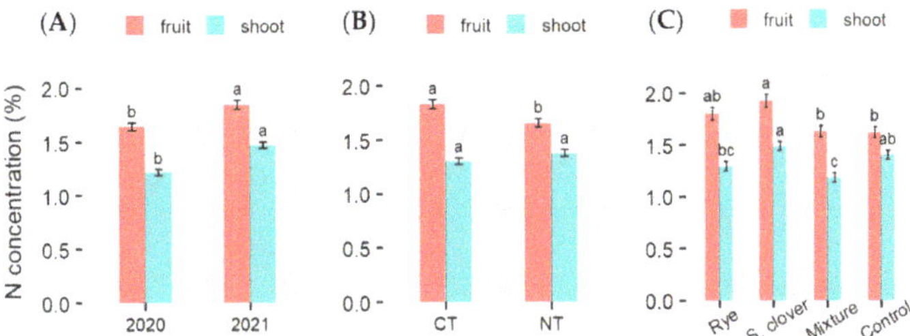

Figure 9. Main effects of tillage (**A**), cover crop (**B**), and year (**C**) on Nitrogen concentration (Nconc) in tomato shoots and fruits. For each variable, different letters indicate significant differences ($p \leq 0.05$) in each main effect. S. clover: Squarrose clover; CT: conventional tillage; NT: no-till. Error bars represent the standard errors of the estimated marginal means.

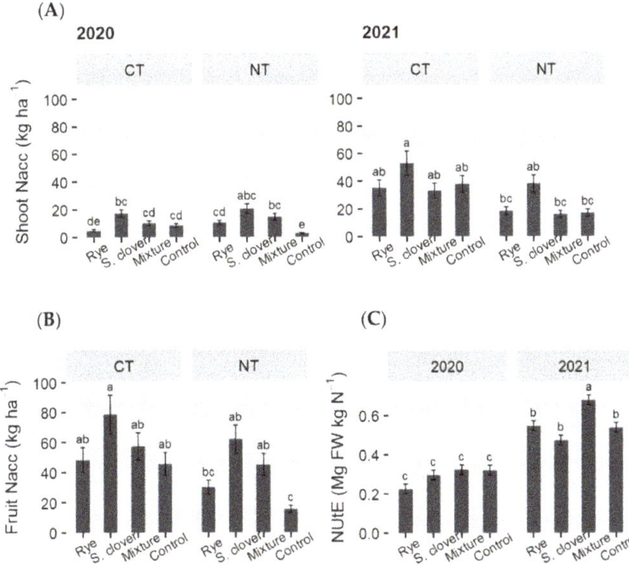

Figure 10. Interaction effects of cover crop, tillage, and year on Nitrogen concentration (Nconc) in tomato shoots (**A**), cover crop and year on Nitrogen accumulation (Nacc) in tomato fruits (**B**), and cover crop and year on Nitrogen utilization efficiency (NUtE) (**C**). Different letters indicate significant differences ($p \leq 0.05$) in each interaction. S. clover: Squarrose clover; CT: conventional tillage; NT: no-till. Error bars represent the standard errors of the estimated marginal means.

Cover crops affected the Nitrogen utilization efficiency (NUtE) only in 2021 with the mixture enhancing the NUtE compared to the monocultures of clover (+43.9%) and rye (+24.2%). The NUtE was lower in 2020 for all treatments, averaging around 0.29 compared to 0.56 Mg of fresh yield kg^{-1} N in 2021.

Statistical analysis outcomes regarding P concentration and accumulation are presented in Table S3 in the Supplementary Material. Tillage and cover crops modified P concentration in tomato shoots as well. Contrary to Nconc, Pconc in shoots was improved by 30% under NT (0.18 vs. 0.24 g P 100 g^{-1}) (Table S4). Regardless of tillage and year, shoot Pconc was the least in squarrose clover (0.16 vs. 0.23 g P 100 g^{-1} in the control in 2020 and 0.15 vs. 0.25 g P 100 g^{-1} in the mixture in 2021) (Table S4). Pconc in fruits seemed indifferent to cover crops and the tillage system adopted but decreased in 2021 (−38.8%) (Table S4). Despite the higher Pconc in tomato shoots under NT, Pacc in shoots and fruits response to tillage was variable with the year and depended mainly on the dry aboveground biomass production: in 2020, plants under NT had higher Pacc in shoots, while in 2021 the contrary was true for fruits too (Figure S2 in the Supplementary Material). Cover crops behaved almost similarly for Pacc in fruits and shoots, and only under NT were they better than the control (Figure S2 in the Supplementary Material). Tomato plants accumulated in fruits between 12.49 and 7.63 kg P ha^{-1} in CT and 11.29 and 3.12 in NT, whereas the uptake in shoots varied between 3.29 and 2.92 in CT and 3.39 and 1.68 in NT.

3.5. Weed Biomass at Tomato Harvest

The response of weed biomass at harvest was influenced by the interaction of the tillage system and year, and cover crop and tillage system, as reported in Table S3 in the Supplementary Material. Weed infestation in CT, averaged over the cover crops, was around 25 times greater in the first year than in the second year, whereas it seemed similar between NT over the years (Figure 11A). The effect of the tillage system on weed biomass was observed only in 2021 with a consistent decrease of almost 92% relative to CT (Figure 11A). Despite the significant effect obtained by the generalized linear model for the interaction between cover crop and the tillage system, the post hoc test returned no significant differences between treatments (Figure 11B). However, it seems that when averaged over the years, the weed-suppressing effect of cover crops tended to be more significant in NT, with the mixture probably reducing weed biomass compared to the control (−65%) (Figure 11B).

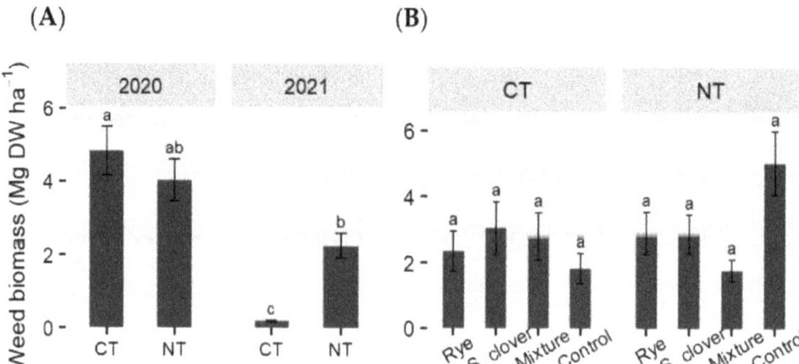

Figure 11. Interaction effects of tillage system and year (**A**) and cover crop and tillage system (**B**) on weed biomass at tomato harvest. Different letters indicate significant differences ($p \leq 0.05$) in each interaction. S. clover: Squarrose clover; CT: conventional tillage; NT: no-till. Error bars represent the standard errors of the estimated marginal means.

4. Discussion

4.1. Cover Crops Performance at Termination Time: Aboveground Dry Biomass, Nutrient Accumulation, and Weed Control

Pure stands of rye, squarrose clover, and their mixture produced a large quantity of residue, with rye and the mixture outperforming the clover monoculture. Over two years, squarrose clover has grown around 6.1 t ha^{-1} of dry matter, significantly higher than in a recent study [5]. Early sowing, i.e., October–November, and late termination dates, i.e., end of May, besides good climatic factors, i.e., at least 595 mm of precipitations, have favored squarrose clover growth and ensured well-performing stands. In addition, all three cover crops have produced stable amounts of biomass yields and almost equivalent N yields across the two years of the experiment, despite differences in temperatures and precipitation distribution between the years. Even though it yielded the same quantity of residues as the most performing monoculture, i.e., rye, we found that the mixture slightly enhanced biomass production relative to the pure stands. The same might be deducted by a LER value superior to 1 (1.10 with a confidence interval between 1.02 and 1.19). Like our findings, mixture "over-yielding" was repeatedly reported when greater biomass was produced than the average monoculture yields [30,31]. On the other hand, "transgressive over-yielding" has been documented when the mixture produced more biomass than the best-performing monoculture [22,32]. The overyielding observed might suggest that squarrose clover–rye mixtures have benefited from positive interspecific or/and lower intraspecific interactions. In our case, the rye component might have accounted for the overyielding of the mix, as it produced more than what ideally could be expected under a half seed rate.

The reasons for the beneficial interspecific interactions in mixtures include facilitation processes resulting from complementarity or selection effects at play according to resource availability [20,32,33]. As Mason et al. [31] described, complementarity effects can occur through the efficient utilization of resources (water, Nitrogen, and light) caused by niche partitioning. In combinations of grasses and forbs, the complementary use of resources was found to increase when resources are limited, while the proportion of biomass gained through the dominance of the high-yielding monoculture increases with resource availability [34]. As a legume, the squarrose clover can fix atmospheric Nitrogen through biological fixation and have a deeper root system than rye which might have allowed rye to access more amounts of N. Moreover, rye is an aggressive N scavenger [23]. Reducing the proportion of rye in the cover crop might have reduced intraspecific competition and increased rye biomass in the mixture, where it accounted for an average of 64% of total biomass.

As expected, N accumulation was the highest in squarrose clover among cover crops, while the mixture was intermediate between rye and clover. Biomass overyielding could be potentially the most significant contributor to the accumulation of N in the mixture, though insufficient to obtain equivalent amounts to squarrose clover [9].

Compared with the control, all three cover crops consistently reduced weed biomass during the fallow period as measured at termination time, with rye outperforming the squarrose clover monoculture. Rye's ability to control weeds has been well established. Rye was reported to provide at least 80% weed control in terms of weed density and occasionally even biomass, primarily of annual weeds, compared to no-cover crops [13,22]. Yet, in our study, the weed suppression of the mixture matched that of pure rye, the most suppressive cover crop. These results are congruent with those of Smith et al. [35], who found no differences between rye and other multispecies mixtures of cover crops in weed suppression despite non-transgressive overyielding. Indeed, including aggressive species such as rye, even at a low rate, such as 20%, in mixtures, can ensure a high level of weed suppression [22,36]. Cover crop weed suppression depends on several mechanisms of which mainly resource preemption and competition are often indicated by cover crop biomass [36–38], soil coverage [39], and/or allelopathy [40]. Instead of the overall biomass it generates, rye has repeatedly been demonstrated to inhibit weeds owing to its comparatively rapid growth, soil covering, N scavenging, and, notoriously, its allelopathic capacity [36,41,42]. Despite some allelopathic effects [43] and the power to smother weeds [18,44], squarrose clover is a slow-growing legume that, by improving N availability in the soil, allows for substantial weed establishment during the season [36,37].

4.2. Effects of Cover Crops and Tillage System on Tomato Plant Growth, Crop Productivity, Nutrient Accumulation, and Residual Weed Abundance

Crop performance was primarily influenced by the interaction of tillage and cover crop and by the general interaction of the single factors with the year. Tomato yields were higher in 2021 because of the interannual variability in weather, particularly the drier summer, the lower pest incidence, and the effectiveness of phytosanitary interventions once pests have emerged.

Cover crops are expected to affect the succeeding cash crop yield positively. Indeed, the least productive tomato systems were those without a cover crop or with rye monoculture, mainly in 2020, as shown by the interaction of cover crop and year. These results were found in both tillage systems despite not being statistically well verified (possibly because of year and field variability). As SPAD readings and N concentration in tomato shoots and fruits have revealed, tomato plants under rye, especially in 2020, and with no-till in general, might have suffered N stress during the season. Low N availability, for which fertilization did not compensate, likely contributed to the reduced growth and yield of tomatoes grown on rye. Dense rye residues, as a cereal with low N concentration and high residue C/N, might have caused a temporary shortage of available Nitrogen through N immobilization, besides releasing allelopathic compounds. For the same reasons, some experiments have documented a decline in vegetable performance with rye residues relative to other or no-cover crops [12,13].

In contrast, squarrose clover might have released N before and after termination, which met some of the nutrient requirements of tomato plants [26,45]. Through the N released, the clover might have also alleviated the weed-crop competition by providing an advantage to tomato plants before weed establishment [46]. These factors were previously behind the enhanced performance of several vegetables under legume mulches [45,47–49]. The effect of the mixture on tomato performance, which was equivalent to that of squarrose clover in 2020, might be due to some N release, lower competition with the weed community, and successful weed control, especially in NT. Increasing the squarrose clover proportion in the mixture may have promoted N release and provided an advantage to the tomato crop in the competition against weeds early in its growth [50]. Besides cover crop species, cover crop aboveground biomass management and weed cultivation may have influenced the residue decomposition rate over the two years. The reduced contact between residue and the soil surface in dead mulches may be responsible for a lower mineralization rate [26], especially in the poor N rye. Interestingly, we found that the no-till system consistently enhanced the P content in tomato shoots, supporting findings from an earlier experiment [6]. Since equivalent plant biomasses were produced between CT and NT in the first year, we argue that one factor responsible for this effect could be the promotion of arbuscular mycorrhizal fungi symbiosis and activity by no-till.

Cover crops and tillage and their interactions with the year affected N and P uptakes driven by tomato biomass and fruit yields. However, despite a lower yield than squarrose clover, the mixture had the highest nitrogen utilization efficiency in 2021. The mixture in that year produced around 678 kg of tomato fresh yield (sum of marketable and green fruits) for each kg of N absorbed from the soil compared to only 322 kg of tomato kg^{-1} N for pure squarrose clover. This output is comparable to past findings, demonstrating an inverse trend between nitrogen utilization efficiency and yield (somehow lower than the squarrose clover monoculture) and nitrogen-use efficiency and N inputs (here indicated by a probable lower N supply by this cover crop) [45,47].

In-season weed suppression might have been provided during tomato growth, especially at earlier stages (i.e., soon after cover crop residue management). However, its magnitude may be affected by residue management and, more generally, the tillage system.

Our results revealed only a tendency (not supported statistically) for all three cover crops, especially for the mixture, to reduce weed biomass under the no-till system. This effect could be ascribed to the physical barrier that cover crops provide, thereby modifying the environmental conditions under the residue necessary for weed emergence and establishment, i.e., blocking light and lowering daily temperature amplitude. The apparent weed control by the mixture of dead mulch could be due to the enhanced release of bioactive compounds in the soil when rye is mixed with clover, slowing weed emergence and resulting in long-season weed suppression [43].

In contrast to earlier works undertaken under Mediterranean conditions in Italy [51,52], we did not observe a significantly lower weed abundance in no-till systems compared to the conventionally tilled systems; we observed instead the contrary in one year. Weed infestation in NT, despite the cover crop weed suppression, was comparable to CT in 2020 and considerably higher in 2021. For the latter, we might think that the improved soil conditions for weed seed germination and inefficient post-transplanting direct weed control (performed as occasional inter-row threshing) could be the reasons behind the increased weed biomass at the harvest of tomato in NT. For 2020, despite weed cultivation, a large weed infestation during tomato growth in CT treatments might have inflicted a crop performance drop and masked the differences with NT. Data on weed population compositions among years and treatment combinations might help unravel other ecological mechanisms behind these different behaviors of weed biomass.

5. Conclusions

The three cover crops compared in this study (rye, squarrose clover and their mixture) are attractive winter cover crops for the Mediterranean region, with fast growth, high productivity, and great weed competition. Mixtures might even benefit from overyielding, generating almost similar amounts to the highest productive cover crop, i.e., rye. However, their effects on cash crops under different tillage systems depend widely on the pedo-climatic and field conditions in which they occur. Cover crops were demonstrated to be of a high value, especially to no-till systems. Using the mixture as dead mulch may provide consistent weed biomass reduction under different infestation levels and possibly weed communities. Nevertheless, the effect of the mixture on tomato performance was variable with the site/year in our study. Squarrose clover, through its Nitrogen fixing capacity, compensated for its lower weed control ability as dead mulch, which ensured the best crop performance even under high weed infestation conditions. A legume also seems a guarantee in conventionally tilled systems under these suboptimal circumstances. Besides the squarrose clover monoculture, a mixture with increased legume proportion could be a trade-off between weed control (before and after cover crop termination) and N supply to the crop. Future research may concentrate on a deeper understanding of how the studied cover crops and the tillage system affect weed management, including weed abundance dynamics, weed community alterations, and weed functional biodiversity promotion. These insights would help us better understand the results obtained herein and develop fine-tuned cover crops that target long-term weed suppression.

Supplementary Materials: The following supporting information can be downloaded at: https://www.mdpi.com/article/10.3390/agronomy13082027/s1, Table S1: Significance of terms (*p*-values) from the linear models fitted for the dependent variables evaluated at cover crop termination. Table S2: Significance of terms (*p*-values) from the linear models fitted for SPAD and tomato plant height at different crop stages. Table S3: Significance of terms (*p*-values) from the linear models fitted for the dependent variables of tomato performance evaluated after cover crop termination. Table S4: Effects (emmean ± standard error) of tillage system, cover crop and year on tomato fruit number, fruit dry matter content, fruit and shoot P concentration (Pconc) and accumulation (Pacc). Figure S1: Effects of the interaction of cover crop and year (A) and cover crop and tillage system (B) on the number of marketable and green fruits, and unmarketable tomato fruits, respectively. Figure S2: Effects of the interaction of tillage system and year (A), cover crop and year (B) and cover crop and tillage system (C) on P accumulation in tomato shoots and fruits.

Author Contributions: Conceptualization, L.A.C. and D.A.; methodology, D.A.; formal analysis, C.F., M.S. and L.A.C.; investigation, M.S., L.A.C. and L.G.T.; resources, A.P. and M.M.; data curation, L.A.C., M.S., D.A., C.F. and L.G.T.; writing—original draft preparation, L.A.C.; writing—review and editing, D.A., C.F., M.S., L.G.T., M.M. and A.P.; visualization, L.A.C.; supervision, D.A. and C.F.; project administration, A.P.; funding acquisition, A.P. and M.M. All authors have read and agreed to the published version of the manuscript.

Funding: This research was funded by the Italian Ministry of Agricultural, Food and Forestry Policies (MiPAAF) under the project MEORBICO (MEccanizzazione dell' ORticoltura BIologica e COnservativa), DM 89238, 19 December 2019.

Data Availability Statement: Data are available upon request.

Acknowledgments: The authors acknowledge all the CiRAA staff involved in the field management, sample processing, and lab analysis. In particular, thanks to Alessandro Pannocchia, Giovanni Melai, Marco Della Croce, Roberta Del Sarto, Nadia Ceccanti, Rosenda Landi, Rosalba Risaliti, and Sabrina Ciampa. Warm thanks to Marco Ginanni, who helped design the trial and the water irrigation plant. We also acknowledge the precious support of Fausto Messina, Niccolò Pignotti, and Romans Caetani in field activities.

Conflicts of Interest: The authors declare no conflict of interest. The funders had no role in the design of the study; in the collection, analyses, or interpretation of data; in the writing of the manuscript; or in the decision to publish the results.

References

1. Palm, C.; Blanco-Canqui, H.; Declerck, F.; Gatere, L.; Grace, P. Conservation Agriculture and Ecosystem Services: An Overview. *Agric. Ecosyst. Environ.* **2014**, *187*, 87–105. [CrossRef]
2. Vincent-Caboud, L.; Peigné, J.; Casagrande, M.; Silva, E.M. Overview of Organic Cover Crop-Based No-Tillage Technique in Europe: Farmers' Practices and Research Challenges. *Agronomy* **2017**, *7*, 42. [CrossRef]
3. Blanco-canqui, H.; Shaver, T.M.; Lindquist, J.L.; Shapiro, C.A.; Elmore, R.W.; Francis, C.A.; Hergert, G.W. Cover Crops and Ecosystem Services: Insights from Studies in Temperate Soils. *Agron. J.* **2015**, *107*, 2449–2474. [CrossRef]
4. Shackelford, G.E.; Kelsey, R.; Dicks, L.V. Effects of Cover Crops on Multiple Ecosystem Services: Ten Meta-Analyses of Data from Arable Farmland in California and the Mediterranean. *Land Use Policy* **2019**, *88*, 104204. [CrossRef]
5. Abou Chehade, L.; Antichi, D.; Martelloni, L.; Frasconi, C.; Sbrana, M.; Mazzoncini, M.; Peruzzi, A. Evaluation of the Agronomic Performance of Organic Processing Tomato as Affected by Different Cover Crop Residues Management. *Agronomy* **2019**, *9*, 504. [CrossRef]
6. Antichi, D.; Sbrana, M.; Martelloni, L.; Abou Chehade, L.; Fontanelli, M.; Raffaelli, M.; Mazzoncini, M.; Peruzzi, A.; Frasconi, C. Agronomic Performances of Organic Field Vegetables Managed with Conservation Agriculture Techniques: A Study from Central Italy. *Agronomy* **2019**, *9*, 810. [CrossRef]
7. Büchi, L.; Gebhard, C.; Liebisch, F.; Sinaj, S.; Ramseier, H.; Charles, R. Accumulation of Biologically Fixed Nitrogen by Legumes Cultivated as Cover Crops in Switzerland. *Plant Soil* **2015**, *393*, 163–175. [CrossRef]
8. Abdalla, M.; Hastings, A.; Cheng, K.; Yue, Q.; Chadwick, D.; Espenberg, M.; Truu, J.; Rees, R.M.; Smith, P. A Critical Review of the Impacts of Cover Crops on Nitrogen Leaching, Net Greenhouse Gas Balance and Crop Productivity. *Glob. Chang. Biol.* **2019**, *25*, 2530–2543. [CrossRef]
9. Thapa, R.; Poffenbarger, H.; Tully, K.L.; Ackroyd, V.J.; Kramer, M.; Mirsky, S.B. Biomass Production and Nitrogen Accumulation by Hairy Vetch–Cereal Rye Mixtures: A Meta-Analysis. *Agron. J.* **2018**, *110*, 1197–1208. [CrossRef]
10. Mirsky, S.B.; Ryan, M.R.; Teasdale, J.R.; Curran, W.S.; Reberg-Horton, C.S.; Spargo, J.T.; Wells, M.S.; Keene, C.L.; Moyer, J.W. Overcoming Weed Management Challenges in Cover Crop—Based Organic Rotational No-Till Soybean Production in the Eastern United States. *Weed Technol.* **2013**, *27*, 193–203. [CrossRef]
11. Wells, M.S.; Smith, A.N.; Grossman, J.M. The Reduction of Plant-Available Nitrogen by Cover Crop Mulches and Subsequent Effects on Soybean Performance and Weed Interference. *Agron. J.* **2013**, *105*, 539–545. [CrossRef]
12. Miville, D.; Leroux, G.D. Rolled Winter Rye—Hairy Vetch Cover Crops for Weed Control in No-till Pumpkin. *Weed Technol.* **2018**, *32*, 251–259. [CrossRef]
13. Campanelli, G.; Testani, E.; Canali, S.; Ciaccia, C.; Trinchera, A. Effects of Cereals as Agro-Ecological Service Crops and No-till on Organic Melon, Weeds and N Dynamics. *Biol. Agric. Hortic.* **2019**, *35*, 275–287. [CrossRef]
14. Vandermeer, J.H. *The Ecology of Intercropping*; Cambridge University Press: Cambridge, UK, 1989.
15. Tilman, D.; Lehman, C.L.; Thomson, K.T. Plant Diversity and Ecosystem Productivity: Theoretical Considerations. *Proc. Natl. Acad. Sci. USA* **1997**, *94*, 1857–1861. [CrossRef]
16. Wortman, S.E.; Francis, C.A.; Bernards, M.L.; Drijber, R.A.; Lindquist, J.L. Optimizing Cover Crop Benefits with Diverse Mixtures and an Alternative Termination Method. *Agron. J.* **2012**, *104*, 1425–1435. [CrossRef]

17. Finney, D.M.; White, C.M.; Kaye, J.P. Biomass Production and Carbon/Nitrogen Ratio Influence Ecosystem Services from Cover Crop Mixtures. *Agron. J.* **2016**, *108*, 39–52. [CrossRef]
18. Ranaldo, M.; Carlesi, S.; Costanzo, A.; Arberi, P.B. Functional Diversity of Cover Crop Mixtures Enhances Biomass Yield and Weed Suppression in a Mediterranean Agroecosystem. *Weed Res.* **2019**, *60*, 96–108. [CrossRef]
19. Elhakeem, A.; Bastiaans, L.; Houben, S.; Couwenberg, T.; Makowski, D.; Werf, W. Van Der Do Cover Crop Mixtures Give Higher and More Stable Yields than Pure Stands ? *Field Crop. Res.* **2021**, *270*, 108217. [CrossRef]
20. Reiss, E.R.; Drinkwater, L.E. Promoting Enhanced Ecosystem Services from Cover Crops Using Intra- and Interspecific Diversity. *Agric. Ecosyst. Environ.* **2022**, *323*, 107586. [CrossRef]
21. Lawson, A.; Cogger, C.; Bary, A.; Fortuna, A. Influence of Seeding Ratio, Planting Date, and Termination Date on Rye-Hairy Vetch Cover Crop Mixture Performance under Organic Management. *PLoS ONE* **2015**, *10*, e0129597. [CrossRef]
22. Hayden, Z.D.; Brainard, D.C.; Henshaw, B.; Ngouajio, M. Winter Annual Weed Suppression in Rye—Vetch Cover Crop Mixtures. *Weed Technol.* **2012**, *26*, 818–825. [CrossRef]
23. Hayden, Z.D.; Ngouajio, M.; Brainard, D.C. Rye—Vetch Mixture Proportion Tradeoffs: Cover Crop Productivity, Nitrogen Accumulation, and Weed Suppression. *Agron. J.* **2014**, *106*, 467. [CrossRef]
24. White, K.E.; Brennan, E.B.; Cavigelli, M.A.; Smith, R.F. Winter Cover Crops Increased Nitrogen Availability and Efficient Use during Eight Years of Intensive Organic Vegetable Production. *PLoS ONE* **2022**, *17*, e0267757. [CrossRef] [PubMed]
25. Fiorini, A.; Remelli, S.; Boselli, R.; Mantovi, P.; Ardenti, F.; Trevisan, M.; Menta, C.; Tabaglio, V. Driving Crop Yield, Soil Organic C Pools, and Soil Biodiversity with Selected Winter Cover Crops under No-Till. *Soil Tillage Res.* **2022**, *217*, 105283. [CrossRef]
26. Radicetti, E.; Mancinelli, R.; Moscetti, R.; Campiglia, E. Management of Winter Cover Crop Residues under Different Tillage Conditions Affects Nitrogen Utilization Efficiency and Yield of Eggplant *(Solanum melanogena* L.) in Mediterranean Environment. *Soil Tillage Res.* **2016**, *155*, 329–338. [CrossRef]
27. SINAB Sistema d'Informazione Nazionale sull'Agricoltura Biologica. Available online: https://www.sinab.it/superfici (accessed on 21 May 2023).
28. Meier, U. *Growth Stages of Mono- and Dicotyledonous Plants: BBCH Monograph*, 2nd ed.; Federal Biological Research Centre for Agriculture and Forestry: Braunschweig, Germany, 2001.
29. Frasconi, C.; Martelloni, L.; Raffaelli, M.; Fontanelli, M.; Abou Chehade, L.; Peruzzi, A.; Antichi, D. A field vegetable transplanter for use in both tilled and no-till soils. *Trans ASABE* **2019**, *62*, 593–602. [CrossRef]
30. Smith, R.G.; Atwood, L.W.; Warren, N.D. Increased Productivity of a Cover Crop Mixture Is Not Associated with Enhanced Agroecosystem Services. *PLoS ONE* **2014**, *9*, e097351. [CrossRef]
31. Mason, N.W.H.; Orwin, K.H.; Lambie, S.; Waugh, D.; Pronger, J.; Perez, C.; Paul, C. Resource-Use Efficiency Drives Overyielding via Enhanced Complementarity. *Oecologia* **2020**, *193*, 995–1010. [CrossRef]
32. Wendling, M.; Büchi, L.; Amossé, C.; Jeangros, B.; Walter, A.; Charles, R. Specific Interactions Leading to Transgressive Overyielding in Cover Crop Mixtures. *Agric. Ecosyst. Environ.* **2017**, *241*, 88–99. [CrossRef]
33. Siebenkäs, A.; Schumacher, J.; Roscher, C. Resource Availability Alters Biodiversity Effects in Experimental Grass-Forb Mixtures Alrun. *PLoS ONE* **2016**, *11*, e0158110. [CrossRef]
34. Blesh, J. Functional Traits in Cover Crop Mixtures: Biological Nitrogen Fixation and Multifunctionality. *J. Appl. Ecol.* **2018**, *55*, 38–48. [CrossRef]
35. Smith, R.G.; Warren, N.D.; Cordeau, S. Are Cover Crop Mixtures Better at Suppressing Weeds than Cover Crop Monocultures? *Weed Sci.* **2020**, *68*, 186–194. [CrossRef]
36. Baraibar, B.; Hunter, M.C.; Schipanski, M.E.; Hamilton, A.; Mortensen, D.A. Weed Suppression in Cover Crop Monocultures and Mixtures. *Weed Sci.* **2018**, *66*, 121–133. [CrossRef]
37. Mckenzie-Gopsill, A.; Mills, A.; Macdonald, A.N.; Wyand, S. The Importance of Species Selection in Cover Crop Mixture Design. *Weed Sci.* **2022**, *70*, 436–447. [CrossRef]
38. Maclaren, C.; Swanepoel, P.; Bennett, J.; Wright, J.; Dehnen-schmutz, K. Cover Crop Biomass Production Is More Important than Diversity for Weed Suppression. *Crop Sci.* **2019**, *59*, 733–748. [CrossRef]
39. Christina, M.; Négrier, A.; Marnotte, P.; Viaud, P.; Auzoux, S.; Techer, P.; Hoarau, E. A Trait-Based Analysis to Assess the Ability of Cover Crops to Control Weeds in a Tropical Island. *Eur. J. Agron.* **2021**, *128*, 126316. [CrossRef]
40. Scavo, A.; Mauromicale, G. Crop Allelopathy for Sustainable Weed Management in Agroecosystems: Knowing the Present with a View to the Future. *Agronomy* **2021**, *11*, 2104. [CrossRef]
41. Boselli, R.; Anders, N.; Fiorini, A.; Ganimede, C.; Faccini, N.; Marocco, A.; Schulz, M.; Tabaglio, V. Improving Weed Control in Sustainable Agro-Ecosystems: Role of Cultivar and Termination Timing of Rye Cover Crop. *Ital. J. Agron.* **2021**, *16*, 1807. [CrossRef]
42. Grint, K.R.; Arneson, N.J.; Oliveira, M.C.; Smith, D.H.; Werle, R. Cereal Rye Cover Crop Terminated at Crop Planting Reduces Early-Season Weed Density and Biomass in Wisconsin Corn—Soybean Production. *Agrosystems Geosci. Environ.* **2022**, *5*, e20245. [CrossRef]
43. Abou Chehade, L.; Puig, C.G.; Souto, C.; Antichi, D.; Mazzoncini, M.; Pedrol, N. Rye *(Secale cereale* L.) and Squarrose Clover *(Trifolium squarrosum* L.) Cover Crops Can Increase Their Allelopathic Potential for Weed Control When Used Mixed as Dead Mulch. *Ital. J. Agron.* **2021**, *16*, 1869. [CrossRef]

44. Adeux, G.; Antichi, D.; Carlesi, S.; Mazzoncini, M.; Munier-jolain, N.; Paolo, B. Cover Crops Promote Crop Productivity but Do Not Enhance Weed Management in Tillage-Based Cropping Systems. *Eur. J. Agron.* **2021**, *123*, 126221. [CrossRef]
45. Campiglia, E.; Mancinelli, R.; Radicetti, E. Influence of No-Tillage and Organic Mulching on Tomato (*Solanum lycopersicum* L.) Production and Nitrogen Use in the Mediterranean Environment of Central Italy. *Sci. Hortic.* **2011**, *130*, 588–598. [CrossRef]
46. Little, N.G.; Ditommaso, A.; Westbrook, A.S.; Ketterings, Q.M.; Mohler, C.L. Effects of Fertility Amendments on Weed Growth and Weed—Crop Competition: A Review. *Weed Sci.* **2021**, *69*, 132–146. [CrossRef]
47. Testani, E.; Montemurro, F.; Ciaccia, C.; Testani, E. Agroecological Practices for Organic Lettuce: Effects on Yield, Nitrogen Status and Nitrogen Utilisation Efficiency Nitrogen Status and Nitrogen Utilisation efficiency. *Biol. Agric. Hortic.* **2019**, *36*, 84–95. [CrossRef]
48. Radicetti, E.; Mancinelli, R.; Campiglia, E. Impact of Managing Cover Crop Residues on the Floristic Composition and Species Diversity of the Weed Community of Pepper Crop (*Capsicum annuum* L.). *Crop Prot.* **2013**, *44*, 109–119. [CrossRef]
49. Manici, L.M.; Caputo, F.; Nicoletti, F.; Leteo, F.; Campanelli, G. The Impact of Legume and Cereal Cover Crops on Rhizosphere Microbial Communities of Subsequent Vegetable Crops for Contrasting Crop Decline. *Biol. Control* **2018**, *120*, 17–25. [CrossRef]
50. Kramberger, B.; Gselman, A.; Kristl, J.; Lesnik, M.; Šuštar, V.; Muršec, M.; Podvršnik, M. Winter Cover Crop: The Effects of Grass—Clover Mixture Proportion and Biomass Management on Maize and the Apparent Residual N in the Soil. *Eur. J. Agron.* **2014**, *55*, 63–71. [CrossRef]
51. Canali, S.; Campanelli, G.; Ciaccia, C.; Leteo, F.; Testani, E.; Montemurro, F. Conservation Tillage Strategy Based on the Roller Crimper Technology for Weed Control in Mediterranean Vegetable Organic Cropping Systems. *Eur. J. Agron.* **2013**, *50*, 11–18. [CrossRef]
52. Ciaccia, C.; Canali, S.; Campanelli, G.; Testani, E.; Montemurro, F.; Leteo, F.; Delate, K. Effect of Roller-Crimper Technology on Weed Management in Organic Zucchini Production in a Mediterranean Climate Zone. *Renew. Agric. Food Syst.* **2016**, *31*, 111–121. [CrossRef]

Article

Effects of Conservation Agriculture Practices on Tomato Yield and Economic Performance

Lorenzo Gagliardi, Mino Sportelli *, Marco Fontanelli, Massimo Sbrana, Sofia Matilde Luglio, Michele Raffaelli and Andrea Peruzzi

Department of Agriculture, Food and Environment, University of Pisa, Via del Borghetto 80, 56124 Pisa, Italy; lorenzo.gagliardi@phd.unipi.it (L.G.); marco.fontanelli@unipi.it (M.F.); massimo.sbrana@phd.unipi.it (M.S.); sofiamatilde.luglio@phd.unipi.it (S.M.L.); michele.raffaelli@unipi.it (M.R.); andrea.peruzzi@unipi.it (A.P.)
* Correspondence: mino.sportelli@phd.unipi.it

Abstract: Conservation agriculture practices, such as reduced tillage and the incorporation of cover crops, play a crucial role in improving the sustainability of organic farming systems. The aim of this two-year field trial was to evaluate five different organic technical itineraries (ST, IN1, IN2, M1, and M2) which differed on soil management practices adopted before processing tomato transplantation and regarding weed control strategies performed. Soil management practices in comparison consisted of conventional deep tillage (ST and M1) or reduced tillage together with the use of a cover crop mixture composed of common vetch and barley (IN1, IN2, and M2). Weed control strategies involved the use of biodegradable mulch together with mechanical weeding (ST and M2), or false seedbed technique and mechanical weeding (IN1, IN2, and M1). Weed biomass at harvest, tomato yield, and the operational and economic performance of each of the technical itineraries was evaluated. No significant differences emerged in terms of weed biomass at harvest between itineraries. Best yield results were obtained tendentially by ST and M2 when biodegradable mulch was used, with values equal to 42.14 and 41.47 Mg ha^{-1} in 2020 and 30.68 and 31.19 Mg ha^{-1} in 2021, respectively. Even though the itineraries where mulch film was used (ST and M2) resulted in significantly onerous processes, they also obtained the highest gross income compared to the other itineraries, with values of 30,998 and 29,900 € ha^{-1} in 2020, and of 16,060 and 15,186 € ha^{-1} in 2021, respectively. These results revealed the importance of using mulching to help cope with critical climatic conditions, such as drought seasons. Further studies are needed to evaluate the yield and economic advantages of both the effect of shallower soil tillage over a longer period in this specific context and the creation of ground cover with cover crops managed as dead mulch.

Keywords: conservation agriculture; reduced tillage; cover crop; false seedbed technique; mechanical weed control; thermal weed control; sustainable agriculture

Citation: Gagliardi, L.; Sportelli, M.; Fontanelli, M.; Sbrana, M.; Luglio, S.M.; Raffaelli, M.; Peruzzi, A. Effects of Conservation Agriculture Practices on Tomato Yield and Economic Performance. *Agronomy* **2023**, *13*, 1704. https://doi.org/10.3390/agronomy13071704

Academic Editor: Antonio Rodríguez-Lizana

Received: 31 May 2023
Revised: 21 June 2023
Accepted: 24 June 2023
Published: 26 June 2023

1. Introduction

Agricultural ecosystems account for 36% of land areas and provide food for over seven billion people. To satisfy the ever-growing demand for food, it is necessary to increase agricultural production [1,2]. Intensive agricultural practices, characterized by excessive fertilization, irrigation, and tillage, have been commonly adopted to achieve this goal, leading to progressive degradation of soil and water quality. These aspects, together with climate change issues, represent a threat to current and future agricultural production [3]. According to some studies, the intensive farming systems currently adopted could lead to a substantial reduction of agricultural production in future climatic conditions [4]. To counter the ongoing environmental decline, European governments have fostered agri-environmental policies, and the use of organic farming has increased, which is supporting the application of sustainable practices by farmers [5]. However, some aspects of organic agriculture are debated, such as the reliance on conventional intensive tillage to prepare soil before crop planting, to incorporate organic fertilizers, cover crops, crop residues, or to control weeds [6]. It is well known that inversion tillage practices, such as moldboard plowing, can have negative effects. These effects include causing the hardening of deeper soil layers, soil erosion, loss of water and organic matter, an increase of greenhouse gases (GHG) emissions, and biodiversity reduction [7]. In addition, these practices usually require high-powered tractors and high fuel consumption. Conventional tillage technologies usually also involve the performance of several operations in the field which, following the repeated machines soil trampling, contribute to soil compaction [8]. The introduction of conservation tillage practices that lead to environmental benefits by aiming to minimize the intensity and/or frequency of tillage operations would help to improve the sustainability of organic farming systems. Reduced tillage represents a common method of conservation tillage. It allows to limit the damage to soil structure decreasing the susceptibility to soil erosion, improves water infiltration, promotes biodiversity conservation and soil biological activity, thus favoring the maintenance of soil fertility [9,10]. Reduced tillage allows also limits the number of air-filled pores in the soil, keeping soil CO_2 emissions at a low level [11,12]. Hu et al. [12] observed that reduced tillage decreases CO_2 emission over conventional tillage by 5.9% for monoculture maize, 7.1% for monoculture wheat, and 6.7% for intercropping. However, there are discordant opinions on conservation tillage effect on N_2O emissions. Some authors argue that these can be increased compared to conventional tillage [13,14], while according to others these are not altered [15,16]. These practices also create economic benefits by allowing farmers to reduce costs for machinery, fuel, and labor, thereby enhancing economic returns in crop production [9]. Filipovic et al. [17] found that reduced tillage allows a decrease in fuel consumption of 1.5–2 times compared to conventional tillage. Nevertheless, some disadvantages are associated with the application of conservation tillage in organic farming in the long-term, such as increased weed pressure, especially from grasses and perennial weeds [6,18,19]. Cooper et al. [20] found an increase in weed incidence in organic systems with conservation tillage equal to 50% compared to systems in which plowing was performed.

Nowadays, there is a growing interest in the introduction of cover crops in crop rotation, particularly in organic farming systems, as they guarantee numerous agro-ecosystem services and are crucial for fertility management [21,22]. Indeed, cover crops can help to increase soil organic carbon, nutrient availability, soil aggregation, water conservation, microbial activity, weed and pest management, and can reduce the risk of soil compaction and erosion [23]. In organic farming systems, cover crops are often managed as green manure before the cash crop is planted; therefore, they are shredded and incorporated into the soil [21]. The choice of which cover crop species to use for green manure depends on the final objective to be achieved. Leguminous species are chosen for their ability to fix and supply large amounts of N, while non-legume species are mostly used to increase soil organic carbon stock, improve soil structure, reduce nutrient leaching, and prevent soil erosion. The adoption of a mixture of legumes and non-legumes can allow more agronomic and environmental benefits to be obtained [24]. Processing tomato is one of the most important vegetable crops in the Mediterranean area and in Italy, with a dedicated area of 65,180 ha and a national average yield of 84 Mg ha^{-1} in 2022 [25]. This crop is demanding in N, and green manuring is becoming increasingly widely accepted as a beneficial method for sustainable processing tomato production [26,27]. The adoption of a cover crop mixture composed of barley and vetch proved to be an effective and sustainable mixture for processing tomatoes production. This mixture provides a biomass with a good C/N ratio to be incorporated in the soil. It allows a stable N accumulation, guaranteeing a good availability of N for the crop while reducing N leaching compared to the use of only vetch as a cover crop [28,29].

In vegetable cropping systems, mulching technique is frequently used. It brings various benefits, such as earlier crop production, higher yields and product quality, more efficient water use, soil erosion protection, pest control, and weeds suppression [30]. In organic farming systems, synthetic herbicides are not allowed, so mulching turns out to be crucial as means of weed control, along with mechanical means, such as weeders, or thermal methods, such as flaming. Films composed of biodegradable polymers currently represent an alternative to the traditionally used non-biodegradable materials. These films can be buried in the soil at the end of the cropping season and microbially degraded, thus reducing the amount of plastic waste left in the ground [31–33]. In the Mediterranean continental climate, Mater-Bi mulch proved to be a sustainable alternative to polyethylene in organic production [34]. Some authors [30,35] found no differences in terms of processing tomatoes yield when grown using Mater-bi or polyethylene film. Cirujeda et al. [36] observed similar results, both in terms of weed control and tomato yield, in farming systems where Mater-bi and polyethylene mulch were used. Comparing plastic mulch (PE mulch), mater-bi mulch, and brush hoe for their weed control effects, authors [37] observed a greater reduction of weed biomass with PE and Mater-bi mulch than with brush hoeing.

To the best of our knowledge, no studies have been conducted on the production of organic processing tomatoes in the Mediterranean area that compare soil management practices at different levels of intensity and different weed control strategies. The aim of this two-years field trial was to evaluate five different organic technical itineraries which differed regarding the soil management practices adopted before processing tomato transplantation and the weed control strategies performed. Soil management practices, in comparison, consisted of conventional deep tillage, or reduced tillage combined with the use of a cover crop mixture composed of common vetch and barley. Weed control strategies involved the use of biodegradable mulch together with mechanical weeding, or false seedbed technique and mechanical weeding. The itineraries were assessed for their impact on processing tomato yield, as well as the operational and economic performance.

2. Materials and Methods

2.1. Site Characteristics

A two-year field experiment (2020–2021) was performed at the Pasquini farm located at Suvereto, Livorno, Italy (43°02′59″ N 10°40′45″ E, 1 m.a.s.l.). The trial was performed on a field with loam soil (38.2% sand, 44.0% silt, and 17.8% clay). Soil organic matter content corresponded to 2.09%, total N was 1.12‰, available P was 4.6 ppm, and pH was 8.2. Electrical conductivity was 51.9 μS cm^{-1}, and cation exchange capacity was 10.7 meq 100 g^{-1}. The farm was managed in accordance with the criteria of organic farming (Reg. CE 834/2007). The area presented the typical Mediterranean climate with seasonal rainfall peaks in spring and fall. Figure 1 shows the monthly total rainfall (mm) and the mean minimum and maximum air temperatures at the experimental site for both years in which the trial was conducted along with the 10-year mean values.

2.2. Experimental Layout and Managament Systems

Two cycles of the same crop were carried out during the experiment involving processing tomato (*Solanum lycopersicum* L.). Processing tomato was planted in a double-row arrangement. The distances were 1.5 m between double-row centers, 0.4 m between rows, 1.0 m between double-rows, and about 0.4 m between tomato plants along the row. In 2020 and 2021, irrigation water was supplied during the vegetable growing season by means of dripline. Fertilization was executed in an identical manner during both years. Before tomato transplantation, 1.5 Mg ha^{-1} of dried manure pellets (NPK 6-15-0) were distributed. At the time of transplantation, 0.15 Mg ha^{-1} of microgranular organic fertilizer (NPK 13-0-0) and 0.1 Mg ha^{-1} of organic-mineral microgranular fertilizer (NPK 7-5-0) were applied through localized distribution. At each irrigation shift, 20 L ha^{-1} of liquid organic fertilizer (NPK 5-0-0) with fertigation were distributed. Pest control treatments were mainly aimed at containing *Tuta absoluta* (Meyrick), which is strongly present in the area where the trial was set up, and at preventing the onset of fungal diseases. Treatments took place mainly in the period between June and July in both 2020 and 2021. The treatments involved the use of copper oxychloride (2 kg ha^{-1}) for the control of fungal diseases and Bacillus thuringiensis (1 kg ha^{-1}) and Spinosad (1 kg ha^{-1}) for the control of *Tuta absoluta*. Five different technical itineraries, which differed mainly in the soil management practices that were adopted before tomato transplanting and the weed control strategies that were carried out, were evaluated. The itineraries were compared according to a randomized complete block design with three replications. Plot size was 75 m^2 (1 × 75 m).

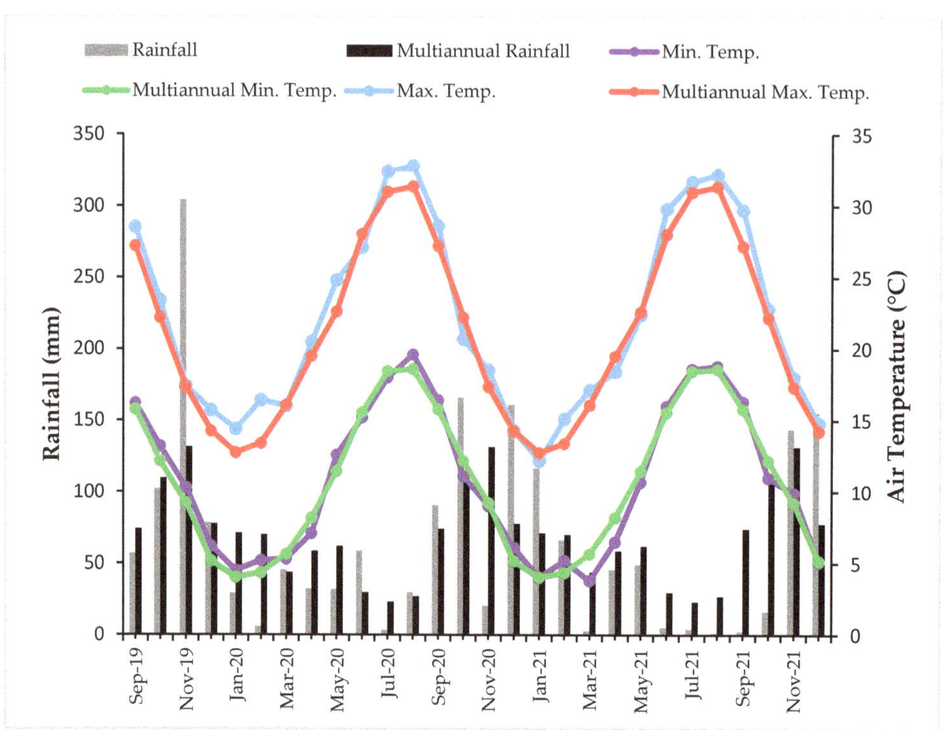

Figure 1. Monthly total rainfall (mm) and mean minimum and maximum air temperatures (°C) at the experimental site for both years in which the experimental trial was conducted along with the 10-year mean values.

Technical Itineraries

Several technical itineraries were tested. There was the standard itinerary adopted by the farmer based on conventional soil tillage before tomato transplanting, the application of biodegradable Mater-bi-based mulch film, and mechanical cultivation between double-rows after transplanting to control weeds (ST). There was an innovative itinerary where a cover crop mixture was sown on soil prepared with reduced tillage, then managed as green manure before transplanting; false seedbed technique was adopted, and mechanical cultivation was performed after transplanting to control weeds (IN1). A second version of the innovative itinerary was used, in which, in addition to the practices described for IN1, after tomato transplanting, weed control was also performed on the rows by means of side-flaming (IN2). For a more accurate evaluation of the strategies used in the standard and innovative itineraries, two "mixed" systems characterized by intermediate strategies between the itineraries ST and IN1 were tested. In the mixed itinerary M1, conventional tillage practices were carried out before transplanting, as in ST. False seedbed technique was adopted, mulch film was not applied, and mechanical cultivation was performed after transplanting to control weeds, as in IN1. This itinerary was tested to evaluate the mulch film effect on weeds and crop yield when conventional tillage is performed before transplanting. In the second mixed itinerary M2, a cover crop mixture managed as green manure was used before transplanting, as in IN1; mulch film was applied, and mechanical cultivation was carried out to control weeds, as in ST. This itinerary allowed for the evaluation of the mulch film effect on weeds and crop yield when reduced tillage practices and green manure cover crops were used. In all the technical itineraries, manual weeding was carried out just before harvesting to facilitate mechanical harvesting of tomatoes.

The conventional tillage carried out before transplanting in ST and M1 consisted of an intervention at a depth of 50 cm with a subsoiler (RTP/T, Nardi, Selci Lama, Italy) equipped with three straight tines that had a working width of 2.5 m. The intervention was performed on 14 February 2020 and 16 February 2021. This operation was followed by the use of a heavy cultivator (CASC7DP, Mipe Viviani, Monteriggioni, Italy) with seven rigid tines arranged in a "v" shape with a working width of 2.5 m at a depth of 30 cm on 21 February 2020 and 22 February 2021. Subsequently, two interventions with a rotary harrow (DRAGO DC., Maschio Gaspardo, Campodarsego, Italy) at a depth of 15 cm were carried out. One intervention was performed to refine the soil on 13 May 2020 and 4 May 2021, the other in proximity of tomato transplanting for surface crust breakage on 19 May 2020 and 14 May 2021. In IN1, IN2, and M2, before cover crop planting, the experimental area was tilled with a 3 m wide combined cultivator (MANDAM, Gliwice, Poland) at a depth of 20 cm. The machine was equipped with nine rigid tines, four couples of inclined discs and a roller for final soil leveling and compacting. This operation was carried out on 14 February 2020 and 16 February 2021. Subsequently, the rotary harrow working at a 15 cm depth was used to establish the seedbed and, at the same time, the sowing of the cover crop was performed with a seed drill (SC MARIA 300, Maschio Gaspardo, Campodarsego, Italy) on 14 February 2020 and 16 February 2021. Cover crop consisted of a mixture of barley (*Hordeum vulgare* L., with a seed rate of 120 kg ha^{-1}) and common vetch (*Vicia sativa* L., with a seed rate of 100 kg ha^{-1}). The high doses of the components of the mixture are motivated by the late sowing periods. Cover crops were then shredded with a flail mower (TORNADO, Maschio Gaspardo, Campodarsego, Italy) and incorporated in the soil using the above-mentioned combined cultivator working at a depth of about 20 cm on 18 May 2020 and 10 May 2021. In all the itineraries tested, the transplantation was carried out with a transplanter (F-MAX/2, Ferrari, Abbiategrasso, Italy) equipped with two transplanting elements, cup rotating distributor, dripline, and film mulch deposition systems on 3 June 2020 and 21 May 2021. In ST and M2, simultaneously with the tomato transplantation, mulch film was also applied. In these itineraries, weed control after transplantation was performed by two interventions with a cultivator developed by the farm owner and equipped with two couples of four spring tines. The operation was executed in the area not covered by the mulch film among double rows on 6 June and 8 July 2020 and 22 June and 8 July 2021. In IN1, IN2, and M1, before transplantation, the false seedbed technique was carried out by performing two interventions with a 2 m wide rolling harrow designed, fully realized, and patented at the University of Pisa [38]. The operation was carried out on 22 and 29 May 2020 and 14 and 18 May 2021. The machine is equipped with spike discs in the front that till the soil at 3–4 cm depth, and cage rolls at the rear of the unit that allow to separate weed seedling roots from the soil. To facilitate the passage of the rolling harrow, previously, an intervention with the rotary harrow was carried out on 19 May 2020 and 14 May 2021, which allows the soil crust to be broken. Subsequently, tomato was transplanted and weed control during the crop growing season was carried out with two interventions with a precision weeder designed and realized at the University of Pisa [38]. The precision weeder was used on 24 June and 8 July 2020 and 22 June and 8 July 2021. This machine was equipped with three weeding units, each of which was equipped with a central goose-foot sweep and two side 'L'-shaped sweeps along with a couple of flexible tines for intra-row selective weed control, with a working width of 3 m. Moreover, in IN2, immediately after the two interventions with the precision weeder, two weed control interventions were carried out on the rows with flaming. The flamer machine (PIRO-TRACK, Maito, Arezzo, Italy) was powered by LPG and consisted of four rod burners each 50 cm wide rotated at 90° with respect to driving direction to perform side-flaming. The operations performed in the five technical itineraries are shown in Table 1.

Table 1. List of operations performed in the five technical itineraries tested during the two-year field experiment.

		ST	IN1	IN2	M1	M2
Soil Tillage	Conventional tillage (subsoiler, heavy cultivator, and rotary harrow).	✓	-	-	✓	-
	Reduced tillage (combined cultivator).	-	✓	✓	-	✓
Cover Crop	Combined harrowing and sowing, shredding and burial with combined cultivator.	-	✓	✓	-	✓
Weed Control	Mulch film application and mechanical weeding.	✓	-	-	-	✓
	False seedbed technique and mechanical weeding after transplanting.	-	✓	✓	✓	-
	Side-flaming.	-	-	✓	-	-

✓ indicates the management strategy implemented in each itinerary regarding Soil Tillage, Cover Crop and Weed Control; ST—standard itinerary; IN1—innovative itinerary 1; IN2—innovative itinerary 2; M1—mixed itinerary 1; M2—mixed itinerary 2.

A 4WD tractor (Case IH, Racine, USA) powered by a 155-kW diesel engine was used for the interventions with subsoiler, heavy cultivator, combined cultivator, rotary harrow, and for the combined intervention of harrowing and cover crop sowing. For cover crop shredding and false seedbed technique, a 4WD tractor (Kubota, Osaka, Japan) powered with a 40.44 kW diesel engine was used. A 2WD FIAT 70/90 tractor (FiatAgri, Torino, Italy) with a 51.47 kW diesel engine was used for mechanical weeding, flame weeding, fertilizing operations, and phytosanitary treatments.

2.3. Data Collection

2.3.1. Agronomic Parameters

Concerning the agronomic aspects considered, cover crops biomass, weed biomass at harvest, and fresh tomato berries weight were surveyed. Sampling of the aboveground biomass of cover crops was carried out in mid-May in both 2020 and 2021 on sample areas of 0.5 m² each. Measurements of weed biomass at harvest and tomato berries weight were carried out on a square frame of 0.25 m², positioned to contain a tomato plant in the center. In both years, weeds and berries surveys were carried out in the first and second week of September. Only berries that reached commercial maturity were collected. Two measurements for each replicate were carried out for the above-mentioned parameters during both years of the trial. The collected samples of cover crops, weeds, and tomato berries were processed and examined at the laboratory of the Agri-environmental Research "Enrico Avanzi" Centre, San Piero a Grado (Pisa), Italy. Both weeds and cover crops biomass were oven-dried at 60 °C until constant weight was reached, and dry biomass was determined. The collected berry samples were examined, processed, weighed, and classified by degree of ripeness at the above-mentioned laboratory.

2.3.2. Operational Performance

Operational parameters, such as working width, working depth, working speed, theoretical field times, turning times, supply times, and idle times were considered. Theoretical field times correspond to the time the machines effectively operate at an optimum working speed and work over their full width of action. Supply times refer to the time required for fuel and technical means refueling. Idle times represent the time necessary for operators to improve the correct positioning of plants during transplantation. For the calculation of the machines forward speeds during the various operations, the travel times for a straight section of 30 m located inside the experimental area were timed. These parameters were used to evaluate, the field time and fuel consumption per unit area of each operation carried out in each of the five itineraries.

2.3.3. Economic Performance

For the evaluation of the technical itineraries economic performance, gross salable production (GSP), average variable costs over the two years (VC), and gross income (GI) were considered. GSP was obtained by multiplying the yield achieved by each technical itinerary by the market price of organic processing tomatoes, which was considered equal to 1170 EUR Mg^{-1} and 1120 EUR Mg^{-1} in 2020 and 2021, respectively [39]. Concerning the average variable costs (VC), costs of labor, fuel, agricultural operations, and technical means were considered. An hourly rate of 20 EUR h^{-1} was used for labor. For each agricultural operation, the cost of labor was calculated knowing the time required and the number of operators employed. For the fuel costs estimation, an average market price of agricultural diesel equal to 0.88 EUR kg^{-1} was considered. The agricultural operations unit costs were estimated as the sum of the variable costs due to the use of tractors and operating machines coupled to tractors for each operation. Fixed costs (depreciation, interest, and miscellaneous expenses related to the useful life of the machines expressed in hours), variable maintenance and repair costs were taken into consideration. Concerning the technical means, costs relating to mulch film, LPG for flame weeding, cover crop seeds, irrigation, fertilization, and phytosanitary treatments were considered. Gross income was obtained from the difference between GSP and total VC.

2.4. Statistical Analysis

Two-way ANOVA test was performed to evaluate the effect of cover crop species, year and interaction among factors on cover crop dry biomass. Two-way ANOVA test was performed also to assess the effect of the technical itinerary, year, and interactions among factors on weed biomass at harvest, and fresh tomato berries weight. ANOVA was performed using SPSS software (IBM SPSS Statistics for Mac, Version 25.0. Armonk, NY, USA: IBM Corp.). Normality distribution was evaluated using the Kolmogorov–Smirnov test, while the Breusch–Pagan test was employed for homoskedasticity. After the ANOVA test, the Bonferroni post hoc test at the 0.05 probability level was performed when necessary.

3. Results

3.1. Dry Biomass of Cover Crops and Weeds

ANOVA revealed that cover crops biomass was affected by species ($p < 0.05$), year ($p < 0.05$), and interaction between these factors ($p < 0.01$). Overall, barley produced more biomass than vetch (4.34 vs. 2.61 Mg ha^{-1}), and, in terms of average biomass produced by the two species per year, better results were obtained in 2021 rather than in 2020 (4.08 vs. 2.87 Mg ha^{-1}). In 2020, vetch produced less biomass than in 2021 and also less than barley in both 2020 and 2021 (0.83 vs. 4.39, 4.92, and 3.76, respectively), while no significant differences emerged between the biomass produced by vetch in 2021 and that of barley in 2020 and 2021 (Figure 2).

Regarding weed dry biomass at harvest, ANOVA revealed that neither the management strategies, nor the year, nor the interaction among factors affected the parameter ($p = 0.665$, $p = 0.453$, $p = 0.277$, respectively).

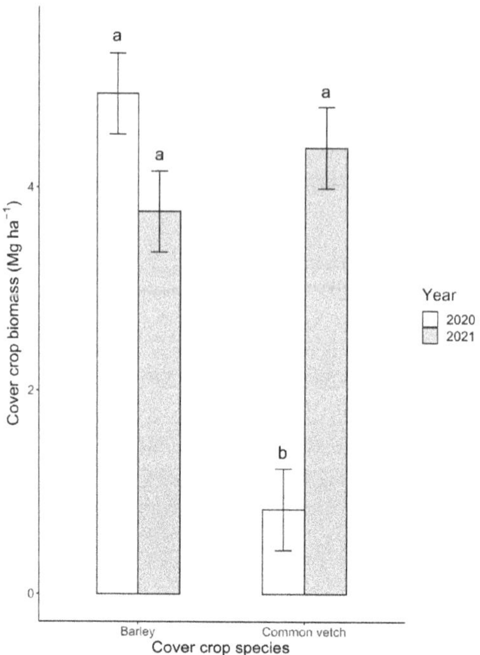

Figure 2. Effect of the interaction among cover crop species and year on cover crop biomass produced. Mean denoted by different letters are significantly different at $p < 0.05$ (Bonferroni test). Error bars indicates upper and lower 95% confidence intervals.

3.2. Fresh Tomato Berries Weight

ANOVA showed that both the technical itinerary ($p < 0.001$) and the interaction among factors ($p < 0.01$) affected the biomass of fresh tomato berries, while the year ($p = 0.470$) did not affect the parameter. In 2020, technical itineraries that produced the highest harvest results were ST and M2, with values equal to 42.14 and 41.47 Mg ha^{-1}, respectively, without showing differences between them. The itineraries in which mulch film was not applied achieved lower results compared to ST and M2, and they were similar to each other. Among these itineraries, the one that obtained the lowest yield in absolute value was IN2, where flame weeding was carried out (13.81 Mg ha^{-1}), followed by IN1 (15.25 Mg ha^{-1}) and M1 (25.12 Mg ha^{-1}). In 2021, the itineraries that produced the highest harvest result in absolute value were still ST and M2 with values slightly above 30 Mg ha^{-1}. However, results obtained with these itineraries were not statistically different from those obtained with IN2, M1, and IN1, which were equal to 24.82, 26.14, and 15.89 Mg ha^{-1}, respectively. It is possible to observe that the IN1 result in absolute values is relevantly lower than those achieved with the other four itineraries (Figure 3).

In 2021, fresh berries weight achieved by ST on average were lower than in 2020. On the other hand, the itineraries in which Mater-bi film was not used (IN1, IN2, and M1) achieved higher results in absolute value compared to 2020. Overall, ST recorded a higher fresh berry weight compared to M1, IN2, and IN1 (36.41 Mg ha^{-1} vs. 25.63, 19.31, and 15.57 Mg ha^{-1}, respectively) and similar to M2. M2 obtained a greater fresh berry weight (35.83 Mg ha^{-1}) than IN1 and a similar weight to IN2 and M1. No differences emerged between M1 and IN2 for this parameter, while IN1 obtained a lower fresh berry weight compared to ST, M1, and M2 (Figure 4).

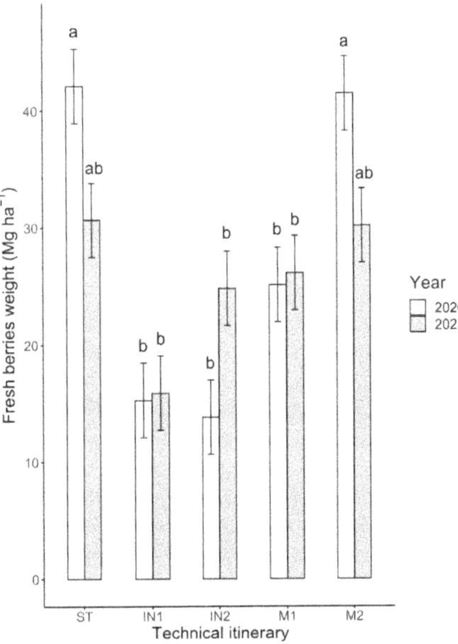

Figure 3. Effect of the interaction among technical itinerary and year on fresh berries weight. Mean denoted by different letters are significantly different at $p < 0.05$ (Bonferroni test). Error bars indicates upper and lower 95% confidence intervals. ST—standard itinerary; IN1—innovative itinerary 1; IN2—innovative itinerary 2; M1—mixed itinerary 1; M2—mixed itinerary 2.

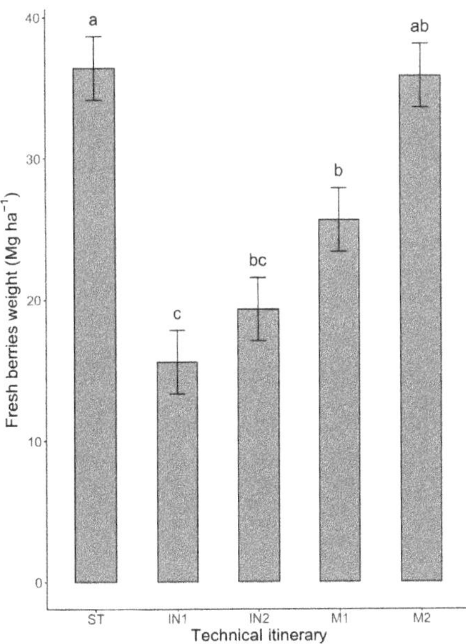

Figure 4. Effect of technical itinerary on fresh berry weight. Mean denoted by different letters are significantly different at $p < 0.05$ (Bonferroni test). Error bars indicates upper and lower 95% confidence intervals. ST—standard itinerary; IN1—innovative itinerary 1; IN2—innovative itinerary 2; M1—mixed itinerary 1; M2—mixed itinerary 2.

3.3. Operative Performances of the Technical Itinearies

ST and M2 resulted in the most onerous technical itineraries in terms of average field time with values of 318.7 and 319.3 h ha^{-1}, respective, which is approximately 65% higher compared to IN1, IN2, and M1. Average field times of IN1, IN2, and M1 were similar in both years, with values of 190, 197.5, and 191.9 h ha^{-1}, respectively (Figure 5).

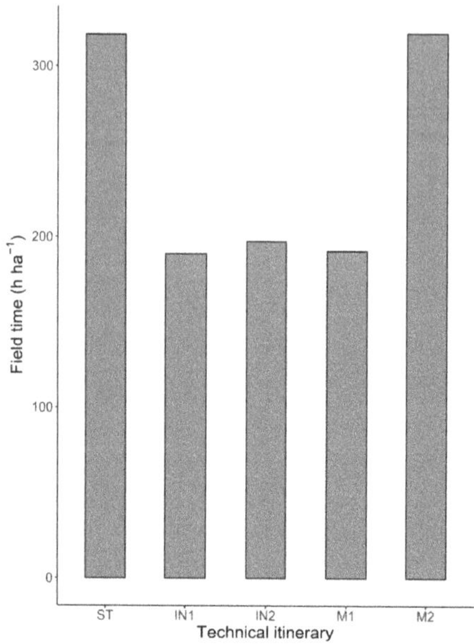

Figure 5. Average total field time of each technical itinerary in 2020 and 2021. ST—standard itinerary; IN1—innovative itinerary 1; IN2—innovative itinerary 2; M1—mixed itinerary 1; M2—mixed itinerary 2.

When mulch film was applied, eight operators were needed for transplanting, five of which took steps to remedy the non-optimal placement of the seedlings and to check adequate film spreading. When transplanting was performed on bare soil, instead, four operators were sufficient and the time to check and optimize seedlings placement were also reduced. Field time required to perform transplanting when mulch film was applied was equal to 200 h ha^{-1}, while on bare soil it was equal to 40 h ha^{-1}, and was therefore five times lower. Thus, transplanting on bare soil leads to a manpower management efficiency that is 2.5 times greater than transplanting on mulch film.

ST and M2 presented also the highest average fuel consumption, with values equal to 323.1 and 300.4 kg ha^{-1}, respectively, with no relevant differences among them. IN1, IN2, and M1 showed similar results that ranged from 225.0 to 258.6 kg ha^{-1} (Figure 6).

Transplanting on film mulch also involved a higher fuel consumption compared to transplanting on bare soil (136.5 vs. 54.6 Kg ha^{-1}) with an increase of 150%. The soil management strategies adopted before transplanting relevantly influenced the technical itineraries fuel consumption. This is evident from the differences in consumption in the comparison of ST and M2, and between IN1 and M1, whose itineraries varies only in terms of soil management practices applied before transplanting. Reducing the depth of tillage offers the advantage of decreasing fuel consumption by approximately 10%. The use of flame weeding led to a slightly higher fuel consumption rather than that of field times compared to IN1 in which weed control was carried out only with mechanical weeding.

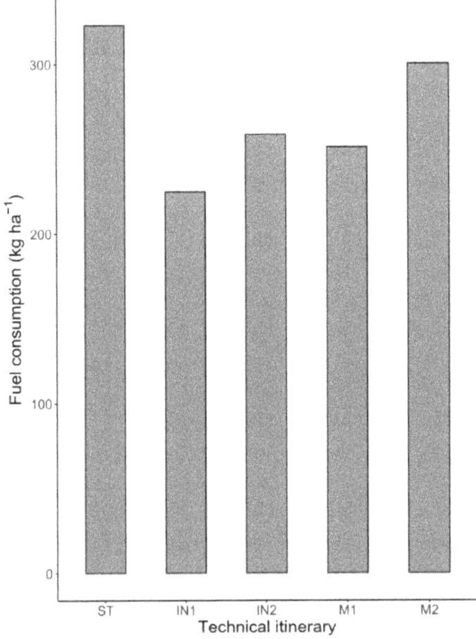

Figure 6. Average total fuel consumption of each technical itinerary in 2020 and 2021. ST—standard itinerary; IN1—innovative itinerary 1; IN2—innovative itinerary 2; M1—mixed itinerary 1; M2—mixed itinerary 2.

3.4. Economic Performances of the Technical Itineraries

The economic performance values of each technical itinerary in terms of gross salable production (GSP), average variable costs (VC), and gross income (GI), are shown in Table 2.

Table 2. Economic performances of the technical itineraries in comparison expressed as gross salable production (GSP), average variable costs (VC), and gross income (IG).

		Unit	ST 2020	ST 2021	IN1 2020	IN1 2021	IN2 2020	IN2 2021	M1 2020	M1 2021	M2 2020	M2 2021
GSP		€ ha^{-1}	49,301	34,364	17,837	17,798	16,153	27,801	29,386	29,276	48,522	33,808
VC	Labor	€ ha^{-1}	6373		3838		3961		3837		6396	
	Fuel	€ ha^{-1}	284		198		227		221		264	
	Machinery	€ ha^{-1}	4585		4639		4905		4710		4580	
	Technical means	€ ha^{-1}	7060		6780		6804		6460		7380	
	Total VC	€ ha^{-1}	18,303		15,456		15,898		15,230		18,622	
GI		€ ha^{-1}	30,998	16,060	2381	2342	254	11,903	14,156	14,046	29,900	15,186

GSP—gross salable production; VC—average variable costs; GI—gross income; ST—standard itinerary; IN1—innovative itinerary 1; IN2—innovative itinerary 2; M1—mixed itinerary 1; M2—mixed itinerary 2.

In 2020, the itineraries that achieved the highest GSP were ST and M2. The itineraries where mulch film was not employed recorded a lower GSP compared to ST and M2, with a reduction of approximately 57%. In 2021, ST and M2 are still the technical itineraries that obtained the highest GSP results. However, in 2021, IN2 and M1 results were not consistently lower than those obtained by ST and M2, with a reduction of approximately 16% on average. GSP achieved by IN1 in 2021 was relevantly lower compared to the other itineraries, as was observed tendentially in 2020.

In terms of total average VC, the most expensive systems were ST and M2 with values of 18,303 and 18,622 EUR ha^{-1} total. Instead, the other systems were similar to each other. Fuel cost was higher in ST in both years. The machinery costs resulted in a slightly higher overall value for IN2 compared to the other itineraries. Soil management technique did not affect machinery costs. Concerning the technical means costs, M2 reported a higher costs compared to the other itineraries, followed by ST, while M1 recorded the lowest costs. The purchase of technical means had the most relevant impact on the total variable cost of itineraries. In contrast, fuel costs had a lower influence on the total variable costs when compared to labor, technical means, and machinery costs. The itineraries connected with higher GI values appeared to be ST and M2 in both years, without showing relevant differences. Finally, it is possible to observe that, compared to the first year of the trial, a reduction in GI of around 50% on average was recorded in the second year for both ST and M2. Instead, values of IN1 and M1 were similar, and that of IN2 resulted in relevantly higher values, with an increase of 45.8%.

4. Discussion

4.1. Cover Crops Biomass Production and Weed Control

The biomass produced by common vetch in 2020 was lower than in 2021 (0.83 vs. 4.39 Mg ha^{-1}) and, compared to barley, lower in both 2020 and 2021 (4.92 and 3.76 Mg ha^{-1}, respectively). Therefore, in 2020, the mixture was mostly composed of barley, and this could be explained by reduced rainfall registered in January and February 2020 compared to 2021, and the resulting lower water availability at that time.

The different weed control strategies tested appeared to have similar effects on spontaneous flora. Considering that weed biomass at harvest was just over 0.4 Mg ha^{-1} on average in both experimental years, it is possible to affirm the high efficiency of all weed control strategies adopted. Results achieved are consistent with those obtained by Raffaelli et al. [38]. Authors compared a conventional weed control system which relied on the use of chemical and mechanical methods with an alternative system based on the use of false seedbed technique and precision weeder. Both management approaches were considered effective, resulting in an average weed biomass at tomato harvest of 0.68 and 0.36 Mg ha^{-1}, respectively. Therefore, mechanical weed control strategies provided good results despite the challenging conditions during weeding. For example, the presence of surface crust, which mainly concerned the first tomato growing season, prevented the weeder's flexible tines from properly controlling weeds in the intra-row space. This was observed also by Cirujeda et al. [37], which stated that the presence of heavy soil or soil crust does not allow for the optimal functioning of tools such as torsion weeders. Overall, the results obtained can also be considered positive in consideration of the limited effectiveness of the false seedbed technique due to surface crust. In the itineraries with film mulch (ST and M2), weeds presence at harvest was attributable to the ability of some plants to pierce the film. Additionally, weeds took advantage of the ecological niches left free by the hole made in the film during transplanting. This has also been observed by other authors [30,40] according to whom biodegradable films can be perforated by some weed species, probably due to their degradation and gradual loss of mechanical consistency over time.

4.2. Processing Tomato Yield

In 2020, ST and M2 obtained the highest fresh berries weight results (42.14 and 41.47 Mg ha^{-1}, respectively), without showing differences between them. Such results are of the same order of magnitude as those obtained by Ronga et al. [41] for processing tomato in an organic farming system. The itineraries without mulch film (IN1, IN2, and M1) achieved lower yields compared to ST and M2. The IN2 itinerary, where flame weeding was implemented, recorded the lowest yield in absolute value (13.81 Mg ha^{-1}). This demonstrates that, in 2020, mechanical weeding was probably more decisive than the intervention of flame weeding. In 2021, ST and M2 were still the itineraries that obtained the highest yield in absolute value, with values slightly above 30 Mg ha^{-1}, partially confirming results obtained in the first year of the trial. However, in this case, yields obtained by IN2, M1, and IN1 (with values of 24.82, 26.14, and 15.89 Mg ha^{-1}, respectively) were not statistically different compared to ST and M2. The higher result obtained by IN2 in absolute value in 2021 compared to IN1 was probably due to the timelier weeding interventions on the row with flaming, although the weed biomass at harvest was not statistically different. On average, in 2021, yield obtained by ST were lower than in 2020 due to the adverse climatic conditions characterized by drought and high temperatures in July and August. Instead, in the itineraries where the Mater-bi film was not employed (IN1, IN2, and M1) the climatic conditions in 2021 probably had fewer adverse impacts. This is evident from the higher results in absolute value obtained compared to 2020. This yield increase is probably due to the lack of surface crust formation and the higher N input available for the crop associated to the higher biomass production by common vetch in 2021. The absence of soil crust could have allowed a higher efficiency of mechanical weeding interventions in 2021, contributing to obtain better yield performances on bare soil (IN1, IN2, and M1). Indeed, mechanical weeding, in addition to controlling weeds and aerating the soil, favors the breaking of soil capillaries, thus preventing water evaporation from the soil [42]. It must also be considered that the adoption of conservative management practices, such as reduced tillage and cover crops, does not always lead to obtaining appreciable benefits in the short term [43]. Several authors observed that the conversion from conventional to conservation tillage practices can lead in the first 4–5 years to some disadvantages in terms of soil properties. These disadvantages include an increase in bulk density and a decrease in soil porosity. However, these can be followed by progressive improvements in terms of soil structure, organic carbon stabilization, availability of nutrients, and, therefore, crop yield [44–46]. In general, ST achieved a higher yield (36.41 Mg ha^{-1}) compared to M1, IN2, and IN1 (25.63, 19.31, and 15.57 Mg ha^{-1}, respectively) and similar to M2 (35.83 Mg ha^{-1}). Therefore, the use of mulch film appeared to be crucial in terms of tomato yield. This is in agreement with Testani et al. [47]. Authors observed a higher tomato yield for the system where Mater-bi film was used than in the system where a cover crop mixture of barley and vetch were managed as green manure. Moreover, it is possible to observe the absence of relevant differences between itineraries based on conservation agriculture practices such as reduced tillage and those with conventional deep tillage. This is especially evident when mulch film was present (ST and M2). This trend showed the importance of ground cover on tomato yield. Ground cover also performed various useful functions, such as the promotion of soil moisture conservation, making it possible to cope with critical climatic conditions, such as drought seasons [30]. This is an important result in favor of adopting conservation agriculture practices. It also supports the progressive abandonment of deep tillage practices which are costly, boost organic matter mineralization, and are related to high GHG emission compared to reduced tillage [7,8].

4.3. Total Field Time and Fuel Consumption of the Technical Itinearies

The higher values recorded by ST and M2 in terms of average total field time (318.7 and 319.3 h ha^{-1}, respectively) and fuel consumption (323.1 and 300.4 kg ha^{-1}, respectively) are mainly related to transplanting operations on mulch film. Transplanting on mulch film requires a longer field time than transplanting on bare soil (200 h ha^{-1} vs. 40 h ha^{-1}). In the first case, eight operators are needed, while in the case of transplanting on bare soil, four operators are sufficient. This results in a manpower management efficiency when transplantation was carried out on bare soil 2.5 times higher compared to the transplantation on mulch film. This is in line with Feldman et al. [48], according to which the use of film mulch can lead to a higher labor requirement for both mulch spreading and planting. Transplanting on film mulch also required a higher fuel consumption compared to transplanting on bare soil, with an increase of 150%. Therefore, these differences in transplantation technique had relevant consequences on these operational parameters. Soil management strategies adopted before transplanting relevantly impacted the technical itineraries' fuel consumption. This is also confirmed by Moitzi et al. [49]. These authors, testing a moldboard plough, a short disc harrow, a universal cultivator, and a subsoiler at different working depths, noticed a rise in fuel consumption as the working depth increased. Authors stated that, as the working depth increased, both drawbar pull and slip increased, and, therefore, fuel consumption rate also increased. This is also in agreement with Pratibha et al. [50], who observed a higher energy input for conventional tillage compared to conservative practices, such as reduced or zero tillage. This result was attributed to a greater depth and number of tillage operations, which resulted in higher machine usage and higher fuel consumption. Thus, a reduction in tillage depth, in addition to the advantages described above, also creates benefits by allowing a reduction in fuel consumption, albeit limited and corresponding to about 10% in the present study.

4.4. Gross Salable Production, Average Variable Costs, and Gross Income

In both 2020 and 2021, ST and M2 were the itineraries that produced the highest results in terms of GSP. This result is in line with the yield trend. Other systems characterized by the lack of mulch film achieved lower GSP compared to ST and M2. The lower GSP of IN1 in 2020 and 2021 and of IN2 in 2020 compared to the other itineraries is mainly related to a lower yield achieved.

The total average VC resulted higher in ST and M2, with values of 18,303 and 18,622 EUR ha^{-1}, while the other itineraries achieved similar results. This is mainly due to the use of mulch film, which was remarkably expensive. Indeed, although Mater-bi film is a more environmentally friendly alternative to polyethylene mulch and allows disposal cost savings through soil incorporation with tillage, its main constrain is its higher cost [51]. Moreover, transplanting on film mulch not only incurred significant expenses in terms of technical means, but also had an impact on fuel consumption and labor costs. In both years, fuel costs were higher in the ST itinerary, highlighting how conventional deep tillage had a greater impact on fuel consumption compared to conservation agriculture practices, such as reduced tillage. In terms of technical means costs, M2 recorded a higher cost compared to the other itineraries due to the use of both cover crops and mulch film. M1 achieved the lowest costs, as in this case, no technical means were used beyond those used in all the other systems. Technical means purchase represented the cost item with the greatest impact on the total variable cost of itineraries. Instead, fuel cost showed a lower incidence to the total variable costs compared to labor, technical means, and machinery costs.

By comparing GI values referring to the results obtained in the first and second year of the trial with those of the GSP and tomato yields, respectively, it is possible to observe a similar trend. This underlines that GI was mainly influenced by yield and GSP, and not by the costs incurred. ST and M2 achieved a higher GI value during both years, without showing relevant differences. This highlights how crucial it is in critical pedo-climatic conditions, such as those in the trial, to maintain soil cover with mulch and thereby positively affect plant microclimate, favor adequate weed control, and conserve soil moisture [30].

5. Conclusions

The present study highlights that, in the particular pedo-climatic conditions in which the study was conducted, there were several limitations in achieving optimal yield performance for organic processing tomato. The best yield results were obtained tendentially by the technical itineraries where permanent coverage with Mater-bi film was used (ST and M2). Therefore, even though this technique resulted in significantly onerous processes in terms of labor, fuel consumption and technical means, it can lead to productive and economic benefits, especially in such critical environmental conditions. Concerning soil management, working depth did not significantly affect tomato performance, either in terms of yield or gross income. Indeed, with the same weed management system, no differences emerged tendentially in tomato performance among itineraries where conventional deep tillage and conservative shallower tillage practices were carried out. This is a significant outcome for the gradual abandonment of deep tillage techniques in favor of conservative agriculture practices, such as reduced tillage. The use of the cover crop mixture composed of barley and common vetch managed as a green manure seemed to be a beneficial method for processing tomato production, as the crop is N demanding. These effects emerged particularly in the second year of the test, in which the higher biomass production by vetch compared to the previous year positively influenced the tomato yield performance in the organic farming systems. By applying conservation agriculture practices, such as reduced tillage and the use of cover crops, improvement of organic farming systems sustainability could be favored. Indeed, these practices can lead to several environmental benefits, such as CO_2 emissions reduction and increase in soil fertility. Further studies are needed to evaluate the effect of shallower soil tillage practices over a longer period in these specific pedo-climatic conditions. Moreover, in this context, it would be useful to evaluate the creation of soil cover with cover crops managed as dead mulch, both in terms of yield and economic advantages.

Author Contributions: Conceptualization, A.P., M.S. (Mino Sportelli) and L.G.; methodology, A.P., M.S. (Mino Sportelli) and L.G.; software, M.S. (Mino Sportelli); validation, A.P. and M.S. (Mino Sportelli); formal analysis, M.S. (Mino Sportelli) and M.S. (Massimo Sbrana); investigation, A.P., S.M.L. and M.S. (Mino Sportelli); resources, A.P.; data curation, A.P., M.S. (Mino Sportelli), M.S. (Massimo Sbrana) and S.M.L.; writing—original draft preparation, L.G.; writing—review and editing, L.G., A.P., M.F. and M.R.; visualization, M.F., M.S. (Massimo Sbrana) and M.R.; supervision, M.S. (Mino Sportelli) and A.P.; project administration, A.P.; funding acquisition, A.P. and M.F. All authors have read and agreed to the published version of the manuscript.

Funding: This research was funded by the Italian Ministry of Agricultural, Forestry and Food Policies (MiPAAF), within the project "Mechanization of organic and conservation horticulture", acronym "MEORBICO", approved by Ministerial Decree 89238 of 19 December 2019.

Data Availability Statement: Not applicable.

Acknowledgments: The authors would like to thank Andrea and Giampaolo Pasquini (owners and managers of Pasquini Farm of Suvereto, Livorno, Italy) for their availability and for hosting the trials. Moreover, the authors want to thank Alessandro Pannocchia and Roberta Del Sarto of the Agri-environmental Research Centre "Enrico Avanzi", San Piero a Grado (Pisa), Italy, very much for their technical help in the conduction of field tests and samples processing. The authors wish also to acknowledge Michel Pirchio for technical support in field data collection.

Conflicts of Interest: The authors declare no conflict of interest.

References

1. Godfray, H.C.J.; Beddington, J.R.; Crute, I.R.; Haddad, L.; Lawrence, D.; Muir, J.F.; Pretty, J.; Robinson, S.; Thomas, S.M.; Toulmin, C. Food Security: The Challenge of Feeding 9 Billion People. *Science* **2010**, *327*, 812–818. [CrossRef]
2. Zalles, V.; Hansen, M.C.; Potapov, P.V.; Stehman, S.V.; Tyukavina, A.; Pickens, A.; Song, X.-P.; Adusei, B.; Okpa, C.; Aguilar, R.; et al. Near Doubling of Brazil's Intensive Row Crop Area since 2000. *Proc. Natl. Acad. Sci. USA* **2019**, *116*, 428–435. [CrossRef]
3. Browne, W.P. *The Failure of National Rural Policy: Institutions and Interests*; Georgetown University Press: Washington, DC, USA, 2001.
4. Oldfield, E.E.; Bradford, M.A.; Wood, S.A. Global Meta-Analysis of the Relationship between Soil Organic Matter and Crop Yields. *Soil* **2019**, *5*, 15–32. [CrossRef]
5. Cisilino, F.; Bodini, A.; Zanoli, A. Rural Development Programs' Impact on Environment: An Ex-Post Evaluation of Organic Faming. *Land Use Policy* **2019**, *85*, 454–462. [CrossRef]
6. Peigné, J.; Ball, B.C.; Roger-Estrade, J.; David, C. Is Conservation Tillage Suitable for Organic Farming? A Review. *Soil Use Manag.* **2007**, *23*, 129–144. [CrossRef]
7. Morris, D.R.; Gilbert, R.A.; Reicosky, D.C.; Gesch, R.W. Oxidation Potentials of Soil Organic Matter in Histosols under Different Tillage Methods. *Soil Sci. Soc. Am. J.* **2004**, *68*, 817–826. [CrossRef]
8. Šarauskis, E.; Kriaučiūnienė, Z.; Romaneckas, K.; Buragienė, S. Impact of Tillage Methods on Environment, Energy and Economy. In *Sustainable Agriculture Reviews 33*; Lichtfouse, E., Ed.; Sustainable Agriculture Reviews; Springer International Publishing: Cham, Switzerland, 2018; Volume 33, pp. 53–97. ISBN 978-3-319-99075-0.
9. Kuhwald, M.; Blaschek, M.; Brunotte, J.; Duttmann, R. Comparing Soil Physical Properties from Continuous Conventional Tillage with Long-Term Reduced Tillage Affected by One-Time Inversion. *Soil Use Manag.* **2017**, *33*, 611–619. [CrossRef]
10. Jaleta, M.; Baudron, F.; Krivokapic-Skoko, B.; Erenstein, O. Agricultural Mechanization and Reduced Tillage: Antagonism or Synergy? *Int. J. Agric. Sustain.* **2019**, *17*, 219–230. [CrossRef]
11. Rutkowska, B.; Szulc, W.; Sosulski, T.; Skowrońska, M.; Szczepaniak, J. Impact of Reduced Tillage on CO_2 Emission from Soil under Maize Cultivation. *Soil Tillage Res.* **2018**, *180*, 21–28. [CrossRef]

12. Hu, F.; Chai, Q.; Yu, A.; Yin, W.; Cui, H.; Gan, Y. Less Carbon Emissions of Wheat–Maize Intercropping under Reduced Tillage in Arid Areas. *Agron. Sustain. Dev.* **2015**, *35*, 701–711. [CrossRef]

13. Mei, K.; Wang, Z.; Huang, H.; Zhang, C.; Shang, X.; Dahlgren, R.A.; Zhang, M.; Xia, F. Stimulation of N_2O Emission by Conservation Tillage Management in Agricultural Lands: A Meta-Analysis. *Soil Tillage Res.* **2018**, *182*, 86–93. [CrossRef]

14. Yuan, J.; Yan, L.; Li, G.; Sadiq, M.; Rahim, N.; Wu, J.; Ma, W.; Xu, G.; Du, M. Effects of Conservation Tillage Strategies on Soil Physicochemical Indicators and N2O Emission under Spring Wheat Monocropping System Conditions. *Sci. Rep.* **2022**, *12*, 7066. [CrossRef]

15. Van Kessel, C.; Venterea, R.; Six, J.; Adviento-Borbe, M.A.; Linquist, B.; Van Groenigen, K.J. Climate, Duration, and N Placement Determine N_2O Emissions in Reduced Tillage Systems: A Meta-Analysis. *Glob. Change Biol.* **2013**, *19*, 33–44. [CrossRef]

16. Feng, J.; Li, F.; Zhou, X.; Xu, C.; Ji, L.; Chen, Z.; Fang, F. Impact of Agronomy Practices on the Effects of Reduced Tillage Systems on CH_4 and N_2O Emissions from Agricultural Fields: A Global Meta-Analysis. *PLoS ONE* **2018**, *13*, e0196703. [CrossRef]

17. Filipovic, D.; Kosutic, S.; Gospodaric, Z.; Zimmer, R.; Banaj, D. The Possibilities of Fuel Savings and the Reduction of CO_2 Emissions in the Soil Tillage in Croatia. *Agric. Ecosyst. Environ.* **2006**, *115*, 290–294. [CrossRef]

18. Salem, H.M.; Valero, C.; Muñoz, M.Á.; Rodríguez, M.G.; Silva, L.L. Short-Term Effects of Four Tillage Practices on Soil Physical Properties, Soil Water Potential, and Maize Yield. *Geoderma* **2015**, *237–238*, 60–70. [CrossRef]

19. Nichols, V.; Verhulst, N.; Cox, R.; Govaerts, B. Weed Dynamics and Conservation Agriculture Principles: A Review. *Field Crop. Res.* **2015**, *183*, 56–68. [CrossRef]

20. Cooper, J.; Baranski, M.; Stewart, G.; Nobel-de Lange, M.; Bàrberi, P.; Fließbach, A.; Peigné, J.; Berner, A.; Brock, C.; Casagrande, M.; et al. Shallow Non-Inversion Tillage in Organic Farming Maintains Crop Yields and Increases Soil C Stocks: A Meta-Analysis. *Agron. Sustain. Dev.* **2016**, *36*, 22. [CrossRef]

21. Benincasa, P.; Tosti, G.; Guiducci, M.; Farneselli, M.; Tei, F. Crop Rotation as a System Approach for Soil Fertility Management in Vegetables. In *Advances in Research on Fertilization Management of Vegetable Crops*; Tei, F., Nicola, S., Benincasa, P., Eds.; Advances in Olericulture; Springer International Publishing: Cham, Switzerland, 2017; pp. 115–148. ISBN 978-3-319-53624-8.

22. Carlesi, S.; Bigongiali, F.; Antichi, D.; Ciaccia, C.; Tittarelli, F.; Canali, S.; Bàrberi, P. Green Manure and Phosphorus Fertilization Affect Weed Community Composition and Crop/Weed Competition in Organic Maize. *Renew. Agric. Food Syst.* **2020**, *35*, 493–502. [CrossRef]

23. Koudahe, K.; Allen, S.C.; Djaman, K. Critical Review of the Impact of Cover Crops on Soil Properties. *Int. Soil Water Conserv. Res.* **2022**, *10*, 343–354. [CrossRef]

24. Hwang, H.Y.; Kim, G.W.; Lee, Y.B.; Kim, P.J.; Kim, S.Y. Improvement of the Value of Green Manure via Mixed Hairy Vetch and Barley Cultivation in Temperate Paddy Soil. *Field Crop. Res.* **2015**, *183*, 138–146. [CrossRef]

25. ISMEA. *Ortaggi—Focus Conserve di Pomodoro. Tendenze e Dinamiche Recenti*; ISMEA: Rome, Italy, 2022.

26. Tribouillois, H.; Cohan, J.-P.; Justes, E. Cover Crop Mixtures Including Legume Produce Ecosystem Services of Nitrate Capture and Green Manuring: Assessment Combining Experimentation and Modelling. *Plant Soil* **2016**, *401*, 347–364. [CrossRef]

27. Schipanski, M.E.; Barbercheck, M.; Douglas, M.R.; Finney, D.M.; Haider, K.; Kaye, J.P.; Kemanian, A.R.; Mortensen, D.A.; Ryan, M.R.; Tooker, J.; et al. A Framework for Evaluating Ecosystem Services Provided by Cover Crops in Agroecosystems. *Agric. Syst.* **2014**, *125*, 12–22. [CrossRef]

28. Farneselli, M.; Tosti, G.; Onofri, A.; Benincasa, P.; Guiducci, M.; Pannacci, E.; Tei, F. Effects of N Sources and Management Strategies on Crop Growth, Yield and Potential N Leaching in Processing Tomato. *Eur. J. Agron.* **2018**, *98*, 46–54. [CrossRef]

29. Tosti, G.; Benincasa, P.; Farneselli, M.; Tei, F.; Guiducci, M. Barley–Hairy Vetch Mixture as Cover Crop for Green Manuring and the Mitigation of N Leaching Risk. *Eur. J. Agron.* **2014**, *54*, 34–39. [CrossRef]

30. Martín-Closas, L.; Bach, M.A.; Pelacho, A.M. Biodegradable Mulching in an Organic Tomato Production System. *Acta Hortic.* **2008**, *767*, 267–274. [CrossRef]

31. Shen, M.; Song, B.; Zeng, G.; Zhang, Y.; Huang, W.; Wen, X.; Tang, W. Are Biodegradable Plastics a Promising Solution to Solve the Global Plastic Pollution? *Environ. Pollut.* **2020**, *263*, 114469. [CrossRef]

32. Shruti, V.C.; Kutralam-Muniasamy, G. Bioplastics: Missing Link in the Era of Microplastics. *Sci. Total Environ.* **2019**, *697*, 134139. [CrossRef]

33. Touchaleaume, F.; Martin-Closas, L.; Angellier-Coussy, H.; Chevillard, A.; Cesar, G.; Gontard, N.; Gastaldi, E. Performance and Environmental Impact of Biodegradable Polymers as Agricultural Mulching Films. *Chemosphere* **2016**, *144*, 433–439. [CrossRef]

34. Sekara, A.; Pokluda, R.; Cozzolino, E.; Del Piano, L.; Cuciniello, A.; Caruso, G. Plant Growth, Yield, and Fruit Quality of Tomato Affected by Biodegradable and Non-Degradable Mulches. *Hortic. Sci.* **2019**, *46*, 138–145. [CrossRef]

35. Macua, J.I.; Jiménez, E.; Suso, M.L.; Gervas, C.; Lahoz, I. The Future of Processing Tomato Crops in The Ebro Valley Lies with the Use of Biodegradable Mulching. *Acta Hortic.* **2013**, *971*, 143–146. [CrossRef]

36. Cirujeda, A.; Aibar, J.; Anzalone, Á.; Martín-Closas, L.; Meco, R.; Moreno, M.M.; Pardo, A.; Pelacho, A.M.; Rojo, F.; Royo-Esnal, A.; et al. Biodegradable Mulch Instead of Polyethylene for Weed Control of Processing Tomato Production. *Agron. Sustain. Dev.* **2012**, *32*, 889–897. [CrossRef]

37. Cirujeda, A.; Aibar, J.; Moreno, M.M.; Zaragoza, C. Effective Mechanical Weed Control in Processing Tomato: Seven Years of Results. *Renew. Agric. Food Syst.* **2013**, *30*, 223–232. [CrossRef]

38. Raffaelli, M.; Fontanelli, M.; Frasconi, C.; Sorelli, F.; Ginanni, M.; Peruzzi, A. Physical Weed Control in Processing Tomatoes in Central Italy. *Renew. Agric. Food Syst.* **2011**, *26*, 95–103. [CrossRef]

39. ISMEA. ISMEA Mercati. Available online: http://www.ismeamercati.it/flex/cm/pages/ServeBLOB.php/L/IT/IDPagina/1881 (accessed on 11 April 2023).
40. Moreno, M.M.; González-Mora, S.; Villena, J.; Campos, J.A.; Moreno, C. Deterioration Pattern of Six Biodegradable, Potentially Low-Environmental Impact Mulches in Field Conditions. *J. Environ. Manag.* **2017**, *200*, 490–501. [CrossRef]
41. Ronga, D.; Gallingani, T.; Zaccardelli, M.; Perrone, D.; Francia, E.; Milc, J.; Pecchioni, N. Carbon Footprint and Energetic Analysis of Tomato Production in the Organic vs. the Conventional Cropping Systems in Southern Italy. *J. Clean. Prod.* **2019**, *220*, 836–845. [CrossRef]
42. Vincent, C.; Panneton, B.; Fleurat-Lessard, F. (Eds.) *Physical Control Methods in Plant Protection*; Softcover reprint of the original 1st ed. 2001; Springer: Berlin/Heidelberg, Germany, 2014; ISBN 978-3-662-04586-2.
43. Castellini, M.; Fornaro, F.; Garofalo, P.; Giglio, L.; Rinaldi, M.; Ventrella, D.; Vitti, C.; Vonella, A. Effects of No-Tillage and Conventional Tillage on Physical and Hydraulic Properties of Fine Textured Soils under Winter Wheat. *Water* **2019**, *11*, 484. [CrossRef]
44. Blanco-Canqui, H.; Ruis, S.J. No-Tillage and Soil Physical Environment. *Geoderma* **2018**, *326*, 164–200. [CrossRef]
45. Strudley, M.; Green, T.; Ascoughii, J. Tillage Effects on Soil Hydraulic Properties in Space and Time: State of the Science. *Soil Tillage Res.* **2008**, *99*, 4–48. [CrossRef]
46. Reichert, J.M.; Da Rosa, V.T.; Vogelmann, E.S.; Da Rosa, D.P.; Horn, R.; Reinert, D.J.; Sattler, A.; Denardin, J.E. Conceptual Framework for Capacity and Intensity Physical Soil Properties Affected by Short and Long-Term (14 Years) Continuous No-Tillage and Controlled Traffic. *Soil Tillage Res.* **2016**, *158*, 123–136. [CrossRef]
47. Testani, E.; Ciaccia, C.; Diacono, M.; Fornasier, F.; Ferrarini, A.; Montemurro, F.; Canali, S. Agroecological Practices Improve Soil Biological Properties in an Organic Vegetable System. *Nutr. Cycl. Agroecosyst.* **2023**, *125*, 471–486. [CrossRef]
48. Feldman, R.S.; Holmes, C.E.; Blomgren, T.A. Use of Fabric and Compost Mulches for Vegetable Production in a Low Tillage, Permanent Bed System: Effects on Crop Yield and Labor. *Am. J. Alt. Ag.* **2000**, *15*, 146–153. [CrossRef]
49. Moitzi, G.; Wagentristl, H.; Refenner, K.; Weingartmann, H.; Piringer, G.; Boxberger, J.; Gronauer, A. Effects of Working Depth and Wheel Slip on Fuel Consumption of Selected Tillage Implements. *Agric. Eng. Int. CIGR J.* **2014**, *16*, 182–190.
50. Pratibha, G.; Srinivas, I.; Rao, K.V.; Raju, B.M.K.; Thyagaraj, C.R.; Korwar, G.R.; Venkateswarlu, B.; Shanker, A.K.; Choudhary, D.K.; Rao, K.S.; et al. Impact of Conservation Agriculture Practices on Energy Use Efficiency and Global Warming Potential in Rainfed Pigeonpea–Castor Systems. *Eur. J. Agron.* **2015**, *66*, 30–40. [CrossRef]
51. Andrade, C.S.; Palha, M.D.G.; Duarte, E. Biodegradable Mulch Films Performance for Autumn-Winter Strawberry Production. *JBR* **2014**, *4*, 193–202. [CrossRef]

Article

How Much Impact Has the Cover Crop Mulch in Mitigating Soil Compaction?—A Field Study in North Italy

Marco Benetti [1], Kaihua Liu [1], Lorenzo Guerrini [1,*], Franco Gasparini [1], Andrea Peruzzi [2] and Luigi Sartori [1]

[1] Dipartimento Territorio e Sistemi Agro-Forestali, Università di Padova, Viale dell'università 16, 35020 Legnaro, Italy

[2] Dipartimento di Scienze Agrarie, Alimentari e Agro-ambientali, Università di Pisa, Via del Borghetto 80, 56124 Pisa, Italy

* Correspondence: lorenzo.guerrini@unipd.it

Abstract: Soil compaction was largely studied in different scenarios with laboratory and field scale experiments, with various soil conditions and traffic intensities. However, a detailed analysis to better understand the protective role of plant residues or cover crop mulch is still required. A field test was conducted in Northeast Italy aiming to fill this gap. Rye was chosen as a winter cover crop, and growth on a controlled traffic random block experimental field. Four different cover crop mulch treatments were compared to study the effects of root systems: roller crimper, flail mower, bare soil control and harvested biomass control. Four different traffic intensities were used to evaluate the multiple passages with 0, 1, 3, 5 traffic events. During traffic events, the mean normal stress was measured. Penetration resistance was then evaluated after trafficking and soil samples were collected. The obtained results showed a 19.3% cone index increase in bare soil compared to flail mower treatment after the first traffic event, while low differences were found in harvested biomass bulk density during the first and third traffic events. Moreover, mean normal stress increased 16.5% on harvested biomass treatment compared to the flail mower. These findings highlight that the cover crop maintains a lower soil penetration resistance during compaction events, helping the subsequent field operations. Furthermore, roller crimper and flail mower cover crop termination impact soil bearing capacity differently due to different soil moisture content. However, the results showed a low contribution of cover crop mulch on mitigating soil compaction effects during the experiment.

Keywords: soil compaction; cover crop; bulk density; mean normal stress; soil cone index; mulch

Citation: Benetti, M.; Liu, K.; Guerrini, L.; Gasparini, F.; Peruzzi, A.; Sartori, L. How Much Impact Has the Cover Crop Mulch in Mitigating Soil Compaction?—A Field Study in North Italy. *Agronomy* **2023**, *13*, 686. https://doi.org/10.3390/agronomy13030686

Academic Editor: Javier M. Gonzalez

Received: 21 January 2023
Revised: 21 February 2023
Accepted: 24 February 2023
Published: 26 February 2023

1. Introduction

Soil compaction is one of the main causes of soil degradation [1]. In fact, soil compaction induces a complex change in soil conditions and behaviour, including loss of soil porosity and pore function [2]. This phenomenon occurs when the load transmitted by vehicles applied a higher stress than the soil bearing capacity, which involves plastic soil deformation [3]. Different approaches can be used to mitigate soil compaction effects on soil and plant growth, such as enhancing soil bearing capacity [4], decreasing soil stress [5] or reducing the trafficked area [6]. During forestry operations, the use of brush mats as a soil protective layer was frequently adopted to decrease compaction and improve trafficability [7]. Conservation agriculture can match the same forestry strategy to mitigate compaction when residues are left on the surface, but the main aim is to mitigate soil erosion. Few experiments studied soil compaction mitigation on the surface residue layer. Ess et al. studied the cover crop effects on mitigating soil compaction with field experiments by analysing roots and surface biomass effects on soil characteristics and behaviour [8]. In addition, field tests with stress state transducers showed residues effects on soil stress compared to bare soil [9], but the quantity, type and conditions of residues was not defined. Other lab and bin experiments showed no differences between different amounts of residues mixed with soil on the compression index [10]. However, a bin experiment with surface residues showed that mean normal stress decreases less than 80 kPa comparing maize residues and bare soil treatment with no statistical differences in bulk density [10]. Other outcomes regarding residues effects on stress mitigation were obtained in a laboratory experiment, but they did not confirm the field test [11]. Recent findings obtained in a laboratory experiment showed that residues effects have subtly increased soil load-bearing capacity (e.g., 15 kPa in apparent precompression stress, 7%). Soil bulk density changes decreased below 100 kPa stress and increased over 400 kPa stress [12]. Few previous studies showed a complete approach to evaluate the residues effects on soil compaction [8]. Indeed, other studies were focused on residues effects only, without analysing the soil conditions due to the lying residues or cover crop growth [11,12]. The aim of our study was to test the effects of residues obtained from differing cover crop management on soil compaction in a real-scale field experiment by considering possible interactions and effects that affect the soil compaction phenomena during cropping operations, which are difficult to observe in laboratory experiments.

2. Materials and Methods

The field experiment was conducted on the University of Padova Experimental farm (Legnaro, Italy, 45°21′4″ N; 11°56′49″ E; 6 m a.s.l.) in 2022 (Figure 1). Temperatures rise from January (min average: −1.5 °C) to July (max average: 27.2 °C). The sub-humid climate receives approximately 850 mm of rainfall annually, with the highest average rainfall in June (100 mm) and October (90 mm). The lowest averages were recorded in January and February (50–60 mm). The soil texture of the experiment field is clay loam and was already analysed in another field experiment [13]. The soil was characterised by a 33.8% sand content, 37.0% silt and 29.2% clay. The organic matter content of the topsoil was 1.81%.

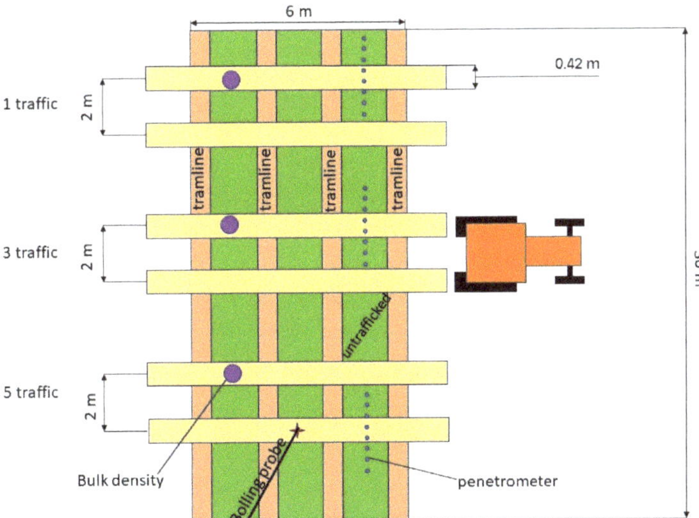

Figure 1. Traffic pattern of the experimental field test for every covering plot. The picture shows how the traffic was managed during the experiment to avoid compaction on the sampling zone through the sampling area of the bulk density, penetrometer to measure cone index and Bolling probe to assess mean normal stress. A total of 16 covering plots were used with 4 different covering treatments and 4 replicates.

The area of the experiment was selected after preliminary soil variability analysis conducted using a high-resolution electromagnetic conductivity meter (CMD mini explorer 6 L, GF instrument, Brno, Czech Republic). Thus, a homogeneous area was selected, avoiding headlands. After this process the field was divided in plots and tilled to prepare a homogeneous seedbed. The field had a 45 cm depth ploughed on 26 October 2021 with a Bordin (Padova, Italy) double furrow plough, and all field operations were carried out following controlled traffic to avoid compaction in the sampling area. A Pegoraro (Lonigo, Italy) PTO powered rotary harrow was used for seedbed preparation, and rye (*Secale cereale* L.) was sown with a common Carraro row drill (Campodarsego, Italy) on 29 October 2021. Seeding rate was 165 kg ha^{-1} (cultivar: Antoninskie). One thousand seeds weight 23.9 g. No fertilizer was applied during the experiment. Bare soil treatments were also tilled, without sowing. On bare soil treatment, two weed controls were conducted with chemical herbicide (glyphosate 1.44 kg ha^{-1} each one).

Four cover crop mulch treatments were considered: roller crimper (Crimped), flail mower (Shredded), bare soil (Bare), harvested canopy (Harvested) (Figures 2 and 3). The cover crop mulch was terminated with one passage of the roller crimper (Crimped treatment) or mounted flail mower (Berti Machine Agricole spa, Caldiero, Italy) (Harvested and shredded treatment) on 26 May 2022, following a random block design. The roller crimper involved in the test was built by modifying an iron smooth-roller. The roller crimper had a total mass of 880 kg and a working width of 3.3 m. The crimping action was achieved using twelve iron plates with 6 mm thickness and 60 mm height. Plates were welded orthogonally to the tangential plane of the roller surface, every 121.7 mm of circumference. The resulting external roller diameter was 585 mm. Mulch was removed by hand just before trafficking in the "Harvested" treatment. Two irrigations were applied with a hose-reel sprinkler irrigator (Irrigazione Veneta srl, Torri di Quartesolo, Italy) four days before traffic due to a drought during spring and summer 2022, the first of 50 mm and second of 15 mm.

<table>
<tr><td>(a)</td><td>(b)</td></tr>
</table>

Figure 2. Different cover crop terminations: (**a**) flail mower, (**b**) roller crimper.

The experimental plots were trafficked simultaneously 0, 1, 3 and 5 times at 4 km h^{-1}, as shown in Figure 1 on the undisturbed sampling zone. The biomass samples were collected before traffic events. The soil samples were collected immediately after traffic events. The two-wheel drive tractor used for trafficking is described in Table 1.

Figure 3. Effect of the four different cover crop managements compared in the experiment: (1) bare soil (Bare); (2) roller crimper (Crimped); (3) harvested canopy (Harvested); (4) flail mower (Shredded).

Table 1. Technical data of the tractor used in the experiment.

Name	Unit	Model
Tractor Model		Fiat 680
Total mass	kg	4310
Front axle	kg	780
Rear axle	kg	3530
Rear tyre	Kleber traker	420/85R30
Front tyre	Vredestein multirill	7.50–16
Front tyre inflation pressure	bar	1.7
Rear tyre inflation pressure	bar	1.45

2.1. Soil Cone Index

Penetration resistance was measured with a cone penetrometer to evaluate the effect of traffic conditions on soil. The cone penetrometer (Penetrologger Eijkelkamp, Geesbek, The Netherlands) was inserted into the soil at a constant speed with a penetration rate of less than 2 cm s^{-1}. The cone used in the penetrometer had a base diameter of 12.83 mm and cone angle of 30°, as specified in ASAE Standards S313.3 and EP 542 (ASAE Standards, 1999, 2001). The penetrometer was mounted on a designed iron frame fixed to the hydraulic piston (Figure 4). The iron frame allows the penetrometer to change the location of measurement horizontally. The hydraulic piston driven by the tractor allows for the uniform insertion speed during measurement. One transect centred on the wheel rut was performed for every level of traffic and cover crop mulch treatment for a total of 64 transects. Every transect had 15 penetration points spaced 5 cm with depth up to 60 cm.

The cone index was calculated as the average of 7 centred on the rut penetration points on 4 defined depth layers: 5–15, 15–25, 25–35, 35–45 cm. After a preliminary data analysis, the deeper layer (35–45 cm depth) was not considered in the statistical analysis due to the presence of plough pan.

Figure 4. A penetrometer was installed on an iron frame with hydraulic ram to insert the cone tip at constant speed in the soil.

2.2. Rut Profile Analysis

Penetrometer data were used to analyse rut profile evolution after 0, 1, 3, 5 passages. The penetrometer was mounted on a hydraulic frame to start penetration every time from the same position. The ultrasonic depth sensor of the penetrometer allows for cone tip depth reading with a 1 cm interval. The value with zero before soil penetration was used to estimate the rut section area. Rut profile areas (cm^2) were calculated using the following Equation (1):

$$\text{Rut profile area} = \sum P_d * D_1 \tag{1}$$

where:

P_d = distance measured with penetrometer ultrasonic sensor (cm)
D_1 = distance between two penetrations (cm)

2.3. Soil Bulk Density and Soil Moisture

Soil samples were collected with a special hydraulic sampler up to 60 cm depth [14]. Frost storage was used to maintain samples during processing. The soil samples were divided into the following depth layers: 5–15, 15–25, 25–35, 35–45 cm. Samples were weighed and oven-dried at 105 °C until constant weight. Bulk densities (g cm^{-3}) were calculated using the following Formula (2):

$$\text{Bulk Density} = \frac{m}{v} \tag{2}$$

where:

m = dry mass of soil collected on undisturbed and defined volume "v" (g)
v = volume of soil sample (cm^3).

The deeper layer (35–45 cm depth) was not considered in the statistical analysis due to the presence of plough pan.

Soil moistures were measured with undisturbed soil samples used to measure bulk density. Volumetric water content (%) was calculated using the following Formula (3):

$$\text{VWC} = 100 * \frac{(M - m)}{(d * v)} \tag{3}$$

where:

M = wet soil mass (g)

m= dry mass of soil collected on undisturbed and defined volume "v" (g)

d = water density (g cm^{-3})

v = volume of soil sample (cm^3).

Additionally in this case, the deeper layer (35–45 cm depth) was not considered in the statistical analysis due to the presence of plough pan.

2.4. Biomass

Biomass samples were collected after trafficking the experimental field, using an iron wire square with 0.4 m side. Biomass was weighed and oven-dried at 105 °C until constant weight to measure dry mass and biomass moisture.

2.5. Mean Normal Stress

Mean normal stress was estimated using the Bolling probe [15,16]. The probes have a deformable cylindrical bulb and can easily be inserted in undisturbed soil at a defined depth with special installation tools. Sampling grids were created to align probe bulbs in order to define the trafficking and measuring area. Bulbs were filled with water and pressurized before trafficking. A Keller (Winterthur, Switzerland) hydraulic pressure sensor provided real time measurements. A laptop computer was connected to the sensor to monitor sampling and datalogging. Bolling probes were installed at 20 and 40 cm depths without damaging the cover crop mulch, as showed in Figure 5. Only maximum values, corresponding to the peak of rear axle traffic, were used. The first two passes were not considered during data analysis to avoid unreliable data due to soil deformation around the bulb zones. The pressure data under different depths were collected after each time compaction (five times in total; only the last three were considered in the analysis).

Figure 5. Installation position of the Bolling probe to measure the mean normal stress with different cover crop mulch treatments.

Mean normal stress (σ_m) was calculated [15,16] from Bolling probe pressure data in order to compare the results in different cover crop mulch treatments. The following is the calculation Formula (4) of the mean normal stress (kPa):

$$\sigma_m = \frac{1+v}{3(1-v)} P_i \tag{4}$$

where:

p_i is the measured stress from the Bolling probe (kPa)

v is the Poisson ratio in the soil matrix

The value of the Poisson ratio for soils is usually between 0.20 and 0.45 [17–21]. We set the Poisson ratio as 0.3 in our study using data available from previous studies [16,19].

2.6. Statistics

The trials were arranged in a randomized block design. In each of the 4 blocks, all the 4 considered cover crop management strategies were tested, in randomized order, as fixed effect factors. The number of transits and depths of measurement were further introduced in the model as random factors. Data were treated with an analysis of variance (ANOVA) model considering the main effects of the 3 tested factors, their interactions, and the experimental blocks. The chosen significant threshold was $p < 0.05$. The Tukey HSD was chosen as the post hoc test.

3. Results

The statistical significance of the results obtained in the experiment are summarized in Table 2.

Table 2. Analysis of variance, with all soil properties and variables taken into account.

Analysis of Variance Summary						
	Cover Crop Mulch	Traffic Events	Depth	Soil Moisture	CCM × TE	CCM × Depth
Bulk Density	<0.001	<0.001	0.04		0.02	
Cone Index	<0.001	<0.001	<0.001	<0.001	0.01	
Volumetric Water Content	<0.001	<0.001	ns		ns	
Rut profile area	0.004	<0.001			ns	
Surface Biomass	<0.001					
Mean Normal Stress	0.004		<0.001			0.001

ns = Not Significant; *p*-values are missing where factors were not involved in statistical analysis. CCM×TE is the interaction between cover crop mulch and traffic events, while CCM×epth is the interaction between cover crop mulch and depth.

3.1. Soil Cone Index

The results of the statistical analysis of the data from the soil cone index obtained in the four compared treatments after different passages of the tractor are shown in Figure 6.

Because the soil was undisturbed, no differences between mulch treatment were found at the zero traffic event. After the first tractor transit, the CI of bare soils becomes statistically significantly higher in comparison to the treatments involving cover crops. Specifically, Bare results in a 19.33% higher CI compared to Shredded. The increasing penetration resistance is higher after the first traffic event, but with a residues effect that resulted in an increase of 62% on Bare and 53% on Harvested compared to an increase of 31% on Crimped and 29% on shredded. The other traffic events recorded a lower increase in penetration resistance. Bare also had higher values after the following traffic events. Harvested treatment showed no significant difference from the other cover-cropped treatments Crimped and Shredded. However, the study of the interaction with traffic highlights that this difference is statistically significant only in the first and fifth traffic events. Differences were also found in the third traffic event, but without statistical significance. Traffic effects show differences between the fifth and other traffic intensities and between zero and other traffic intensities only. No differences were found between the first and third traffic events (Figure 6).

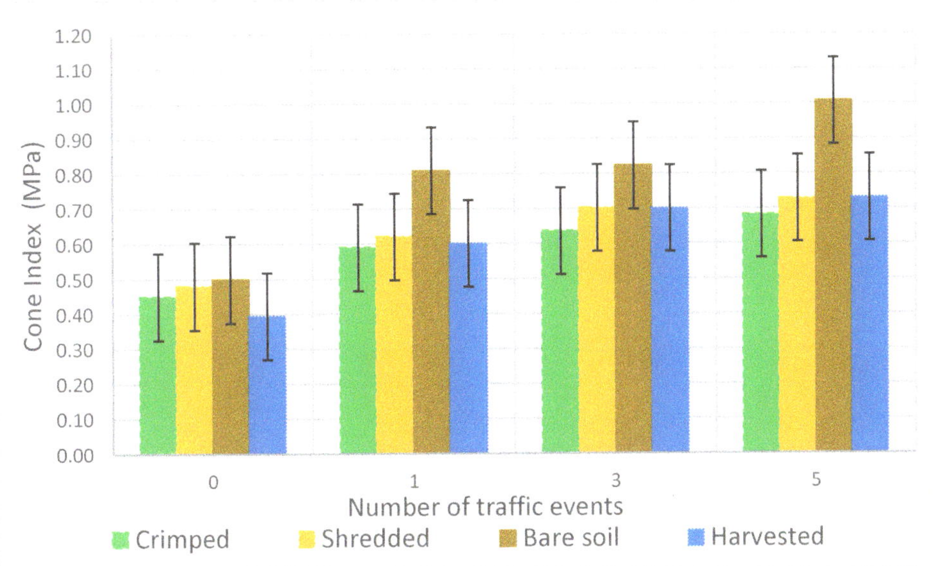

Figure 6. Cone index means after different numbers of traffic events and cover crop mulch treatments (Harvested; Bare soil; Crimped; Shredded). Residual standard errors were used on bar error.

3.2. Rut Profile Analysis

Rut profile analysis showed statistically significant differences between Shredded and Harvested-Bare mulch treatments. Bare and Harvested resulted in the higher rut profile area compared to the other treatments. Significant differences were also found between traffic treatments, as shown in Table 3. The increase in the number of traffic events was followed by an increase in the rut profile area. No interaction effect was found among the cover crop mulch treatments and number of traffic events.

Table 3. Result of statistical analysis on Rut area profiles (cm^2).

Treatment	Rut Profile Area	Standard Error		Treatment	Rut Profile Area	Standard Error	
Crimped	1405.94	55.13	ab	5	1639.06	50.56	a
Shredded	1328.75	54.31	b	3	1495.63	59.43	b
Bare	1507.19	54.35	a	1	1365.00	38.84	bc
Harvested	1498.44	66.80	a	0	1240.63	36.97	c
alpha: 00.5							

Treatments with the same letter are not significantly different.

3.3. Soil Bulk Density and Soil Moisture

The results of the statistical analysis of bulk density data obtained in the four compared treatments after each passage of the tractor are shown in Figure 7.

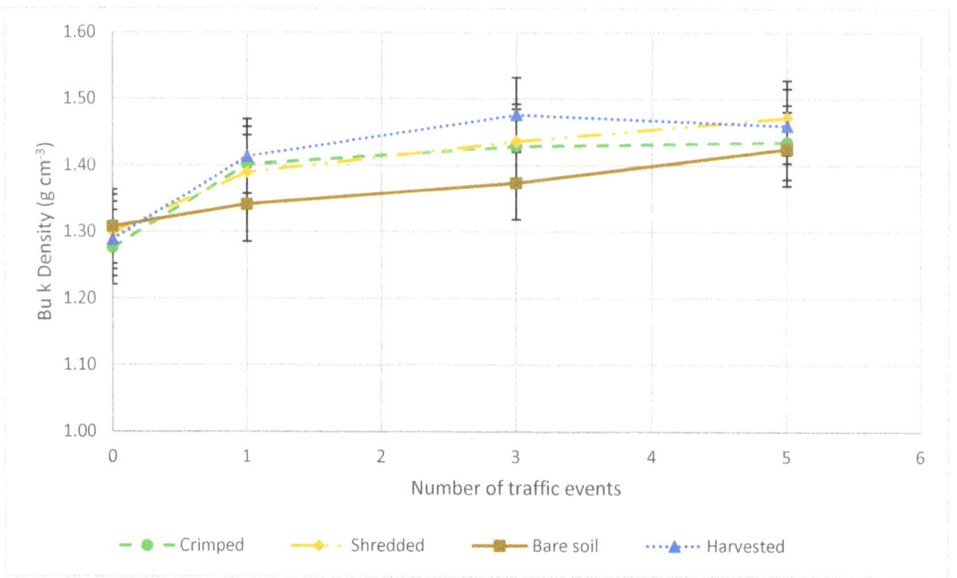

Figure 7. Bulk density means resulted from different number of traffic events and cover crop mulch treatments (Harvested; Bare soil; Crimped; Shredded). Residual standard errors were used on the bar error.

Bulk density increased significantly until the third traffic event. Zero traffic was statistically lower than other traffic events. One traffic event was statistically higher than the zero traffic event, but was statistically lower than three or more traffic intensities. No differences were found between the third and fifth passages. Cover crop mulch treatment was characterized by significantly lower bulk density on bare soil treatment compared to Harvested and Shredded. Crimped mulch treatment cannot be considered statistically different from the other mulch treatments. Statistically higher BD values were found on removed mulch treatment (Harvested) after the first and third passages in comparison with the other treatments when considering the interaction effect between traffic and cover crop mulch treatments. Moreover, BD also did not show differences between mulch treatment at zero and five traffic events. Statistical analysis also showed differences in bulk density in the soil layers between 5 and 15 cm depth and 15 and 25 cm depth. No differences were found between the deeper layer and the other layers.

Volumetric soil water content did not show any statistical difference at the different depths where it was analyzed. Interaction effects between traffic and mulch treatment were not detected during statistical analysis. Harvested and Shredded treatments resulted in significantly higher water content in comparison to Bare and Crimped treatments. No differences were found between Harvested and Shredded, nor between Bare and Crimped. Only zero traffic was significantly lower than the other traffic events. No differences in water content were found between the first, third and fifth passages. Removed mulch treatment (Harvested) showed a higher level of moisture at every traffic intensity.

3.4. Biomass

The results of the statistical analysis of mulch dry biomass are shown in Table 4.

Table 4. Result of statistical analysis on mulch dry biomass.

Treatment	Mulch Dry Biomass (Mg ha^{-1})		Standard Error
Crimped	11.24	a	4.18
Shredded	12.63	a	3.08
Bare	1.14	b	0.61
Harvested	3.54	b	4.49

Treatments with the same letter are not significantly different. Groups according to probability of means differences and alpha level (0.05).

Significant differences were found between the covered treatments (Crimped and Shredded) and not-covered treatments (Harvested and Bare). There were no statistical differences found between covered and not covered. Crimped treatment showed higher variability, probably caused by early bedding of cereal rye in a random position before crimping.

3.5. Mean Normal Stress

The results of the statistical analysis of soil mean normal stress determined at 20 and 40 cm obtained in the four compared treatments after each passage of the tractor are shown in Figure 8.

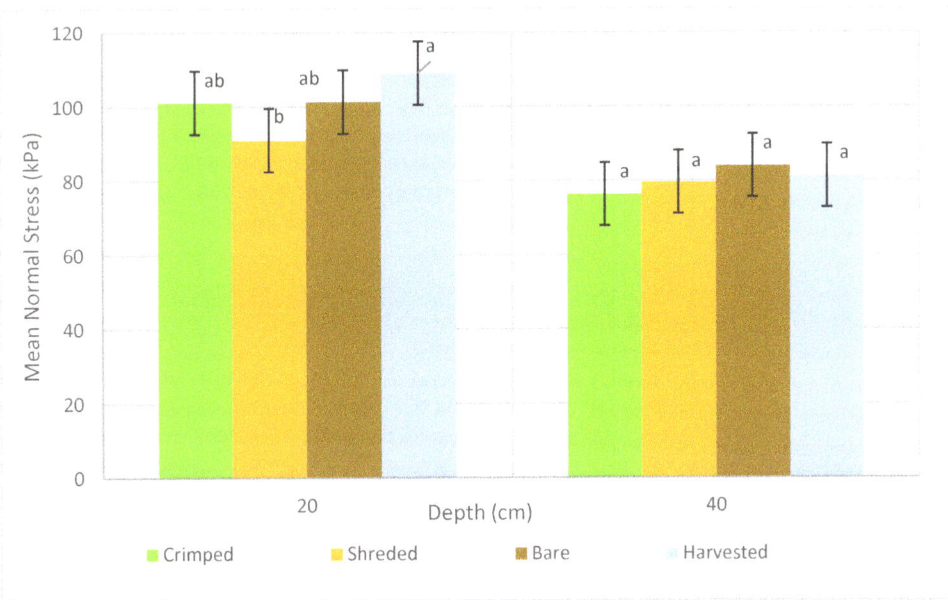

Figure 8. Results of the statistical analysis of mean normal stress (kPa). Residual standard errors were used on the bar error. Treatments with the same letter are not significantly different.

Significant differences were found between 20 and 40 cm depth mean normal stress. No statistical differences were found between treatment at 40 cm depth; only Harvested and Shredded can be considered different with $p < 0.05$. Strong variability was found in treatment Crimped at 20 cm depth, probably due to the spatial variability of rye biomass distribution caused by early crop bedding, before roller crimper passage. This effect could be mitigated by the flail mower action in treatment Shredded, resulting in less variability of stress data.

4. Discussion

4.1. Effect of Cover Crop Mulch

The effect of cover crop mulch on soil compaction could be divided into three different actions:

First, the cover crop presence, growth and termination method affect soil moisture. The higher volumetric water content found in treatment (Harvested, Shredded) can be explained by the increased water retention gained through cover crop root penetration, while their exudates improved soil pore stability, which was consistent with the study by Angers and Caron [22]. In contrast to our initial hypothesis, Crimped treatment shows low volumetric water content in comparison to other mulched treatments, probably due to incomplete termination immediately after crimping. The cover crop died slowly in the days after crimping and, in the meantime, water transpiration from the soil continued, as established in previous research [23]. This phenomenon probably did not happen in Harvested and Shredded treatments where termination was achieved with a flail mower. This tool causes the immediate death of the plant by cutting and shredding the stem, decreasing soil moisture loss. Furthermore, we hypothesized that the lower variability found on Shredded treatment mulch biomass was explained by the spreading effect of the flail mower compared to Crimped treatment. The rolling action during crimping, on Crimped treatment, probably did not modify the biomass distribution, maintaining the mulch rooted to the ground

Second, the root exudation and exploration change soil structure. The higher cone index on bare soil could be explained by the lack of cover crop root growing action, which stabilizes soil aggregates, mitigating the internal slaking of soil structure [24–26]. The internal slaking and the wetting and drying cycle cause the formation of a surface crust that could increase the susceptibility to soil compaction [25], resulting in increased penetration resistance, as observed in zero traffic; there was no difference in the cone index between all treatments, and there was an increase in the cone index in "Bare" after the traffic events. Indeed, the Bare treatment did not have a different soil moisture level compared to Crimped, but did record a higher cone index. The higher cone index in Bare could explain the slower increasing in bulk density compared to the cover cropped treatment, due to the phenomena described above.

Third are the surface residues. Indeed, "Shredded" resulted in 16.54% lower mean normal stress compared to "Harvested" treatment, and this finding could be explained by the surface residues effects. Moreover, this residues effects result was higher than a previous experiment on the effect of residues on soil compaction [10]. The differences recorded between "Shredded" and "Crimped" treatments suggest a different behaviour of the two cover crop termination methods considered in this experiment. This effect can be also divided to obtain a deep analysis on the direct residues effects on force transfer during traffic and on the effect of the termination method on soil bearing capacity, as analysed in the previous point. In this experiment, the strong variability found in "Crimped" limited the analysis of the residues effect.

All of this action at different magnitudes affects the soil compaction, but the main effects could be considered the change in soil structure. This could be explained by bulk density results during traffic events. Indeed, treatment with higher moisture content resulted in higher bulk density because higher soil moisture content increased the risk of soil compaction. The significant—but lower—difference in soil moisture (<3%) could only partially explain the difference with lower moisture Bare treatment because the Crimped treatment without different VWC from B is not different from the other mulching treatments. The secondary action of the residues could also be explained by the same results. In fact, the Shredded treatment did not have a different bulk density from Harvested, despite the significant decreasing in mean soil stress measured and compared to Harvested treatment. Moreover, no additive effects were measured on Crimped treatment, where both enhanced conditions, low water content and residue mulch can be found.

4.2. Effect of Repeated Traffic

All of the soil analyses highlight an increased susceptibility to soil compaction during multiple trafficking, especially comparing all the trafficked treatments with the zero traffic event. The VWC at zero traffic was significantly lower than after the other traffic events, probably due to a concentration effect caused by the loss of air-filled macroporosity. Moreover, the cone index had an average increase of 30.41% after the first traffic event and of 9.11% between the third and fifth traffic events. The rut profile area increased about 9.11% on the first wheeling and maintained a constant average increase of 8.86% for every traffic treatment. Cone index and rut profile analysis showed no difference between the first and third passages. Otherwise, bulk density significantly increased (7.21%) between zero and one passages and also between one and three traffic events (2.95%). These findings were aligned to the result obtained in other studies [27–29]. The repeated traffic effect on soil compaction was more relevant after five traffic events and less influenced by the residues effect. Indeed, no differences were found between different mulch treatments on soil bulk density after five traffic events.

5. Conclusions

Cover crop mulch can, in some cases, dissipate part of the machine load during traffic, but this effect could already be negligible with 4 Mg of tractor traffic. The residues effect changes with cover crop termination methods and affects soil susceptibility to compaction in different ways and magnitudes, first on soil water content. The use of a roller crimper could cause a slow cover crop termination that seems to determine an increase in the amount of water used by the cover crop during its life cycle and a decrease in soil water availability. Furthermore, the cover crop roots improve soil water retention, resulting in higher moisture content if termination occurs early in the season and according to the termination method. Moreover, the termination method affects the residue bearing capacity, and in some cases, counterbalances the lower soil bearing capacity with lower soil stress due to higher moisture content. Finally, the use of a cover crop can affect the soil structure, decreasing the penetration resistance and resulting in an easier soil penetration during the following operations, such as no-till planting or tillage. The findings of this study confirmed the results obtained in previous research, but underlined the variability of residues effects due to different soil conditions and cover crop termination methods. However, taking into account the limited load adsorbed by the surface residues layer, further studies are needed to increase the knowledge on how residues interact on soil compaction with light loads, as happens on a no-till planter closing wheel, where a compacting action was needed and could be disturbed by different types of residues.

Author Contributions: Conceptualization, M.B.; methodology, M.B. and F.G.; software, M.B.; validation, M.B., F.G. and L.S.; formal analysis, M.B., F.G. and K.L.; data curation, M.B. and L.G.; writing—original draft preparation, M.B.; writing—review and editing, L.G. and A.P.; supervision, A.P. and L.S.; project administration, L.S.; funding acquisition, L.S. All authors have read and agreed to the published version of the manuscript.

Funding: This research was completed during a PhD scholarship founded by IBF Servizi s.p.a.

Data Availability Statement: Data presented in this study are available on request from the corresponding author.

Acknowledgments: The authors acknowledge Roberto, Angelica, Marco, Francesco and all the workers of the experimental farm for the support given during the experimental activity.

Conflicts of Interest: The authors declare no conflict of interest.

References

1. Osman, K.T. Physical Deterioration of Soil BT. In *Soil Degradation, Conservation and Remediation*; Osman, K.T., Ed.; Springer: Dordrecht, The Netherlands, 2014; pp. 45–67, ISBN 978-94-007-7590-9.
2. Soane, B.D.; van Ouwerkerk, C. Soil Compaction Problems in World Agriculture. In *Soil Compaction in Crop Production-Developments in Agricultural Engineering 11*; Elsevier: Amsterdam, The Netherlands, 1994; pp. 1–21.
3. Keller, T.; Lamandé, M.; Schjønning, P.; Dexter, A.R. Analysis of soil compression curves from uniaxial confined compression tests. *Geoderma* **2011**, *163*, 13–23. [CrossRef]
4. Batey, T. Soil compaction and soil management—A review. *Soil Use Manag.* **2009**, *25*, 335–345. [CrossRef]
5. ten Damme, L.; Stettler, M.; Pinet, F.; Vervaet, P.; Keller, T.; Munkholm, L.J.; Lamandé, M. Construction of modern wide, low-inflation pressure tyres per se does not affect soil stress. *Soil Tillage Res.* **2020**, *204*, 104708. [CrossRef]
6. Antille, D.L.; Peets, S.; Galambošová, J.; Botta, G.F.; Rataj, V.; Macak, M.; Tullberg, J.N.; Chamen, W.C.T.; White, D.R.; Misiewicz, P.A.; et al. Review: Soil compaction and controlled traffic farming in arable and grass cropping systems. *Agron. Res.* **2019**, *17*, 653–682. [CrossRef]
7. Poltorak, B.J.; Labelle, E.R.; Jaeger, D. Soil displacement during ground-based mechanized forest operations using mixed-wood brush mats. *Soil Tillage Res.* **2018**, *179*, 96–104. [CrossRef]
8. Ess, D.R.; Vaughan, D.H.; Perumpral, J.V. Crop residue and root effects on soil compaction. *Trans. ASAE* **1998**, *41*, 1271–1275. [CrossRef]
9. Holthusen, D.; Brandt, A.A.; Reichert, J.M.; Horn, R.; Fleige, H.; Zink, A. Soil functions and in situ stress distribution in subtropical soils as affected by land use, vehicle type, tire inflation pressure and plant residue removal. *Soil Tillage Res.* **2018**, *184*, 78–92. [CrossRef]
10. Gupta, S.C.; Schneider, E.C.; Larson, W.E.; Hadas, A. Influence of corn residue on compression and compaction behavior of soils. *Soil Sci. Soc. Am. J.* **1987**, *51*, 207–212. [CrossRef]
11. Reichert, J.M.; Brandt, A.A.; Rodrigues, M.F.; Reinert, D.J.; Braida, J.A. Load dissipation by corn residue on tilled soil in laboratory and field-wheeling conditions. *J. Sci. Food Agric.* **2016**, *96*, 2705–2714. [CrossRef]
12. Cherubin, M.R.; Franchi, M.R.A.; de Lima, R.P.; de Moraes, M.T.; Luz, F.B. da Sugarcane straw effects on soil compaction susceptibility. *Soil Tillage Res.* **2021**, *212*, 105066. [CrossRef]
13. Piccoli, I.; Sartori, F.; Polese, R.; Berti, A. Crop yield after 5 decades of contrasting residue management. *Nutr. Cycl. Agroecosystems* **2020**, *117*, 231–241. [CrossRef]
14. Dal Ferro, N.; Piccoli, I.; Berti, A.; Polese, R.; Morari, F. Organic carbon storage potential in deep agricultural soil layers: Evidence from long-term experiments in northeast Italy. *Agric. Ecosyst. Environ.* **2020**, *300*, 106967. [CrossRef]
15. Bolling, I. *Bodenverdichtung und Triebkraftverhalten bei Reifen-Neue Meß-und Rechenmethoden*; Lehrstuhl für Landmaschinen, Technische Universität: München, Germany, 1987.
16. Berli, M.; Eggers, C.G.; Accorsi, M.L.; Or, D. Theoretical Analysis of Fluid Inclusions for In Situ Soil Stress and Deformation Measurements. *Soil Sci. Soc. Am. J.* **2006**, *70*, 1441–1452. [CrossRef]
17. Défossez, P.; Richard, G.; Boizard, H.; O'Sullivan, M.F. Modeling change in soil compaction due to agricultural traffic as function of soil water content. *Geoderma* **2003**, *116*, 89–105. [CrossRef]
18. Keller, T.; Ruiz, S.; Stettler, M.; Berli, M. Determining Soil Stress beneath a Tire: Measurements and Simulations. *Soil Sci. Soc. Am. J.* **2016**, *80*, 541–553. [CrossRef]
19. de Lima, R.P.; Keller, T. Impact of sample dimensions, soil-cylinder wall friction and elastic properties of soil on stress field and bulk density in uniaxial compression tests. *Soil Tillage Res.* **2019**, *189*, 15–24. [CrossRef]
20. Kirby, J.M. Soil stress measurement: Part I. Transducer in a uniform stress field. *J. Agric. Eng. Res.* **1999**, *72*, 151–160. [CrossRef]
21. Naderi-Boldaji, M.; Alimardani, R.; Hemmat, A.; Sharifi, A.; Keyhani, A.; Tekeste, M.Z.; Keller, T. 3D finite element simulation of a single-tip horizontal penetrometer-soil interaction. Part II: Soil bin verification of the model in a clay-loam soil. *Soil Tillage Res.* **2014**, *144*, 211–219. [CrossRef]
22. Angers, D.A.; Caron, J. Plant-induced changes in soil structure: Processes and feedbacks. *Biogeochemistry* **1998**, *42*, 55–72. [CrossRef]
23. Mischler, R.A.; Curran, W.S.; Duiker, S.W.; Hyde, J.A. Use of a Rolled-rye Cover Crop for Weed Suppression in No-Till Soybeans. *Weed Technol.* **2010**, *24*, 253–261. [CrossRef]
24. Chen, G.; Weil, R.R. Root growth and yield of maize as affected by soil compaction and cover crops. *Soil Tillage Res.* **2011**, *117*, 17–27. [CrossRef]
25. Kooistra, M.J.; Tovey, N.K. Effects of Compaction on Soil Microstructure. In *Developments in Agricultural Engineering*; Elsevier: Amsterdam, The Netherlands, 1994.
26. Roberson, E.B.; Firestone, M.K.; Sarig, S. Cover Crop Management of Polysaccharide-Mediated Aggregation in an Orchard Soil. *Soil Sci. Soc. Am. J.* **1991**, *55*, 734–739. [CrossRef]
27. Seehusen, T.; Riggert, R.; Fleige, H.; Horn, R.; Riley, H. Soil compaction and stress propagation after different wheeling intensities on a silt soil in South-East Norway. *Acta Agric. Scand. Sect. B Soil Plant Sci.* **2019**, *69*, 343–355. [CrossRef]

28. ten Damme, L.; Schjønning, P.; Munkholm, L.J.; Green, O.; Nielsen, S.K.; Lamandé, M. Soil structure response to field traffic: Effects of traction and repeated wheeling. *Soil Tillage Res.* **2021**, *213*, 105128. [CrossRef]

29. Pulido-Moncada, M.; Munkholm, L.J.; Schjønning, P. Wheel load, repeated wheeling, and traction effects on subsoil compaction in northern Europe. *Soil Tillage Res.* **2019**, *186*, 300–309. [CrossRef]

Article

Spectral Response of Camelina (*Camelina sativa* (L.) Crantz) to Different Nitrogen Fertilization Regimes under Mediterranean Conditions

Clarissa Clemente [1,†], Leonardo Ercolini [2,†], Alessandro Rossi [1], Lara Foschi [1], Nicola Grossi [1], Luciana G. Angelini [1], Silvia Tavarini [1,*] and Nicola Silvestri [1]

[1] Department of Agriculture, Food and Environment, University of Pisa, Via del Borghetto 80, 56124 Pisa, Italy; clarissa.clemente@phd.unipi.it (C.C.); alessandro.rossi@agr.unipi.it (A.R.); lara.foschi@unipi.it (L.F.); nicola.grossi@unipi.it (N.G.); luciana.angelini@unipi.it (L.G.A.); nicola.silvestri@unipi.it (N.S.)

[2] Centro di Ricerche Agro-Ambientali "Enrico Avanzi", Via Vecchia di Marina 6, 56122 Pisa, Italy; leonardo.ercolini@avanzi.unipi.it

* Correspondence: silvia.tavarini@unipi.it

† These authors contributed equally to this work.

Abstract: Knowledge about the spectral response of camelina under different regimes of nitrogen (N) fertilization is very scarce. Therefore, 2-year open-field trials were carried out in the 2021 and 2022 growing seasons with the aim of evaluating the spectral response of spring camelina to four different N fertilization regimes by using remote (UAV) and proximal (leaf-clip Dualex) sensing techniques. The tested treatments were: (i) control: no N application (T0); (ii) top dressing: 60 kg N ha^{-1} before stem elongation (T1); basal dressing: 60 kg N ha^{-1} at sowing (T2); basal + top dressing combination: 60 kg N ha^{-1} at sowing + 60 kg N ha^{-1} before stem elongation (T3). Camelina seed yield and N use efficiency were strongly affected by fertilization regimes, with the best results obtained at T2. A reduction in plant development and seed yield was detected in 2022, probably due to the rise in air temperatures. A significant effect of both growing season and N fertilization was observed on the photosynthetic pigments content with the T1 highest values in 2022. The highest seed oil content was achieved at T1, while the protein content increased with increasing N, with the best values at T3. Positive and significant correlations were observed among several vegetation indices obtained through UAV flights (NDVI, MRS705, FGCC) and seed yield, as well as between FGCC and leaf N concentration. Overall, these findings demonstrate the feasibility of utilizing remote sensing techniques from UAVs for predicting seed yield in camelina.

Keywords: sustainable agriculture; UAV-derived vegetation indices; camelina; nitrogen use efficiency; seed yield; oil and protein content

Citation: Clemente, C.; Ercolini, L.; Rossi, A.; Foschi, L.; Grossi, N.; Angelini, L.G.; Tavarini, S.; Silvestri, N. Spectral Response of Camelina (*Camelina sativa* (L.) Crantz) to Different Nitrogen Fertilization Regimes under Mediterranean Conditions. *Agronomy* **2023**, *13*, 1539. https://doi.org/10.3390/agronomy13061539

Academic Editor: Gniewko Niedbała

Received: 6 May 2023
Revised: 23 May 2023
Accepted: 30 May 2023
Published: 31 May 2023

1. Introduction

Nowadays, the great potential arising from technological progress seems to be able to meet the challenges of modern agriculture aimed at increasing the sustainability and quality of cropping systems [1]. The application of new monitoring techniques to manage intensive agriculture is crucial. Indeed, modern technologies allow farmers to keep plants healthy and achieve high and stable yields, and, at the same time, reduce environmental impacts. Particular attention is generally paid to nitrogen (N) fertilization, as N is an essential mineral nutrient known to enhance photosynthetic capacity and improve crop yield [2,3]. In agriculture, nitrogen use efficiency (NUE) is generally quite low, and nitrogen losses, mainly through leaching (nitrates) and gas emissions (nitrous oxide and ammonia), cause serious environmental concerns such as water contamination and climate change [4,5]. Therefore, dynamic monitoring of crop nutrition status is a key point to rationalize N management by ensuring higher yield and food quality and minimizing the negative environmental impacts due to fertilizer losses [6,7]. However, traditional methods based on N analysis of plant tissues are often expensive and time consuming because of the large number of samples required. Moreover, they do not provide real-time information about the nutritional status of the crop, and thus, they do not allow timely management of nitrogen fertilization [8]. Conversely, remote and proximal sensing are able to provide non-destructive and real-time information on leaf N and can be considered a viable alternative to laboratory analyses [8–12]. Indeed, many authors reported that foliar chlorophyll meters are a good proxy of plant photosynthetic activity, which in turn is affected by nitrogen content in addition to environmental stresses such as drought, salinity, disease, and pests [13–15].

Moreover, multispectral and hyperspectral imagery captured through unmanned aerial vehicles (UAVs) can be used to calculate several vegetation indices (VIs) widely used for the prediction of crop yields and the evaluation of nutritional status and stress conditions of plants [16]. They are calculated by considering the reflectance of a single (or a combination of two or more) wavelengths [17–19]. In this way, maps based on the variable rate distribution of fertilizers are created to match nitrogen supply with crop needs, both in space and time, combining crop yield, N efficiency, and compliance with the environment.

In this context, the literature about *Camelina sativa* (L.) Crantz's response to different nitrogen fertilization rates—as well as on the use of remote/proximal sensing for evaluating its N status—is scarce.

Camelina, an oilseed crop belonging to the Brassicaceae family, is mainly used in biofuel production, thanks to its ability to reduce greenhouse gas emissions. Recently, it has gained renewed interest due to its numerous favorable agronomic traits and multiple uses in food, feed, and bio-based industry [20,21]. It shows a considerably high seed oil content (up to 49% dry matter) with an unusual fatty acid profile of which α-linolenic acid (C18:3, ALA) accounts for almost 28–50%, while linoleic acid (C18:2, LA) accounts for approximately 15–23% of all FAs [21,22]. Its seeds are also characterized by a good crude protein content (28–32%) and by the presence of high amounts of bioactive compounds with high antioxidant activity, such as phenolic acids, flavonoids, tocopherols, and xanthophylls [22]. Moreover, camelina seed cake, a co-product deriving from oil extraction, can be used as an ingredient in livestock feed diets due to its high protein content (up to 45% DM) with a favorable amino acid profile and energetic value [23–25]. This end-use of camelina cake represents the main driving force for the expansion of this crop in Europe, where the current shortage in the domestic production of vegetable proteins and oils for feed use has imposed a strong dependence on foreign imports.

Camelina is a short-season crop with a crop cycle varying from 90 to 250 days, with spring and autumn sowing [26]. It is resistant to drought, cold, pests, and diseases and has low agricultural input requirements, thus showing broad environmental adaptability, including in poor and marginal soils [26,27]. Furthermore, the ability of camelina to adapt well to the agricultural equipment available on farms and the possibility of being introduced as the main crop, cover crop, and relay crop make it a winning and sustainable diversification strategy for European cropping systems. However, camelina is characterized by lower yield stability compared to other *Brassica* crops [28]. This fact, together with the lack of adequate agronomic knowledge, often represents the main obstacle to its introduction and diffusion into current cropping systems.

Therefore, it is of primary importance to optimize the N management in order to increase crop yield and improve production quality. Unfortunately, the literature is scarce about camelina's response to different nitrogen fertilization rates and, even more, about the use of remote/proximal sensing in evaluating its nitrogen status.

This study was conducted to explore the use of digital technologies on *Camelina sativa* (L.) Crantz under different N availabilities in order to: (i) optimize N fertilization and (ii) evaluate the suitability of remote and proximal sensing in monitoring nutritional status. In particular, the effects of different N fertilizer regimes (rate and timing) on crop growth (seed yield and yield components) and quality (protein and oil seed content) were assessed, along with traditional (destructive diagnosis) and non-destructive methods (Dualex portable sensor and UAVs equipped with RGB and high-definition multispectral cameras).

2. Materials and Methods

2.1. Experimental Setup and Site Description

Two-year experimental field trials were conducted during 2021 and 2022 growing seasons at the Experimental Center of the Department of Agriculture, Food, and Environment (DAFE) of the University of Pisa, San Piero a Grado, Pisa, Italy (43°40′ N; 10°19′ E; 5 m above sea level). In both years of cultivation, winter wheat (*Triticum turgidum* L. subsp. *durum* (Desf.) Husn.) preceded camelina. An integrated management system was adopted with conventional tillage practices and mineral fertilization. Pre-planting phosphorus and potassium fertilizations were performed at a rate of 80 kg P_2O_5 ha^{-1} as triple superphosphate and 50 kg K_2O ha^{-1} as potassium sulfate. Both years, spring sowing of camelina was performed by a Wintersteiger plot drill on 0.15 m spaced rows and a depth of 0.01 m on 8 March. The seeding rate, in both years, was around 6.5 kg ha^{-1}, considering percent seed germination as well as 1000-seed weight (TSW), in order to reach a target of 500 plants m^{-2}. For these experiments, the commercial variety Calena was used, and different nitrogen levels (applied as ammonium nitrate, NH_4NO_3) and timing (basal and top dressing) were compared. The tested treatments were: T0 = control: no N application; T1 = top dressing: 60 kg N ha^{-1} before stem elongation; T2 = basal dressing: 60 kg N ha^{-1} at sowing; T3 = basal + top dressing combination: 60 kg N ha^{-1} at sowing + 60 kg N ha^{-1} before stem elongation. A completely randomized block design was adopted with three replications for each treatment (plot size: 8 m × 6 m). No chemical treatments for pests and diseases were necessary during the whole experiment, and weeds were controlled by hand weeding.

Daily meteorological data (temperature and rainfall) were measured through an automatic meteorological station located near the experimental site. The climate in the area is Mediterranean (Csa according to Köppen classification), characterized by hot and dry summers, with rainfall concentrated during winter. Long-term (20 years) rainfall in the March–June period was 212 mm, while in the same period, temperature varied from a minimum of 5.8 °C (1st decade of March) to a maximum of 26.6 °C (3rd decade of June).

Soils on which trials were carried out are classified as Typic Xerofluvens according to USDA soil taxonomy. In both growing seasons, soil samples were collected at 0–30 cm depth at the beginning of the trial (spring 2021 and spring 2022) for physical and chemical analysis. According to USDA soil texture classes, in both years (2021 and 2022), field trials were characterized by loamy soils with a sub-alkaline pH, good soil organic matter (SOM) content, medium–low levels of total nitrogen content and available phosphorus, and medium–high levels of exchangeable potassium (Table 1). Soil pH determination was performed on a 1:2.5 soil:water suspension following McLean procedure [29]. Electrical conductivity (EC) was measured at 20 °C by using a GLP-31 Crison conductimeter (52.93 electrode) (Montepaone s.r.l., San Mauro Torinese, Torino, Italy). Soil total nitrogen was evaluated using the macro-Kjeldahl digestion procedure [30], available phosphorus by colorimetric analysis using the Olsen method [31], and the cation exchange capacity (CEC) following Mehlich method [32]. The exchangeable K was determined using the Thomas method [33]. Soil organic matter was calculated by multiplying by 1.724 the soil organic carbon concentration, measured using the modified Walkley–Black wet combustion method [34].

2.2. Crop Growth Cycle Monitoring

For each growing season, the crop growth cycle was monitored, and the main phenological phases of camelina, including basal rosette (BBCH 19), beginning of stem elongation (BBCH 31), full flowering (BBCH 65), and seed maturation (BBCH 89), were determined using the extended BBCH scale described by Martinelli and Galasso [35]. In addition, number of days and accumulated growing degree days (GDD) for both years of cultivation (2021 and 2022) were calculated using daily maximum air temperature (Tmax), daily minimum air temperature (Tmin), and base temperature (Tbase) of 4 °C [36], as follows:

$$GDD = \Sigma \left[(Tmax + Tmin)/2 - Tbase \right] \tag{1}$$

Table 1. Physical and chemical characteristics of the soil in experimental spring trials (2021 and 2022).

Characteristics	Spring 2021	Spring 2022
Clay (<0.002 mm, %)	15.1	14.6
Silt (0.05–0.002 mm, %)	35.9	42.0
Sand (2–0.05 mm, %)	48.9	43.4
pH (H_2O 1:2.5 soil:water suspension; McLean method)	8.2	8.1
Total Nitrogen (Kjeldahl method, g kg^{-1})	0.9	1.2
Organic matter (Walkley–Black method, %)	2.1	2.8
Available phosphorus (Olsen method, mg kg^{-1})	3.4	8.0
Exchangeable potassium (Thomas method, mg kg^{-1})	65.0	146.0
EC (μS cm^{-1})	53.7	78.0
CEC (Method $BaCl_2$, pH 8.1, meq 100 g^{-1})	13.7	8.4

2.3. Non-Destructive Analysis and Destructive Sampling

Proximal and remote sensing methods, in combination with destructive methods, were carried out when camelina plants were at full flowering stage (BBCH 65) (21 and 12 May, in 2021 and 2022, respectively). In particular, aerial surveys were carried out on two sites of camelina cultivation using the DJI MAVIC 2 PRO drone equipped with a Hasselblad L1D-20C camera and a PARROT SEQUOIA multispectral camera (Parrot Drones S.A.S., Paris, France) with four spectral bands (Green, Red, Red-edge, NIR). The flights were performed at an altitude of 15 m, with an overlap of 80%, and a spatial resolution in terms of pixels on the ground equal to 3.5 mm for the RGB camera and 14 mm for the multispectral camera. The multispectral sensor was calibrated using a calibration panel (provided by the sensor producer), which must be scanned after take-off and before landing to correct the acquired data based on the incident light conditions during the flight. The survey activities were preceded by the identification on the ground, with the use of the total station, of 13 ground control points (GCPs), with known coordinates, useful for the georeferencing of all the surveys in a common reference system (ETRF2000 UTM 32N). These points were then materialized with clearly visible "targets" from drones, useful for the georeferencing of the models and orthophotos obtained from the processing of aerial images. Data obtained from UAV flights were used for the calculation of some vegetation indices (VIs) reported in Table 2.

After aerial surveys, three leaves randomly chosen from three plants for each plot were clipped with a field-portable leaf-clip sensor Dualex® Force-A (Orsay, Cedex, France) for the evaluation of chlorophyll content (μg cm^{-2}), flavonols (Abs unit), anthocyanins (Abs unit), and nitrogen balance index (NBI). This instrument performs simple, fast, and non-destructive measurements acquiring information from the UV absorbance measurement of the leaf epidermis by double excitation of chlorophyll fluorescence [37]. Two measurements for each leaf per plant and per plot were performed on both the adaxial and abaxial sides, with 72 total measurements detected. Chlorophyll and anthocyanin values were obtained as average between adaxial and abaxial sides, while the flavonol value was the sum, and NBI was the ratio between chlorophyll and flavonol values.

Table 2. Vegetation indices (VIs) evaluated in this study.

Vegetation Indices	Formula
Normalized difference vegetation index (NDVI) [38]	NDVI = (NIR − RED)/(NIR + RED)
Ratio vegetation index (RVI) [39]	RVI = RED/NIR
Transformed normalized difference vegetation index [40]	TNDVI = (NIR − RED)/(NIR + RED) + 0.5]$^{1/2}$
Renormalize difference vegetation index (RDVI) [41]	RDVI = (NIR − RED)/(NIR + RED)$^{1/2}$
Improved modified chlorophyll absorption ratio index (MCARI2) [42]	MCARI 2 = 1.5 × [2.5 × (NIR − RED) − 1.3 × (NIR − GREEN)]/{[2NIR + 1]2 − [6NIR − 5 × (RED)$^{1/2}$] − 0.5}
Modified red edge simple ratio index (MSR705) [43]	MSR 705 = (NIR/RED − 1)/(NIR/RED + 1)$^{1/2}$
Soil adjustment vegetation index (SAVI) [44]	SAVI = (NIR − RED) × (1 × L)/(NIR + RED + L); L = 0.5

In order to validate the collected data non-destructively, three fresh leaf discs (Ø 0.8 cm) from each leaf per three plants per plot were sampled immediately after Dualex® Force-A measurements and processed for chlorophyll and carotenoid content according to Lichtentahler and Buschmann [45]. The extraction was carried out in 4 mL of 80% acetone by placing the samples in a cold chamber to reduce the volatility of acetone and in the dark so as not to interfere with chlorophyll activity. Finally, destructive plant samplings (~0.075 m^2) were performed for N determination in the different plant organs (stems, leaves, and inflorescences). Fresh weight was measured, and plants subsequently allowed to dry in a ventilated oven (70 °C) for dry weight determination and evaluated for their moisture content. Soil samples were collected in the same area of destructive plant samplings to analyze soil moisture and nitrate concentration (NO_3^-). The soil moisture was calculated with the gravimetric method (samples were weighed twice before and after oven drying at 60 °C until constant weight). NO_3^- content was determined on the same samples taken to monitor soil moisture by pooling the first three (0–30 cm) and the last two (30–50 cm) layers. The ion chromatography method was used (Dx-500 ion chromatograph; Dionex, Sunnyvale, CA, USA).

2.4. Image Processing

The RGB and multispectral images of experimental field plots acquired from the UAVs were orthorectified and mosaicked using Agisoft Photoscan Professional Edition 1.1.6 (Agisoft LLC, St. Petersburg, Russia) in order to correct the possible perspective deformations of the single frames. Finally, subsets of the orthomosaics were created by isolating the various test plots using Erdas Imagine software by Hexagon AB. The RGB images of the textures were then processed using Canopeo, which analyzes the images using color values in the red (R), green (G), and blue (B) system and classifies all the pixels according to three selection parameters: R/G, B/G, and the green excess index (GEI = 2G-R-B) [46–49]. The primary image was transformed into a binary image depending on whether or not the pixels met the selection criteria (R/G < 0.95, B/G < 0.95, and GEI > 20, default values). The final result of the image processing is, therefore, a percentage index of FGCC (fraction green canopy cover), which expresses the number of pixels complying with the aforementioned parameters and which are therefore classified as green, with respect to the number of total pixels. The multispectral images were processed using the Erdas Imagine 2020 software, a tool for geospatial digital image processing developed by Hexagon AB [50].

2.5. Crop Production and Yield Component

Camelina plants were harvested by manually sampling 7 rows from the central portion of each plot for a length of 1.5 m (sampling area = 1.58 m^2) at full seed ripening, when more than 90% of siliques were dried and turned brown, and most seeds were reddish-brown in color (seed moisture ≤ 12%, BBCH 89). Spring-sown camelina were harvested on 24 and 20 June, in 2021 and 2022, respectively. Plant density, plant height, branching, aboveground biomass dry weight (stems, empty siliques, and seeds), seed yield, and yield components (number of siliques, number of seeds per silique, 1000-seeds weight) were evaluated. To assess seed yield, the plants were threshed by a fixed machine, using sieves suitable for small seeds. Fresh weight was measured, and plants subsequently allowed to dry in a ventilated oven (70 °C) for dry weight determination and evaluated for their moisture content. The harvest index (HI), an important trait associated with crop yields and the successful partitioning of photosynthates to harvestable products, was calculated by dividing dry seed weight by the dry weight of total aboveground biomass.

Thousand seed weight (TSW, g DW) was also assessed at the Seed Research and Testing Laboratory of DAFE on representative seed samples deriving from each plot according to the International Rules for Seed Testing [51]. The content of N (determined by macro-Kjeldahl digestion procedure) in the different plant organs (stems, straw, and seeds) was also evaluated. Fresh weight was measured, and plant organs subsequently allowed to dry in a ventilated oven (70 °C) for dry weight determination and evaluated for their moisture content. Moreover, Agronomic Efficiency of applied Nitrogen (*AE*, kg seed yield per kg N applied) was calculated as the difference between the seed yield of treatments with N and the achene yield of the non-fertilized control, divided by the *N* rate applied. Apparent recovery efficiency (ARE) is defined as the difference in N accumulation between plots receiving fertilizer and unfertilized plots and is in proportion to the amount of N fertilizer applied. *AE* and *ARE* indices were then calculated with the following equations [52,53]:

$$AE = \frac{fertilized\,plot\,seed\,yield\left[kg\,ha^{-1}\right] - unfertilized\,plot\,seed\,yield\left[kg\,ha^{-1}\right]}{N\,fertilization\left[kg\,N\,ha^{-1}\right]} \quad (2)$$

$$ARE = \frac{fertilized\,plot\,N\,uptake\left[kg\,N\,ha^{-1}\right] - unfertilized\,plot\,N\,uptake\left[kg\,N\,ha^{-1}\right]}{N\,fertilization\left[kg\,N\,ha^{-1}\right]} \quad (3)$$

where *N* uptake was computed by multiplying biomass yield per *N* tissue concentration.

2.6. Seed Quality

Seed oil and protein contents were determined on representative samples from each plot. Oil content was determined using a Soxhlet extractor apparatus (mod. R 306) from Behr Labor-Technik (Düsseldorf, Germany). Briefly, about 30 g of camelina seeds were finely ground in a coffee grinder for 40 s. An aliquot of 1.5 g of ground material was exactly weighed in a cellulose extraction thimble (22 mm × 80 mm) from Axiva Sichem Biotech (Delhi, India). The thimble was successively inserted in a glass extractor, and oil extraction was carried out for two hours in a Soxhlet extractor using 60 mL of n-hexane as an organic solvent. The extract containing the oily fraction was then filtered with anhydrous sodium sulfate in a 100 mL flat bottom flask and removed under reduced pressure at 30 °C in a rotary evaporator. The residual oil was dried under a gentle nitrogen flow for 5 min with the flask in a water bath (50–55 °C), exactly weighed, transferred by means of 5 mL of n-hexane/i-propanol 4/1 (*v/v*) in a 10 mL Teflon screw-cap glass tube, and stored at −18 °C. Solvents used were of analytical grade. Oil yield (kg DM ha^{-1}) was obtained by multiplying seed yield by seed oil content of each individual replicate. Seed crude protein content was calculated by multiplying the total nitrogen percentage by 6.25, and total N content was determined by means of the mini-Kjeldahl method on different plant organs (stems, empty siliques, and seeds).

2.7. Statistical Analysis

Data were subjected to statistical analysis using the software COSTAT cohort V6.201 (2002). Before ANOVA, Levene's test was used to verify homoscedasticity, and Shapiro–Wilk test was used to check the normal distribution of residuals. A two-way ANOVA for evaluating the effects of growing season (2021 vs. 2022) and the different N fertilization regimes (control, T0, T1, T2, and T3) on agronomic and qualitative parameters were performed. Means were compared using Tukey's HSD test when ANOVA F-test was significant at $p \leq 0.05$. For each year, Pearson's correlation coefficients were computed between VIs and Dualex chlorophyll and seed yield, leaf chlorophyll, and leaf N concentration.

3. Results

3.1. Meteorological Data and Crop Phenology

Weather conditions recorded over the study timespan are reported in Figure 1. The average minimum and maximum temperatures referring to the camelina growing period (from March to June) were very similar for both years, with a minimum of 8.7 °C and 9.6 °C, and a maximum of 21.4 °C and 21.8 °C in spring 2021 and 2022, respectively. In the first growing season, the increase in temperature occurred in the third decade week of April 2021, while in the second growing season, temperatures started to rise significantly from the second decade of April 2022. A shortening of about 10 days in the vegetative growth phase of the second year of experimentation was induced by the differences in the trend of average temperatures (Table 3). In the first growing season (spring 2021), cumulative rainfall and the GDD from sowing to seed maturity were equal to 117.9 mm and 1195.2, respectively, while in the second growing season (2022), they were 140.8 mm and 1241.9 GDD, respectively. A few days of cycle length discern the two spring sowings; in fact, the plants completed their growth cycle in 109 and 105 days in 2021 and 2022, respectively (Table 3).

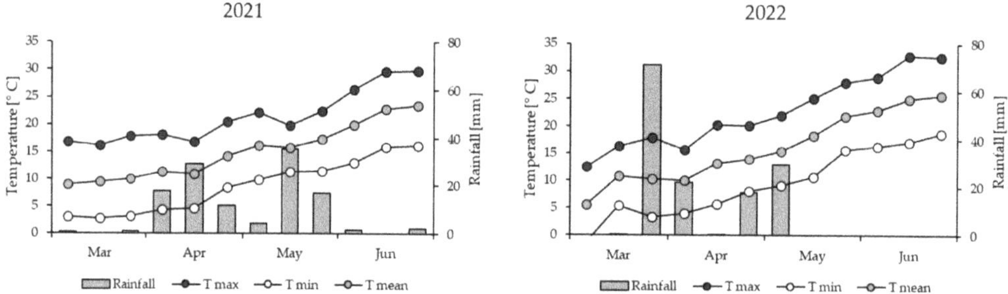

Figure 1. Monthly rainfall (mm) and mean minimum and maximum air temperatures (°C) during the spring camelina cultivation in 2021 and 2022 at San Piero a Grado, Pisa.

Table 3. Length (days), growing degree days (GDD) (°C d), and cumulative rainfall (mm) corresponding to vegetative and reproductive cycles of camelina sown in spring 2021 and 2022. Data are averaged over all N fertilizer treatments.

Growing Season	Cycle Length (Days)		GDD (°C d)		Cumulative Rainfall (mm)	
	Vegetative †	Reproductive ††	Vegetative	Reproductive	Vegetative	Reproductive
Spring 2021	75	34	633.3	561.8	99.4	18.5
Spring 2022	66	39	531.8	710.1	140.8	0

† Vegetative cycle corresponds to the period from sowing till start of flowering. †† Reproductive cycle corresponds to the period from start of flowering till seed maturity.

3.2. Vegetation Indices (VIs)

Results of the statistical two-way analysis (F-Test, ANOVA), related to the effects of year (Y) and N fertilization regimes (F) and their interaction (Y × F), are reported in Table 4. The effect of the year (Y) was significant for the indices NDVI, RVI, TNDVI, WvVI, MSR705, SAVI, and FGGC calculated on the images acquired from the drone. While the effect of N fertilization (F) was statistically appreciable only for NDVI, MSR705, and FGGC, the interaction between year and fertilization was significant only for NDVI.

3.3. Dualex Detection and Destructive Samplings

The effects of year (Y) and N fertilization (F) and their reciprocal interactions were evaluated on Dualex parameters and on destructive samplings. In particular, flavonols and NBI values detected by the Dualex leaf-cleap sensor were significantly affected by the year, and only flavonols were also affected by the Y × F interaction (Table 5). Specifically, the flavonols showed an average value of the different N fertilizers higher than T0 in the first year (+17.4%). T0, in the second year, instead, showed the highest value (+21.34%) compared to the average of all other N fertilizers, except for T1. Furthermore, the NBI detected in 2021 showed an average value higher (+12.4%) than in 2022.

Table 4. Effect of year (Y) and N fertilization (F) and their interaction on different vegetation indices calculated on image acquisition by UAV.

Main Effects	Vegetation Indices (Vis)							
	NDVI	RVI	TNDVI	WvVi	Mcari2	MSR705	SAVI	FGGC
Y	***	***	***	***	ns	***	***	*
F	***	ns	ns	ns	ns	***	ns	***
Y × F	**	ns	ns	ns	ns	ns	ns	ns

The significance of variability factors according to the F-test: ns, not significant; *, significant at $p \leq 0.05$; **, significant at $p \leq 0.01$; ***, significant at $p \leq 0.001$ level.

Regarding destructive samplings, a significant effect of both Y and F and their reciprocal interaction was also observed in the content of chlorophylls and carotenoids deriving from spectrophotometric assays with some exceptions, as reported in Table 5. Specifically, chlorophyll a content was higher at the top dressing in the second year: T2 (+36.2%) and T3 (+32.3%). T0 and T2 in 2022, despite having lower values, were, on average, higher (+39%) than all fertilizers in 2021. The concentration of chlorophyll b, on the other hand, was, on average, higher in the second year (+41%), with T1 and T2 exhibiting higher average values (+12.9%) than T0 and T3. The concentration of carotenoids was, on average, higher in T0 and T2 in 2022 (+13.9%) than in T1 and T3 and compared to the average in 2021 (+69%).

Soil moisture and nitrate concentration were only affected by the year, while total aboveground biomass produced by the crop was significantly affected by both Y and F, as well as by their reciprocal interaction (Table 5). In particular, soil moisture percentage was higher in 2022 (+17%) compared to 2021, and nitrate concentration was also higher in the second year of the experimentation (+42.5%) than in the first one.

3.4. Dry Matter Accumulation and Nitrogen Uptake

At full flowering, the total aboveground biomass was significantly higher in 2021 (+43%) than in 2022 (Figure 2A). Considering the dry matter partitioned in the different organs, camelina plants produced more stems and inflorescences in 2021 than in 2022, except for leaf biomass production. In fact, T2 (2021) and T3 (2022) produced more leaves than the other treatments. Among these, all N fertilizer treatments were more productive than the control in two years (Figure 2A). At seed maturity, camelina plants produced more total aboveground biomass at harvest when they were treated with top-dressing N fertilizer (T1) and with basal + top dressing combination (T3) in 2021 (+41%) and 2022 (+23.5%) compared to the other respective treatments. Nevertheless, 2021 showed clearly higher average values of biomass produced than those of 2022 (+45.5%) (Figure 2B).

Total N uptake by camelina plants when they were at full flowering, was higher in 2021 (+33%), with a better distribution of N uptake among organs (leaves, stems, and inflorescences) than in 2022 (Figure 2C). At seed maturity, in the first year of cultivation, camelina plants treated with N fertilization showed an average N straw uptake higher (+30%) than in the second one, although the control showed an opposite trend. Additionally, N seed uptake was also higher in 2021 (+37.4%) compared to 2022 if considering the different N fertilizer levels (Figure 2D).

Table 5. Results of two-way ANOVA for Dualex parameters, plant leaf, and soil analysis carried out when camelina plants were at full flowering stage (BBCH 65).

Main Effects	Factor Level	Dualex				Plant Leaf Analysis					Soil Analysis	
		Chls (μg cm^{-2})	Anth (Abs Units)	Flav (Abs Units)	NBI	Chl_a (μg cm^{-2})	Chl_b (μg cm^{-2})	Chl_$a + b$ (μg cm^{-2})	Chl_a/b	Car (μg cm^{-2})	M (%)	NO$_3^-$
Year (Y)	2021	24.9 ± 0.4	0.117 ± 0.002	1.15 ± 0.03 [b]	21.8 ± 0.4 [a]	9.6 ± 0.3 [b]	4.3 ± 0.1 [b]	13.8 ± 0.5 [b]	2.3 ± 0.01	1.2 ± 0.01 [b]	12.7 ± 0.2 [b]	23.9 ± 4.7 [b]
	2022	24.1 ± 0.4	0.110 ± 0.003	1.29 ± 0.05 [a]	19.1 ± 0.9 [b]	17.0 ± 0.5 [a]	7.3 ± 0.2 [a]	24.4 ± 0.7 [a]	2.3 ± 0.1	3.7 ± 0.1 [a]	15.3 ± 0.2 [a]	41.6 ± 7.5 [a]
N fertilizer thesis (F)	T0	23.3 ± 0.7	0.114 ± 0.002	1.24 ± 0.10	19.4 ± 1.7	12.4 ± 1.7 [b]	5.0 ± 0.7 [b]	17.4 ± 2.3 [b]	2.5 ± 0.1 [a]	2.5 ± 0.6	14.1 ± 0.6	22.1 ± 6.1
	T1	25.1 ± 0.3	0.113 ± 0.005	1.18 ± 0.04	21.4 ± 0.9	14.7 ± 2.2 [a]	6.4 ± 0.8 [a]	21.1 ± 3.0 [a]	2.3 ± 0.1 [b]	2.7 ± 0.7	14.5 ± 0.7	40.0 ± 13.3
	T2	24.8 ± 0.5	0.115 ± 0.004	1.29 ± 0.05	19.5 ± 1.1	13.2 ± 1.1 [ab]	6.0 ± 0.6 [a]	19.2 ± 1.7 [ab]	2.2 ± 0.1 [b]	2.3 ± 0.5	14.1 ± 0.6	21.6 ± 4.3
	T3	24.9 ± 0.6	0.113 ± 0.003	1.16 ± 0.02	21.5 ± 0.3	12.9 ± 1.9 [b]	5.8 ± 0.7 [b]	18.7 ± 2.6 [b]	2.2 ± 0.1 [b]	2.3 ± 0.5	13.3 ± 0.6	47.3 ± 8.5
Significance												
Y		ns	ns	**	**	***	***	***	ns	***	***	*
F		ns	ns	ns	ns	**	***	**	**	ns	ns	ns
Y × F		ns	ns	**	ns	***	ns	***	**	*	ns	ns

Means ± standard error followed by identical letters are not significantly different for $p < 0.05$, according to Tukey's HSD post-hoc test. The significance of variability factors according to the F-test: ns, not significant; *, significant at $p \leq 0.05$; **, significant at $p \leq 0.01$; ***, significant at $p \leq 0.001$ level. Chls = chlorophylls; Anth = anthocyanins; Flav = flavonols; NBI = nitrogen balance index; Car = carotenoids; M = moisture content; NO$_3^-$ = nitrate concentration.

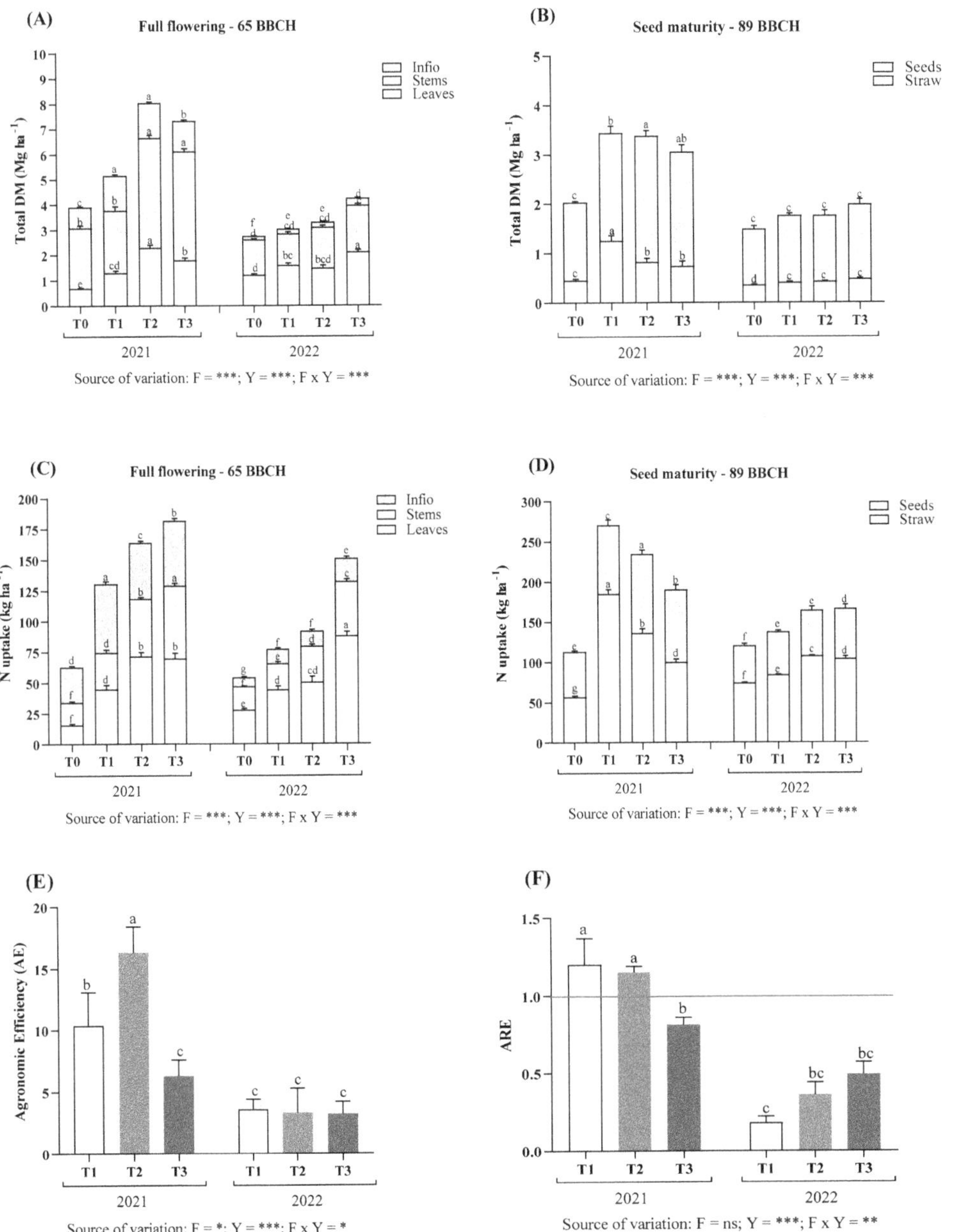

Figure 2. Effect of different N fertilization regimes and year of cultivation on total DM accumulation (**A,B**), N uptake of camelina (**C,D**) both at full flowering stage (65 BBCH) and seed maturity (89 BBCH), and on agronomic efficiency (AE) (**E**) and apparent recovery efficiency (ARE) (**F**). Means ± standard error followed by identical letters are not significantly different for $p \leq 0.05$, according to Tukey's HSD post-hoc test. Significance is indicated as follows: *, $p \leq 0.05$; **, significant at $p \leq 0.01$; ***, significant at $p \leq 0.001$. Red line in sub-image F: values >1 indicate that N up-taken by crop was higher than that applied as fertilizer.

Agronomic efficiency (AE), described as the economic production obtained per unit of nitrogen applied, was significantly highest in 2021 and when the plants were treated with N-fertilizer at sowing (T2) followed by top-dressing N-fertilizer (T1) (Figure 2E). Furthermore, the apparent recovery efficiency (ARE), commonly defined as the difference in nutrient uptake in aboveground parts of the plant between the fertilized and unfertilized crop relative to the quantity of nutrients applied, was significantly higher in 2021 when plants were treated with N-fertilizer both at sowing and top-dressing (Figure 2F).

3.5. Yield, Yield Components, and Seed Quality

The effects of year (Y) and N fertilization (F) and their reciprocal interaction were evaluated on different agronomic and seed qualitative characteristics, as reported in Table 6. The results obtained showed that plant height was significantly influenced by both Y and F, with the highest values recorded in camelina plants in the first year of experimentation, with an average of 13 cm taller compared to the second year. No differences in plant height were observed between the T0 and T1 treatments, although both were smaller (-9.2 cm) than the T2 and T3. Furthermore, a significant effect of the year (Y) was highlighted, while N fertilization was significant only for plant density, branching, and number of siliques per plant. For plant density, a statistically significant interaction between the year and N fertilization was found. In addition, except for plant density, a significant decrease in branching (-0.8 stems per plant), siliques per plant (-11.1 siliques), seeds per silique (-0.8 seeds), and TSW (-3.7%) were observed in the second growing season (2022). In N fertilized plots, the application of $60 + 60$ kg N ha^{-1} (T3) increased branching on average of $+16.6\%$, $+33.6\%$, and $+48.5\%$ compared to T2, T0, and T1, respectively, while no statistical differences with the application of 60 kg N ha^{-1} (T2) and control (T0) was observed. In the first year, only T2 had a higher plant density compared to the untreated control, while in the second year, no significant differences emerged between N fertilizer treatments and control (Figure 3A). Considering the seed yield, a significant effect of both year of cultivation (Y) and N fertilizer treatment (F) and their reciprocal interaction was observed (Table 6). Camelina crop produced a higher seed yield in the first year ($+38\%$) with respect to the second one. All the fertilizer treatments showed higher seed yield (1.9 Mg ha^{-1} as the average value) compared to the control (1.4 Mg ha^{-1}) (Table 6). By the Y x F interaction, the N fertilizer showed, only in 2021, the highest seed yield (2.4 Mg ha^{-1}) compared to the control and all the treatments in 2022 (Figure 3B).

No statistical difference between the two years of cultivation was observed in the oil content, although it was affected by N fertilization, with the highest value achieved when camelina plants were top-dressing fertilized (T1), as reported in Table 6 and Figure 3C. Regarding oil yield, a significant effect of both Y and F, as well as of their interaction (Y $\times$ F), was observed (Table 6). In particular, camelina produced a higher oil yield in 2021 ($+38\%$) compared to 2022, with an average value of the different N fertilizer treatments higher ($+29\%$) than the control (Figure 3D). Finally, a significant effect of both Y and F was observed on the crude protein content (Table 6). Specifically, a slight increase ($+6\%$) was observed in 2022 compared to 2021, with the highest value obtained when camelina plants were treated with basal + top dressing N fertilizer (T3) (25.3%), followed by T1 (24.8%) and T2 (24.4%).

Table 6. Main effects of the year (Y), N fertilization (F), and their interaction on camelina biometric characteristics, seed yield, yield components, and qualitative traits.

Agronomic and Seed Quality Features	Main Effects						Significance		
	Year (Y)		N Fertilizer Regimes (F)						
	2021	2022	T0	T1	T2	T3	Y	F	YxF
Plant height (cm)	77.1 ± 1.4 [a]	64.1 ± 2.0 [b]	64.3 ± 4.2 [b]	67.6 ± 3.2 [b]	75.5 ± 2.6 [a]	74.5 ± 2.8 [a]	***	***	ns
Branching (no. stems $plant^{-1}$)	6.8 ± 0.3 [a]	6.0 ± 0.3 [b]	5.9 ± 0.3 [bc]	6.7 ± 0.2 [b]	5.3 ± 0.2 [c]	7.8 ± 0.3 [a]	**	***	ns
Plant density (plants m^{-2})	153.5 ± 14.2 [b]	235.5 ± 7.2 [a]	168.9 ± 35.1 [b]	175.9 ± 23.0 [b]	217.4 ± 14.3 [a]	215.9 ± 9.5 [a]	***	**	***
no. Silique $plant^{-1}$	224.2 ± 13.3 [a]	132.0 ± 3.7 [b]	193.3 ± 30.7 [a]	180.4 ± 20.2 [ab]	141.4 ± 18.8 [b]	197.1 ± 23.1 [a]	***	*	ns
no. Seeds $silique^{-1}$	8.2 ± 0.2 [a]	7.4 ± 0.1 [b]	8.0 ± 0.4	7.6 ± 0.2	8.1 ± 0.2	7.6 ± 0.4	*	ns	ns
Seed yield (Mg ha^{-1} DW)	2.2 ± 0.1 [a]	1.3 ± 0.1 [b]	1.4 ± 0.1 [b]	1.8 ± 0.2 [a]	1.9 ± 0.3 [a]	1.9 ± 0.2 [a]	***	***	*
TSW (g)	1.1 ± 0.02 [a]	1.1 ± 0.02 [b]	1.0 ± 0.02	1.1 ± 0.03	1.1 ± 0.02	1.1 ± 0.02	*	ns	ns
Oil content (% dry weight)	34.2 ± 1.6	34.7 ± 0.97	34.4 ± 1.66 [b]	38.8 ± 1.66 [a]	30.7 ± 0.46 [c]	33.9 ± 1.05 [bc]	ns	**	***
Oil yield (Mg ha^{-1})	0.7 ± 0.04 [a]	0.5 ± 0.02 [b]	0.5 ± 0.04 [b]	0.7 ± 0.08 [a]	0.6 ± 0.09 [a]	0.7 ± 0.07 [a]	***	***	*
Crude protein (%)	23.7 ± 0.3 [b]	25.2 ± 0.4 [a]	23.3 ± 0.3 [b]	24.8 ± 0.3 [ab]	24.4 ± 0.3 [ab]	25.3 ± 0.5 [a]	*	**	ns

Values $\pm$ standard error followed by identical letters are not significantly different for $p < 0.05$, according to Tukey's HSD post-hoc test. The significance of variability factors according to the F-test: ns, not significant; *, significant at $p \leq 0.05$; **, significant at $p \leq 0.01$; ***, significant at $p \leq 0.001$ level.

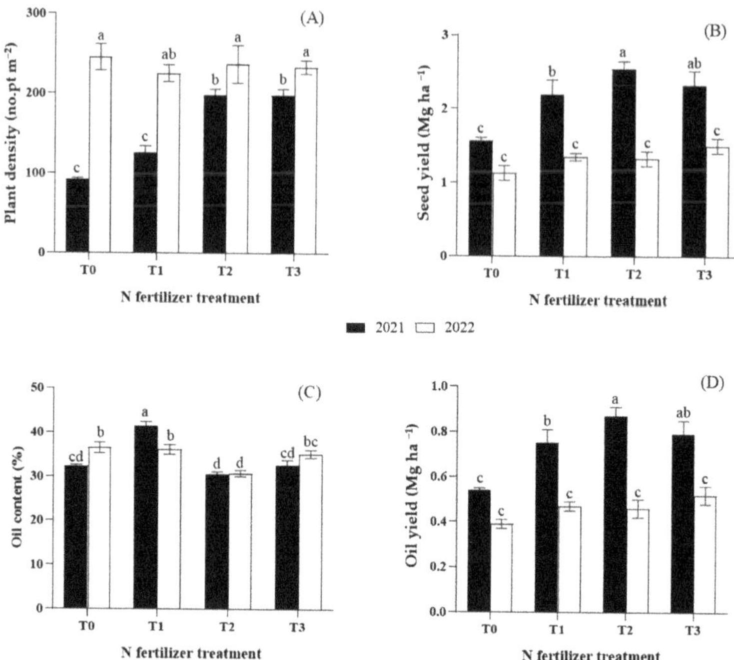

Figure 3. Effect of the year and N fertilization interaction (Y × F) on camelina plant density (**A**), seed yield (**B**), oil content (**C**), and oil yield (**D**). Means followed by the same letter are not significantly different at $p < 0.05$ based on Tukey's HSD test. T0 = control; T1 = 0 + 60 N kg ha^{-1}; T2 = 60 N kg ha^{-1} + 0; T3 = 60 N + 60 N kg ha^{-1}.

3.6. Correlation Analysis

The vegetation indices (VIs) calculated on the spectral data acquired from the drone and chlorophyll obtained with Dualex leaf-clip sensor were correlated with the seed yield, leaf laboratory chlorophyll content, and leaf N concentration (Table 7). During the first year of experimentation (2021), a significantly high level of positive correlation was obtained by NDVI, TNDVI, MSR705, and FGCC indices (0.81, 0.73, 0.72, and 0.74, respectively), followed by WvVi (r = 0.67) and SAVI (0.64), while only RVI showed negative correlation (−0.69). The MCARI2 index, on the other hand, was not significantly correlated to the seed yield. In the second year of experimentation (2022), however, only MSR705 (r = 0.80) and FGCC (r = 0.82) indices had a good level of correlation, followed by NDVI (r = 0.53). The other indices were not significantly correlated with camelina seed yield. The only indices significantly correlated with seed yield in both years of experimental trials (Figure 4) were NDVI (Figure 4A), MSR705 (Figure 4B), and FGCC (Figure 4C), with the latter presenting higher significance values on average. The correlation analysis between the same vegetation indices previously mentioned and the chlorophyll leaf content obtained by the spectrophotometric method did not reveal significantly good correlations for any of the parameters analyzed (Table 7). Low correlation levels in the first year of camelina cultivation (2021) were recorded on NDVI and MCARI2, which, however, were not statistically acceptable (r < 0.52). In the first growing season, leaf N concentration was significantly and positively correlated with NDVI (r = 0.68) and FGCC (r = 0.72), while in the second growing season, a strong positive correlation was found with MSR705 (r = 0.90) and FGCC (r = 0.83). However, only FGCC was found to have a significant correlation with foliar N at full flowering in both years of the trials (Figure 4D). A significant correlation was highlighted by Dualex with seed yield in the first year of cultivation (2021) (r = 0.50) and with leaf chlorophyll content in 2021 (r = 0.51). A non-significant correlation was confirmed, instead, in 2022. During the first growing season, a significant positive correlation was found between leaf N concentration and Dualex chlorophyll; however, this correlation was not found during the second growing season.

Table 7. Pearson's correlation coefficient (r) between VIs, obtained by UAV images, and chlorophyll obtained by Dualex leaf-clip sensor, in relation to camelina seed yield, chlorophyll leaf content, and foliar N % in the two growing seasons (2021 and 2022).

Parameters	Growing Season	VIs by UAV Images								Dualex
		NDVI	RVI	TNDVI	WvVi	MCARI2	MSR705	SAVI	FGCC	Chls
Seed yield	2021	0.81 ***	−0.69 **	0.73 **	0.67 **	0.43 ns	0.72 **	0.64 *	0.74 **	0.50 *
	2022	0.53 *	0.25 ns	−0.12 ns	−0.18 ns	−0.03 ns	0.80 ***	−0.19 ns	0.82 ***	0.29 ns
Leaf Chl content	2021	0.51 *	−0.37 ns	0.43 ns	0.36 ns	0.59 *	0.35 ns	0.32 ns	0.37 ns	0.51 ***
	2022	0.49 ns	−0.02 ns	0.07 ns	0.08 ns	0.47 ns	0.22 ns	0.05 ns	0.13 ns	0.17 ns
Leaf N%	2021	0.68 *	−0.56 ns	0.58 ns	0.61 ns	0.24 ns	0.50 ns	0.43 ns	0.72 *	0.92 ***
	2022	−0.07 ns	−0.18 ns	0.44 ns	0.28 ns	0.14 ns	0.90 **	0.23 ns	0.83 **	0.03 ns

The significance of variability factors according to the F-test: ns, not significant; *, significant at $p \leq 0.05$; **, significant at $p \leq 0.01$; ***, significant at $p \leq 0.001$ level.

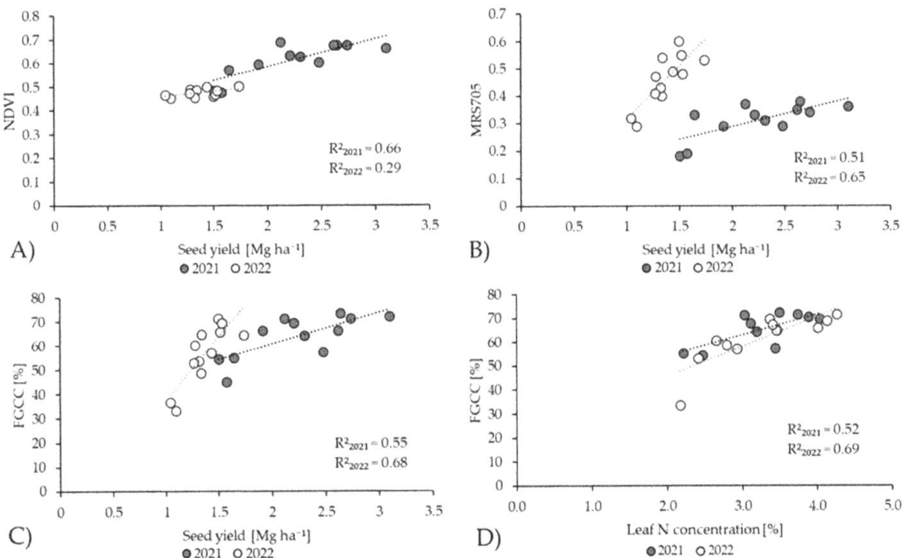

Figure 4. Linear relationship of NDVI (**A**), MRS705 (**B**), FGCC (**C**) with camelina seed yield, and FGCC (**D**) with leaf nitrogen concentration in the two growing seasons (spring 2021 and 2022).

4. Discussion

The use of remote sensing has proved to be a valid tool for estimating the seed production of camelina. In fact, the NDVI, MSR 705 indices, and FGCC% showed good levels of correlation with the recorded seed yields. In line with what has been reported in other works [54–56], the NDVI was significantly correlated with the grain yield in a linear positive way, with a Pearson coefficient equal to 0.81 in the first year of the trial and 0.53 in the second. While the MSR705 index, although not significantly correlated with the leaf chlorophyll content, as reported in other works [43,57,58], is instead significantly correlated with the grain yield of the crop in 2021 (r = 0.72), more than in 2022 (0.80). The FGCC%, derived from Canopeo measurements, is a widely used tool for estimating biomass production [59,60], but there are still few works that use this data to predict grain production [61]. In this trial, a good correlation was found between FGCC and seed yield, in agreement with what was reported by other authors [61,62]. Considering the low cost necessary to calculate this data, from an application point of view, it would seem to be the best among those proven to be reliable. In our study, FGCC was the only VIs capable of providing an indication of camelina nutritional status since it was correlated with leaf N concentration. In the literature, it is reported that the use of FGCC for yield prediction can be obtained with greater precision during the reproductive phase [63], in line with our results. However, for agronomic purposes, the most useful phase for the determination of nutritional stress is during vegetative development in order to adequately calibrate top-dressing fertilization. In this regard, further studies are needed to verify whether the identified index (FGCC) can give reliable indications of camelina nutritional status even during early vegetative phases.

In our trials, meteorological differences among growing seasons were not remarkable; however, in the second year, the increase in air temperatures during spring occurred almost 10 days earlier compared to 2021. It is likely that the rise in temperatures caused the reduction in vegetative development length, with negative effects on the accumulation of vegetative biomass. Consequently, the decrease in photosynthesizing biomass probably contributed to the reduction in seed yield observed in the second growing season. Anyhow, the overall duration of plant development and GDD accumulation are comparable to those already reported in the literature for spring-sown camelina in Central Italy [26]. In the first-year trial, all yield components of camelina gave better results, and this was probably partially due to the rainfall that occurred during the reproductive phase. Therefore, water availability in this phase could have contributed to reducing potential water deficit and improving yield component performances. It has been observed that the occurrence of post-anthesis water stress conditions can lead to negative effects on yield components in various crops, such as rapeseed [64], chickpea [65], and wheat [66]. Secchi et al. [67] reported that in *Brassica napus* L., water stress during pod formation has a higher negative effect on the number and weight of seeds. In camelina, it has been observed that the absence of water availability or irrigation can induce a decrease in the number of siliques and branching per plant [68]. Similarly, in our trials, the number of siliques per branch showed a higher drop in the second year in comparison with the other yield components. Although there may have been an effect of early temperature rise and post-anthesis water stress during the second growing season, it is not possible to establish to what extent each of these factors influenced seed yield or individual yield components. These two stress events occurred in different periods of plant development, and, to the best of our knowledge, no research describes the sequence and timing of yield component formation in camelina. In this regard, further investigations on the production physiology of camelina are needed. In both years of our study, the increase in N dose from 60 to 120 kg N ha^{-1} did not lead to a significant increase in the seed yield, although it had a positive effect on the branching level. However, it is reported that camelina can be responsive to N inputs up to 300 kg N ha^{-1} [69]. Probably, in our study, when the environmental conditions were less favorable, as in the second growing season, the effect of nitrogen fertilization on total aboveground biomass, seed yield, and yield components was deleted. Coherently to Jankowsky et al. [70], in our study, the agronomic efficiency of fertilization in camelina varies between the two years, dropping dramatically during the second growing season. Bronson et al. [71] reported that in camelina, fertilizer agronomic efficiency can be influenced by soil water level, so the AE differences observed in our study between the two growing seasons could be attributed to different rainfall patterns. In addition, AE was not influenced by the N application rate, differently from that observed by Jankowsky et al. [70], who reported that fertilization agronomic efficiency started to decrease from the distribution of 80 kg N ha^{-1}. Similar to our findings, Allen et al. [72], in a 5-year trial, reported that in the Brassicaceae taxon, ARE (named nitrogen recovery index) varied among growing seasons, and for camelina, an average value of 0.3 was detected. Similarly, Afchar et al. [73] observed an increase in the apparent efficiency of nitrogen use in the growing seasons characterized by more favorable conditions for rainfed camelina growth. Contrarywise, in Mahli et al. [74], environmental factors (site–year) did not affect ARE (named percent recovery of applied N). During the first year, high values of ARE were reached, even exceeding the threshold value of 1.0, indicating a condition in which N removals exceeded the N distribution. On the contrary, in the second year, environmental conditions reduced plant development, at the same time decreasing N uptake deriving from fertilization in all fertilized plots, and probably leaving some unused N in the soil, likely to be lost in the environment via leaching. A reduction in root uptake during the second year is confirmed by soil analysis carried out at flowering, which showed an increase in the presence of N and moisture observed compared to the first growing season. This further underlines the necessity of identifying new methodologies to assess the crop nutritional status, also in terms of early seed yield prediction, to properly calibrate top-dressing fertilization. Finally, the oil content measured in our trials resulted in slightly lower values compared to those measured for spring-sown camelina in Central Italy [26], but nonetheless, in line with values (range 28 to 49% DM) obtained from camelina spring cultivation in Central Europe [75]. As expected, while the highest seed oil content was achieved at T1, the protein content increased with increasing N with the highest values at T3. With regard to the protein content (%), even if in previous studies values up to 32% [21] have been found, our results showed an average value of around 25%, which, however, was good and comparable.

5. Conclusions

Our results confirmed the low productive stability of camelina in the Mediterranean areas and highlighted how the effect of nitrogenous fertilization is dependent on meteorological conditions, especially for spring sowings.

Aside from these evaluations, the further value of this study is for demonstrating the possibility of applying remote and proximal techniques to camelina cultivation, with particular attention to N management. This aspect is of fundamental importance for optimizing agronomic management and for increasing cropping system sustainability and quality. Among the different N fertilization rates and timing tested in the present study, satisfactory seed yields and high NUE, as well as good seed oil content, were reached with the application of 60 kg N ha^{-1} before sowing in spring camelina.

We find that UAV-derived vegetation indices during flowering are a good seed yield predictor in camelina. In particular, FGCC was the only VIs capable of providing an indication of camelina nutritional status. Anyway, further studies are needed to verify whether the UAV-derived VIs can give reliable indications on camelina nutritional status even during early vegetative phases. Although preliminary, these results can help in developing proper strategies for N application rates for camelina and for site-specific recommendations for its cultivation in Central Italy.

Author Contributions: Conceptualization, S.T., N.S. and L.G.A.; methodology, S.T., N.S. and L.G.A.; software, L.F., N.G., C.C. and L.E.; formal analysis, C.C., A.R. and L.E.; investigation, C.C., A.R. and L.E.; resources, S.T., N.S. and L.G.A.; data curation, C.C., A.R., L.E., L.F. and N.G.; writing—original draft preparation, all authors; writing—review and editing, all authors; visualization, S.T., C.C., A.R. and L.E.; supervision, S.T., N.S. and L.G.A.; project administration, S.T., N.S. and L.G.A.; funding acquisition, S.T., N.S. and L.G.A. All authors have read and agreed to the published version of the manuscript.

Funding: This research was partially funded by the Research Project of National Relevance—PRIN2017 LZ3CHF_001 "ARGENTO—Agronomic and genetic improvement od camelina for sustainable poultry feeding and healthy food products". Furthermore, this study was carried out partly within the (i) Agritech National Research Center and received funding from the European Union Next-Generation EU (PIANO NAZIONALE DI RIPRESA E RESILIENZA (PNRR)—MISSIONE 4 COMPONENTE 2, INVESTIMENTO 1.4—D.D. 1032 17/06/2022, CN00000022) and within (ii) the PE NEST—Network 4 Energy Sustainable Transition, SPOKE 3 Bioenergy and New Biofuels for a Sustainable Future, funding from the European Union Next-Generation EU (PIANO NAZIONALE DI RIPRESA E RESILIENZA— PNRR)—Mission 4 Component 2 Investment 1.3—D.D. 341 del 15/03/2022, PE0000021). This manuscript reflects only the authors' views and opinions; neither the European Union nor the European Commission can be considered responsible for them.

Data Availability Statement: The data presented in this study are available on request from the corresponding author.

Conflicts of Interest: The authors declare no conflict of interest. The funders had no role in the design of the study; in the collection, analyses, or interpretation of data; in the writing of the manuscript, or in the decision to publish the results.

References

1. Khan, N.; Ray, R.L.; Sargani, G.R.; Ihtisham, M.; Khayyam, M.; Ismail, S. Current Progress and Future Prospects of Agriculture Technology: Gateway to Sustainable Agriculture. *Sustainability* **2021**, *13*, 4883. [CrossRef]
2. Feng, W.; Guo, B.B.; Wang, Z.J.; He, L.; Song, X.; Wang, Y.H.; Guo, T.C. Measuring leaf nitrogen concentration in-winter wheat using double-peak spectral reflection remote sensing data. *Field Crops Res.* **2012**, *159*, 43–52. [CrossRef]
3. Nasar, J.; Khan, W.; Khan, M.Z.; Gitari, H.I.; Gbolayori, H.F.; Moussa, A.A.; Mandozai, A.; Rizwan, N.; Anwari, G.; Maroof, S.M. Photosynthetic Activities and Photosynthetic Nitrogen Use Efficiency of Maize Crop Under Different Planting Patterns and Nitrogen Fertilization. *J. Soil Sci. Plant Nutr.* **2021**, *21*, 2274–2284. [CrossRef]
4. Mahmud, K.; Panday, D.; Mergoum, A.; Missaoui, A. Nitrogen Losses and Potential Mitigation Strategies for a Sustainable Agroecosystem. *Sustainability* **2021**, *13*, 2400. [CrossRef]
5. Li, Z.; Cui, S.; Zhang, Q.; Xu, G.; Feng, Q.; Chen, C.; Li, Y. Optimizing wheat yield, water, and nitrogen use efficiency with water and nitrogen inputs in China: A synthesis and life cycle assessment. *Front. Plant Sci.* **2022**, *13*, 930484. [CrossRef]
6. He, L.; Zhang, H.Y.; Zhang, Y.S.; Song, X.; Feng, W.; Kang, G.Z.; Wang, C.Y.; Guo, T.C. Estimating canopy leaf nitrogen concentration in winter wheat based on multi-angular hyperspectral remote sensing. *Eur. J. Agron.* **2016**, *73*, 170–185. [CrossRef]
7. Esteves, E.; Locatelli, G.; Bou, N.A.; Ferrarezi, R.S. Sap Analysis: A Powerful Tool for Monitoring Plant Nutrition. *Horticulturae* **2021**, *7*, 426. [CrossRef]
8. Guo, B.B.; Qi, S.L.; Heng, Y.R.; Duan, J.Z.; Zhang, H.Y.; Wu, Y.P.; Feng, W.; Xie, Y.X.; Zhu, Y.J. Remotely assessing leaf N uptake in winter wheat based on canopy hyperspectral red-edge absorption. *Eur. J. Agron.* **2017**, *82*, 113–124. [CrossRef]
9. Luisa España-Boquera, M.; Cárdenas-Navarro, R.; López-Pérez, L.; Castellanos-Morales, V.; Lobit, P. Estimating the nitrogen concentration of strawberry plants from its spectral response. *Commun. Soil Sci. Plant Anal.* **2006**, *37*, 2447–2459. [CrossRef]
10. Sishodia, R.P.; Ray, R.L.; Singh, S.K. Applications of Remote Sensing in Precision Agriculture: A Review. *Remote Sens.* **2020**, *12*, 3136. [CrossRef]
11. Weiss, M.; Jacob, F.; Duveiller, G. Remote sensing for agricultural applications: A meta-review. *Remote Sens. Environ.* **2020**, *236*, 111402. [CrossRef]
12. Mezera, J.; Lukas, V.; Horniaček, I.; Smutný, V.; Elbl, J. Comparison of Proximal and Remote Sensing for the Diagnosis of Crop Status in Site-Specific Crop Management. *Sensors* **2022**, *22*, 19. [CrossRef]
13. Berger, K.; Verrelst, J.; Féret, J.B.; Wang, Z.; Wocher, M.; Strathmann, M.; Danner, M.; Mauser, W.; Hank, T. Crop nitrogen monitoring: Recent progress and principal developments in the context of imaging spectroscopy missions. *Remote Sens. Environ.* **2020**, *242*, 111758. [CrossRef]
14. Zhang, J.; Zhang, D.; Cai, Z.; Wang, L.; Wang, J.; Sun, L.; Fan, X.; Shen, S.; Zhao, J. Spectral technology and multispectral imaging for estimating the photosynthetic pigments and SPAD of the Chinese cabbage based on machine learning. *Comput. Electron. Agric.* **2022**, *195*, 106814. [CrossRef]
15. Kamarianakis, Z.; Panagiotakis, S. Design and Implementation of a Low-Cost Chlorophyll Content Meter. *Sensors* **2023**, *23*, 2699. [CrossRef]
16. Zarco-Tejada, P.J.; Berjon, A.; Lopez-Lozano, R.; Miller, J.R.; Martin, P.; Cachorro, V.; Gonzalez, M.R.; de Frutos, A. Assessing vineyard condition with hyperspectral indices: Leaf and canopy reflectance simulation in a row-structured discontinuous canopy. *Remote Sens. Environ.* **2005**, *99*, 271–287. [CrossRef]
17. Fang, H.; Liang, S. Reference Module in Earth Systems and Environmental Sciences. In *Encyclopedia of Ecology*; Elsevier Reference Collection; Elsevier: Amsterdam, The Netherlands, 2014; pp. 2139–2148. ISBN 978-0-12-409548-9.
18. Liu, S.; Li, L.; Fan, H.; Guo, X.; Wang, S.; Lu, J. Real-time and multi-stage recommendations for nitrogen fertilizer topdressing rates in winter oilseed rape based on canopy hyperspectral data. *Ind. Crops Prod.* **2020**, *154*, 112699. [CrossRef]
19. Amiri, M.; Pourghasemi, H.R. Mapping the NDVI and monitoring of its changes using Google Earth Engine and Sentinel-2 images. In *Computers in Earth and Environmental Sciences in: Artificial Intelligence and Advanced Technologies*; Elsevier: Amsterdam, The Netherlands, 2022; pp. 127–136.
20. Berti, M.; Gesch, R.; Eynck, C.; Anderson, J.; Cermak, S. Camelina uses, genetics, genomics, production, and management. *Ind. Crop Prod.* **2016**, *94*, 690–697. [CrossRef]
21. Zanetti, F.; Alberghini, B.; Marjanović, A.; Jeromela, A.M.; Grahovac, N.; Rajković, D.; Kiprovski, B.; Monti, A. Camelina, an ancient oilseed crop actively contributing to the rural renaissance in Europe: A review. *Agron. Sustain. Dev.* **2021**, *41*, 2. [CrossRef]
22. Kurasiak-Popowska, D.; Stuper-Szablewska, K. The phytochemical quality of *Camelina sativa* seed and oil. *Soil Plant Sci.* **2019**, *70*, 39–47. [CrossRef]

23. Oryschak, M.A.; Christianson, C.B.; Beltranena, E. *Camelina sativa* cake for broiler chickens: Effects of increasing dietary inclusion on clinical signs of toxicity, feed disappearance, and nutrient digestibility. *Transl. Anim. Sci.* **2020**, *4*, 1263–1277. [CrossRef]

24. Juodka, R.; Nainienė, R.; Juškienė, V.; Juška, R.; Leikus, R.; Kadžienė, G.; Stankevičienė, D. Camelina (*Camelina sativa* (L.) Crantz) as Feedstuffs in Meat Type Poultry Diet: A Source of Protein and n-3 Fatty Acids. *Animals* **2022**, *12*, 295. [CrossRef]

25. Cullere, M.; Singh, Y.; Pellattiero, E.; Berzuini, S.; Galasso, I.; Clemente, C.; Dalle Zotte, A. Effect of the dietary inclusion of Camelina sativa cake into quail diet on live performance, carcass traits and meat quality. *Poult. Sci.* **2023**, *102*, 102650. [CrossRef]

26. Angelini, L.G.; Abou Chehade, L.; Foschi, L.; Tavarini, S. Performance and potentiality of camelina (*Camelina sativa* L. Crantz) genotypes in response to sowing date under Mediterranean Environment. *Agronomy* **2020**, *10*, 1929. [CrossRef]

27. Vollmann, L.J.; Moritz, T.; Kargl, C.; Baumgartner, S.; Wagentristl, H. Agronomic evaluation of camelina genotypes selected for seed quality characteristics. *Ind. Crop Prod.* **2007**, *26*, 270–277. [CrossRef]

28. Jankowsky, K.; Sokolski, M. Spring camelina: Effect of mineral fertilization on the energy efficiency of biomass production. *Energy* **2021**, *220*, 119731. [CrossRef]

29. McLean, E.O. Soil pH and lime requirement. In *Methods of Soil Analysis, Part 2: Chemical and Microbiological Properties*; Page, A.L., Miller, R.H., Keeney, D.R., Eds.; American Society of Agronomy, Inc.: Madison, WI, USA, 1982; pp. 199–224.

30. Bremner, J.M.; Mulvaney, C.S. Nitrogen total. In *Methods of Soil Analysis, Part 2: Chemical and Microbiological Properties*; Page, A.L., Miller, R.H., Keeney, D.R., Eds.; American Society of Agronomy, Inc.: Madison, WI, USA, 1982; pp. 595–624.

31. Olsen, S.R.; Sommers, L.E. Phosphorus. In *Methods of Soil Analysis, Part 2: Chemical and Microbiological Properties*, 2nd ed.; Page, A.L., Miller, R.H., Keeney, D.R., Eds.; Agronomy Monograph 9; American Society of Agronomy, Inc.: Madison, WI, USA, 1982; pp. 403–430.

32. Mehlich, A. Determination of cation- and anion-exchange properties of soils. *Soil Sci.* **1948**, *66*, 429–446. [CrossRef]

33. Thomas, G.W. Exchangeable cations. In *Methods of Soil Analysis, Part 2: Chemical and Microbiological Properties*; Page, A.L., Miller, R.H., Keeney, D.R., Eds.; American Society of Agronomy, Inc.: Madison, WI, USA, 1982; pp. 159–165.

34. Nelson, P.W.; Sommers, C.E. Total Carbon, organic Carbon and organic matter. In *Methods of Soil Analysis, Part 2: Chemical and Microbiological Properties*; Page, A.L., Miller, R.H., Keeney, D.R., Eds.; American Society of Agronomy, Inc.: Madison, WI, USA, 1982; pp. 539–579.

35. Martinelli, T.; Galasso, I. Phenological growth stages of *Camelina sativa* according to the extended BBCH scale. *Ann. Appl. Biol.* **2011**, *158*, 87–94. [CrossRef]

36. Gesch, R.W. Influence of genotype and sowing date on camelina growth and yield in the north central U.S. *Ind. Crop Prod.* **2014**, *54*, 209–215. [CrossRef]

37. Cerovic, Z.G.; Masdoumier, G.; Ghozlen, N.B.; Latouche, G. A new optical leaf-clip meter for simultaneous non-destructive assessment of leaf chlorophyll and epidermal flavonoids. *Physiol. Plant.* **2012**, *146*, 251–260. [CrossRef]

38. Xie, Q.; Dash, J.; Huang, W.; Peng, D.; Qin, Q.; Mortimer, H.; Casa, R.; Pignatti, S.; Laneve, G.; Pasucci, S.; et al. Vegetation indices combining the red and red-edge spectral information for leaf area index retrieval. *IEEE J. Sel. Top. Appl. Earth Obs. Remote Sens.* **2018**, *11*, 1482–1493. [CrossRef]

39. Pearson, R.L.; Miller, L.D. Remote mapping of standing crop biomass for the estimation of the productivity of the short-grass prairie, Pawnee National Grasslands, Colorado. In Proceedings of the 8th International Symposium on Remote Sensing of Environment, Ann Arbor, MI, USA, 2–6 October 1972; pp. 1357–1381.

40. Bannari, A.; Asalhi, H.; Teillet, P.M. Transformed difference vegetation index (TDVI) vor vegetation cover mapping. In Proceedings of the IEEE International Geoscience and Remote Sensing Symposium, Toronto, ON, Canada, 24–28 June 2002; Volume 5, pp. 3053–3055.

41. Sun, H.Y.; Liu, C.M.; Zhang, X.Y.; Shen, Y.J.; Zhang, Y.Q. Effects of irrigation on water balance yield and WUE of winter wheat in the North China Plan. *Agric. Water Manag.* **2006**, *85*, 211–218. [CrossRef]

42. Badura, G.P.; Bachmann, C.M.; Tyler, A.C.; Goldsmith, S.; Eon, R.S.; Lapszynsky, C.S. A novel approach for deriving LAI of salt marsh vegetation using structure from motion and multiangular spectra. *IEEE J. Sel. Top. Appl. Earth Obs. Remote Sens.* **2019**, *12*, 599–613. [CrossRef]

43. Sims, D.A.; Gamon, J.A. Relationship between leaf pigment content and spectral reflectance across a wide range of species, leaf structures and developmental stages. *Remote Sens. Environ.* **2002**, *81*, 331–354. [CrossRef]

44. Mokhtari, A.; Noory, H.; Vazifedoust, M.; Bahrami, M. Estimating net irrigation requirement of winter theat using model and satellite based single and basal crop coefficient. *Agric. Water Manag.* **2018**, *208*, 95–106. [CrossRef]

45. Lichtentahler, H.K.; Buschmann, C. Chlorophylls and carotenoids: Measurement and characterization by UV-VIS spectroscopy. *Curr. Protoc. Food Anal. Chem.* **2001**, *1*, F4. 3.1–F4. 3.8. [CrossRef]

46. Paruelo, J.M.; Lauenroth, W.K.; Roset, P.A. Estimating aboveground plant biomass using a photographic technique. *Range Ecol. Manag.* **2000**, *53*, 190–193. [CrossRef]

47. Richardson, A.D.; Jenkis, J.P.; Braswell, B.H.; Hollinger, D.Y.; Ollinger, S.V.; Smith, M.L. Use of digital webcam images to track spring green-up in a deciduous broadleaf forest. *Oecologia* **2007**, *152*, 323–334. [CrossRef]

48. Chen, L.; Zhang, J.G.; Su, H.F.; Guo, W. Weed identification method based on probabilistic neural network in the corn seedlings field. In Proceedings of the 2010 International Conference on Machine Learning and Cybernetics, Haifa, Istrael, 11–14 July 2010; Volume 3, pp. 1528–1531.

49. Liang, L.; Schwartz, M.D.; Fei, S. Photographic assessment of temperate forest understory phenology in relation to springtime meteorological drivers. *Int. J. Biometeorol.* **2012**, *56*, 343–355. [CrossRef]

50. Nelson, S.A.; Khorram, S. *Image Processing and Data Analysis with ERDAS IMAGINE®*; CRC Press: Boca Raton, FL, USA, 2018.

51. ISTA (The International Seed Testing Association). *International Rules for Seed Testing*; ISTA: Bassersdorf, Switzerland, 2015; Volume 215.

52. Congreves, K.A.; Otchere, O.; Ferland, D.; Fazadfar, S.; Williams, S.; Arcand, M.M. Nitrogen use efficiency definitions of today and tomorrow. *Front. Plant Sci.* **2021**, *12*, 637108. [CrossRef]

53. Fageria, N.K.; Baligar, V.C. Enhancing nitrogen use efficiency in crop plants. *Adv. Agron.* **2005**, *88*, 97–185.

54. Lukas, V.; Huňady, I.; Kintl, A.; Mezera, J.; Hammerschmiedt, T.; Sobotková, J.; Brtnický, M.; Elbl, J. Using UAV to Identify the Optimal Vegetation Index for Yield Prediction of Oil Seed Rape (*Brassica napus* L.) at the Flowering Stage. *Remote Sens.* **2022**, *14*, 4953. [CrossRef]

55. Janousek, J.; Jambor, V.; Marcoň, P.; Dohnal, P.; Synková, H.; Fiala, P. Using UAV-based photogrammetry to obtain correlation between the vegetation indices and chemical analysis of agricultural crops. *Remote Sens.* **2021**, *13*, 1878. [CrossRef]

56. Tucker, C.J.; Holben, B.N.; Elgin Jr, J.H.; Mcmurtrey III, J.E. Relationship of spectral data to grain yield variation. *Photogramm. Eng. Remote Sens.* **1980**, *46*, 657–666.

57. Curran, P.J.; Dungan, J.L.; Peterson, D.L. Estimating the foliar biochemical concentration of leaves with reflectance spectrometry: Testing the Kokaly and Clark methodologies. *Remote Sens. Environ.* **2001**, *76*, 349–359. [CrossRef]

58. Zhao, D.; Li, J.; Qi, J. Hyperspectral characteristic analysis of a developing cotton canopy under different nitrogen treatments. *Agronomy* **2004**, *24*, 463–471. [CrossRef]

59. Jáuregui, J.M.; Delbino, F.G.; Bonvini, M.I.B.; Berhongaray, G. Determining yield of forage crops using the Canopeo mobile phone app. *J. N. Z. Grassl.* **2019**, *81*, 41–46. [CrossRef]

60. Chung, Y.S.; Choi, S.C.; Silva, R.R.; Kang, J.W.; Eom, J.H.; Kim, C. Case study: Estimation of sorghum biomass using digital image analysis with Canopeo. *Biomass Bioenergy* **2017**, *105*, 207–210. [CrossRef]

61. Reed, V.; Arnall, D.B.; Finch, B.; Bigatao Souza, J.L. Predicting winter wheat grain yield using fractional green canopy cover (FGCC). *Int. J. Agron.* **2021**, *2021*, 1443191. [CrossRef]

62. Goodwin, A.W.; Lindsey, L.E.; Harrison, S.K.; Paul, P.A. Estimating wheat yield with normalized difference vegetation index and fractional green canopy cover. *Crop Forage Turfgrass Manag.* **2018**, *4*, 1–6. [CrossRef]

63. Schmitz, P.K.; Kandel, H.J. Using canopy measurements to predict soybean seed yield. *Agronomy* **2021**, *13*, 3260. [CrossRef]

64. Champolivier, L.; Merrien, A. Effects of water stress applied at different growth stages to *Brassica napus* L. var. *oleifera on yield, yield components and seed quality. Eur. J. Agron.* **1996**, *5*, 153–160.

65. Leport, L.; Turner, N.C.; Davies, S.L.; Siddique, K.H.M. Variation in pod production and abortion among chickpea cultivation under terminal drought. *Eur. J. Agron.* **2006**, *24*, 236–246. [CrossRef]

66. Ercoli, L.; Lulli, L.; Mariotti, M.; Masoni, A.; Arduini, I. Post-anthesis dry matter and nitrogen dynamics in durum wheat as affected by nitrogen supply and soil water availability. *Eur. J. Agron.* **2008**, *28*, 138–147. [CrossRef]

67. Secchi, M.A.; Fernandez, J.A.; Stamm, M.J.; Durret, T.; Vara Prasad, P.V.; Messina, C.D.; Ciampitti, I.A. Effects of heat and drought on canola (*Brassica napus* L.) yield, oil, and protein: A meta-analysis. *Field Crops Res.* **2023**, *293*, 108848. [CrossRef]

68. Ahmed, Z.; Liu, J.; Waraich, E.A.; Yan, Y.; Qi, Z.; Gui, D.; Zeng, F.; Tariq, A.; Shareed, M.; Iqbal, H.; et al. Differential physio-biochemical and yield responses of *Camelina sativa* L. under varying irrigation water regimes in semi-arid climatic conditions. *PLoS ONE* **2020**, *15*, e0242441. [CrossRef]

69. Solis, A.; Vidal, I.; Paulino, L.; Johnson, B.L.; Berti, M.T. Camelina seed yield responses to nitrogen, sulphur, and phosphorus fertilizer in South Central Chile. *Ind. Crops Prod.* **2013**, *44*, 132–138. [CrossRef]

70. Jankowsky, K.J.; Sokolski, M.; Kordan, B. Camelina: Yield and quality response to nitrogen and sulfur fertilization in Poland. *Ind. Crops Prod.* **2019**, *141*, 111776. [CrossRef]

71. Bronson, K.F.; Hunsaker, D.J.; Thorp, K.R. Nitrogen fertilizer and irrigation effects on seed yield and oil in camelina. *Agron. J.* **2019**, *111*, 1712–1719. [CrossRef]

72. Allen, B.L.; Lenssen, A.W.; Sainju, U.M.; Caesar-Ton That, T.; Evans, R.G. Nitrogen use in durum and selected Brassicaceae oilseed in two-year rotations. *Agron. J.* **2019**, *106*, 821–830. [CrossRef]

73. Afshar, R.K.; Lin, R.; Mohammed, Y.A.; Chen, C. Agronomic effects of urease and nitrification inhibitors on ammonia volatilization and nitrogen utilization in a dryland farming system: Field and laboratory investigation. *J. Clean. Prod.* **2018**, *172*, 4130–4139. [CrossRef]

74. Mahli, S.S.; Johnson, E.N.; Hall, L.M.; May, W.E.; Phelps, S.; Nybo, B. Effects of nitrogen fertilizer application on seed yield, N uptake, and seed quality of *Camelina sativa. Can. J. Soil Sci.* **2014**, *94*, 35–47.

75. Krzyżaniak, M.; Stolarski, M.J.; Tworkowski, J.; Puttick, D.; Eynck, C.; Załuski, D.; Kwiatkowski, J. Yield and seed composition of 10 spring camelina genotypes cultivated in the temperate climate of Central Europe. *Ind. Crops Prod.* **2019**, *138*, 111443. [CrossRef]

Article

Variety and Sowing Date Affect Seed Yield and Chemical Composition of Linseed Grown under Organic Production System in a Semiarid Mediterranean Environment

Alessandro Rossi, Clarissa Clemente, Silvia Tavarini * and Luciana G. Angelini

Department of Agriculture, Food and Environment (DAFE), University of Pisa, 56124 Pisa, Italy
* Correspondence: silvia.tavarini@unipi.it; Tel.: +39-050-221-89-48

Abstract: The use of suitable species and varieties in organic cropping systems is essential for improving resource use efficiency, biodiversity, and agroecosystem resilience. Within the SIC-OLEAT project, a 2-year field trial was carried out in two contrasting environments of Central Italy, with the aim to hypothesize a production path for linseed inclusion within organic farming. The effects of location, genotype and sowing date on crop phenology, agronomic performances, and qualitative traits were evaluated. Generally, linseed showed good agronomic traits that make it suitable to be introduced in organic systems. Autumn sowing coupled with milder and wetter conditions seemed to be more favorable for linseed cultivation, allowing a higher seed yield (2.1 vs. 1.3 Mg ha^{-1}) and oil content (47.2 vs. 45.2%). From multivariate analysis, the superior genotypes were Kaolin > Szafir > Galaad, and among these Kaolin had the highest production stability. On the contrary, Libra was the lowest performing one and the most unstable. These findings underline the importance of a site-specific approach for choosing the most suitable variety, since both sowing date and location are meteoclimatic-related factors. Definitively, our results demonstrated that linseed might be a valuable autumn alternative for organic cropping system diversification, contributing to the local production of vegetable oils and proteins.

Keywords: agroecosystem diversification; organic management; seed yield; yield components; PUFAs

Citation: Rossi, A.; Clemente, C.; Tavarini, S.; Angelini, L.G. Variety and Sowing Date Affect Seed Yield and Chemical Composition of Linseed Grown under Organic Production System in a Semiarid Mediterranean Environment. *Agronomy* **2023**, *13*, 45. https://doi.org/10.3390/agronomy13010045

Academic Editor: Christian Frasconi

Received: 16 November 2022
Revised: 16 December 2022
Accepted: 20 December 2022
Published: 22 December 2022

1. Introduction

European cropping systems are often characterized by short rotations or even monocropping, leading to environmental issues such as soil degradation, water eutrophication, and air pollution including greenhouse gas emissions, which contribute to climate change and biodiversity loss [1–3]. A transition in Europe to biodiversity-based cropping systems relying more on ecosystem services, along with the development of local and short value chains, is a major path to face the challenges of balancing production with environmental preservation [4]. The diversification of cropping systems, with species and varieties suited to the specificities of the cultivation environment, is a key action to confer flexibility to agricultural rotations, increasing within-field biodiversity, improving soil health, helping to stabilize yields and, ultimately, to guarantee farming profitability and supporting more diversified agrifood systems at large [5,6]. Multipurpose oilseed crops can represent a reasonable choice to help farmers in increasing cropping system reliability, due to low input requirements and high stress tolerance [7–9]. At the same time, these crops can offer new market opportunities to European farmers and support the development of a bio-based economy and sustainable farming system designed in the EU. New/rediscovered/multipurpose oilseed crops include linseed (*Linum usitatissimus* L.), safflower (*Carthamus tinctorius* L.), hemp (*Cannabis sativa* L.), Ethiopian rape (*Brassica carinata* A. Brown),camelina (*Camelina sativa* (L.) Crantz), and others.

Aside from the centuries-old utilization mainly as fiber crop and paint component [10], nowadays linseed is gaining the attention of both scientific and industrial audiences in view of its unique fatty acid composition, agronomic advantages, provision of new rotation options, and the ability to be grown on lands where commodity crops are not currently economically viable [11]. Linseed is a valuable source of vegetable oil, which can be used as functional food or converted to biofuels and other biobased products through various chemical transformations [12–14]. Linseed oil has an interesting lipid-fraction composition with high concentration of polyunsaturated fatty acids (PUFAs), mainly represented by α-linolinic acid (ALA) [15]. ALA is considered, also at a legislative level, as an important constituent for a healthy diet, being included in the list of permitted health claims according to the Commission Regulation (EU) n.432/2012 as: "ALA contributes to the maintenance of normal blood cholesterol levels". In addition to the PUFA content, other molecules contained in linseed, such as lignans and tocopherols, have been shown to have anti-inflammatory and antioxidant properties [16]. Linseed oil and linseed by-products such as seedcake, can be further used as feed raw material, for obtaining functional foods derived both from monogastric and polygastric animals [17,18], which, in addition to the health value, constitute an important market niche that can economically enhance the organic farming sector [19]. In the 2010–2020 period, the main linseed producers worldwide were Canada, Kazakhstan, and the Russian Federation [20], due to cooler climates more suited to flax cultivation; on the contrary, in Southern European countries characterized by a Mediterranean climate, linseed still represents a crop with marginal spread, in favor of herbaceous oilseed crops for which the production-tradition and agronomic practices are well consolidated, such as sunflower (*Helianthus annuus* L.), rapeseed (*Brassica napus* L.) and soy (*Glycine max* (L.) Merr). The lesser linseed spread probably depends on its lower intrinsic productivity compared to the other above-mentioned crops as well as to its environmental needs (i.e., water availability and mild temperatures during flowering and seed filling) especially when accomplished as spring crop [21,22].

In Tuscan cropping systems, with particular reference to the organic ones, the crop species choice often falls on rainfed winter cereals or spring rainfed crops such as sunflower. In this context linseed could represent a promising winter or spring crop to be introduce in organic systems, due to its low input requirements. However, a site-specific assessment of crop suitability/adaptability is needed, as linseed can be sensitive to late-spring cold or to high temperature and drought occurring during flowering and seed filling, causing a significant reduction in seed yield and quality [23].

The study, carried out within the SIC-OLEAT project, aimed to assess the adaptability of different linseed varieties—as vegetable oil and protein sources—to the pedoclimatic conditions of the Tuscany region (Central Italy) and to organic production systems, through a site-specific approach. Consequently, a 2-year field trial was carried out in two contrasting environments, representative of the northern and southern coastal areas, respectively, comparing five commercial linseed varieties in autumn and spring sowing with the aim to hypothesize a production path for the inclusion of this new crop within organic rainfed cropping systems. The effects of the environment, genotype and sowing date (autumn and spring) on the crop phenology, agronomic performances, seed and oil production, fatty acid composition, and protein content have been evaluated.

2. Materials and Methods

2.1. Experimental Setup and Plant Material

Field-plot trials were conducted in the 2020 and 2021 growing seasons at the Experimental Center of DAFE, located in the lower Arno River plain (San Piero a Grado—SPG, Pisa province, 43°40′29″, 10°18′47″) in northern Tuscany, and at the Tuscany Region Agricultural Center (TeReTo), located at Alberese (ALB, Grosseto province, 42°41′38″, 10°08′29″), in southern Tuscany.

In each site, five commercial linseed varieties (Galaad, Libra, Sideral, Szafir, Kaolin) were compared in spring and autumn sowings within an organic system. Among the tested varieties, Galaad and Sideral were winter varieties, while the other ones were "alternative" types, suitable for both autumn and spring sowing. The spring sowing was accomplished in the 2020 growing season (7 April and 13 February 2020, in SPG and ALB, respectively) and the plants closed their cycle in the summer of the same year (on 20 July 2020 and 26 June 2020 at SPG and ALB, respectively). The autumn sowings were performed in the autumn 2020 (10 October and 11 November 2020, in SPG and ALB, respectively) and the crops were ready to be harvested in the summer of 2021 (30 June 2021 and 21 June 2021 at SPG and ALB, respectively).

Winter wheat (*Triticum turgidum* L. subsp. *durum* (Desf.) Husn.) preceded the linseed, assuming a rainfed cereal-based cropping system. In each site and for each sowing date, plots of the different genotypes were arranged in a randomized complete block design with four replications with sowing time and cultivar as variability factors. The plot area was 18 m^2 (6 × 3 m) and a seed rate of 45 kg ha^{-1} was adopted, with a raw-spacing of 15 cm. Commercial organic fertilizer (2.8% N; 2,5% P_2O_5; 3% K_2O) was applied in each plot at the rate of 700 kg ha^{-1}. Weed management was performed manually and no insect or pathogen attack was observed.

In both sites, soil physical and chemical properties were evaluated at a 30 cm depth at the beginning of each trial. Soil total nitrogen was evaluated using the macro-Kjeldahl digestion procedure [24]. Soil organic carbon was determined using the modified Walkley–Black wet combustion method, then soil organic matter was estimated by multiplying soil organic carbon concentration by 1.724 [25]. According to USDA soil texture classes, in both locations sandy loam soils were used for spring sowings, while autumn sowings were realized in loamy soils, both at SPG and ALB. In both locations and sowing dates, soils were characterized by a medium level of total nitrogen and average organic matter content (Table 1). Soil texture differences observed in ALB were due to the different fields used for experimentation, pointing out the high soil variability that characterizes the Tuscan territory [26].

Table 1. Physical and chemical characteristics of the soil at the experimental sites (0–0.3 m).

	SPG—Spring	SPG–Autumn	ALB—Spring	ALB–Autumn
Sand (%)	55.4	50.9	82.5	48.1
Silt (%)	32.2	38.0	2.4	35.3
Clay (%)	12.4	11.1	15.2	16.6
pH	8.0	8.2	8.0	7.9
Total Nitrogen (‰)	1.1	1.2	1.0	0.8
Organic matter (%)	1.9	2.9	1.7	1.1
Available phosphorus (mg kg^{-1})	8.9	5.8	9.3	28.4
Electrical conductivity (meq 100^{-1}g)	6.5	10.0	39.3	13.7

Daily meteorological data (rainfall and temperature), of both growing seasons and long-term period (30 years) trends were obtained via automatic meteorological stations located near each experimental site. The long-term data record was then used for the computation of the standardized precipitation–evapotranspiration index (SPEI) on a 3 (SPEI_3) and 6 (SPEI_6) months basis, as proposed by Vicente-Serrano et al. [27]. SPEI, being calculated using both rainfall and temperature, is considered more reliable compared to other indexes for the assessment of the extent and duration of drought periods [28], even in the climatic conditions present in Central Italy [29]. Furthermore, SPEI index could assume a not-negligible importance in assessing whether the observed meteorological data reflect the normal weather pattern of the area or an extraordinary weather condition. Afterwards, the reference crop evapotranspiration (ET0) for each site and each growing season, was obtained by the Hargreaves–Saemani formula in the form reported by Allen et al. [30].

2.2. Plant Sampling and Measurements

For each genotype and along the two growing seasons, dates of emergence, stem elongation, flowering, and seed maturity (harvest) were recorded according to Smith and Froment [31].

Cycle length was calculated as the number of days from sowing to harvest and the accumulated growing degree days (GDD) were calculated for each growing season considering a 5 °C T_{base}.

When the seeds had fully ripened (seed moisture < 12%, BBCH 89), the plants were manually harvested on a sampling area of 1 m^2 in the inner part of each plot, collecting all aboveground biomass. Harvests were performed on 20 July 2020 and 26 June 2020 for spring crops at SPG and ALB, respectively, and on 30 June 2021 and 21 June 2021 for autumn crops at SPG and ALB. Within the collected biomass, sub-samples of 15 plants were randomly selected, and, plant density, plant height, number of primary branches, number of fertile and sterile capsules and yield components were measured. Afterwards, all plants were separated by the different organs (seeds, capsules, straw) using a fixed-point trasher for fresh and dry weight determinations. Dry weights were measured after oven-drying (60 °C) to constant weight. Thousand seed weight (TSW) was assessed according to ISTA [32], and dry matter harvest index (HI) was calculated as the dry seed yield/total above-ground dry biomass × 100.

2.3. Oil and Protein Content and Fatty Acid Composition

Oil and protein percentages were obtained using a FOSS-NIRS DA1659 analyser (FOSS, Hillerød, Denmark).

Fatty acid composition was determined by gas chromatography of free and bound fatty acids following their conversion into fatty acid methyl-esters, according to Tavarini et al. [33]. Fatty acid methyl-esters were obtained by pouring 0.2 g of finely ground sample and 3 mL of 10% methanolic solution into 20 mL vials and mixed for 60 s. An amount of 0.5 mg of internal standard (nonadecanoic acid) was added to the mix. After 8 h, 1 mL of hexane were poured into the vial and the mixed for 1 min. The mix was then centrifuged for 10 min at 5000 rpm to separate the layers and the hexane fraction was injected to gas-chromatographic analysis, using a GC2010 Shimadzu gas chromatograph (Shimadzu, Columbia, MD, USA) equipped with a flame-ionization detector and a high-polar fused-silica capillary column (Chrompack CP-Sil88 Varian, 152 Middleburg, Netherlands). Hydrogen was used as carrier gas at 1 mL min^{-1} flow. The injector temperature was set at 270 °C and the detector temperature was set at 300 °C. Individual fatty acid methyl-esters were identified by comparison with a standard mixture of a 52-component fatty acids methyl-esters mix (Nu-Chek Prep Inc., Elysian, MN, USA).

2.4. Statistical Analysis

Results were subjected to 3-way ANOVA (Genotype × Sowing Date × Location), and means were compared by Tukey's HSD test when the ANOVA F-test was significant at the 0.05 probability level. Pearson's correlation coefficients were estimated to determine the relationship between all traits analysed. For statistical analysis, CO-STAT cohort V6.201 (2002) was used. An additional statistical analysis was carried out to evaluate the significant interaction between genotype and environment (G × E) for seed yield to define the varieties characterized by more yield stability across environments. To explain the G × E interaction, the multivariate analysis was performed graphically based on the AMMI and GGE biplot using R studio (a simplified version of R statistical software) developed by the R Core Team. The metan package, an open-source R package designed to provide an efficient and reproducible workflow for the analysis of MET (multi-environment trials) data, was used. A stable version of metan is available on CRAN (https://CRAN.R-project. org/package=metan; accessed on 10 December 2022). A GUI package of R studio was used for GGE biplots while the agricolae package was used for AMMI, involving two concepts, the biplot concepts and the GGE concept [34]. The GGE biplots and AMMI are graphical images to exemplify G × E interaction and genotype ranking based on mean and stability. The graph generated is based on multi-environment evaluation (which-won-where pattern), genotype evaluation (ranking genotype and mean versus stability), and tested environment raking (discriminative versus representative). The biplots were based on singular-value partitioning (SVP) = 1, transformed (transform = 0), environment-centered (G + GE, centering = 2), and standard deviation-standardized (scaling = 0).

3. Results

3.1. Weather Conditions and Crop Phenology

In both sites, the long-term rainfall pattern was similar, with precipitation mainly concentrated during the October-February period, but on average, higher annual cumulative rainfall was measured at SPG (890 mm), compared to ALB (685 mm). Long-term average temperature almost overlaps among sites; however, ALB had a wider temperature range, with lower average minimum (-1.1 °C on annual base) and higher average maximum temperature ($+1.9$ °C on annual base) compared to SPG.

Wider temperature range was also observed at ALB during the two growing seasons, in the first one (February 2020–July 2020), T_{max} and T_{min} were, respectively, $+2.2$ °C and -2.1 °C compared to SPG, while in the second season (October 2020–June 2021) T_{max} and T_{min} at ALB were $+2.1$ °C and -1.2 °C compared to SPG. During both growing seasons, cumulative rainfall was higher at SPG compared to ALB, as reported in Figures 1 and 2.

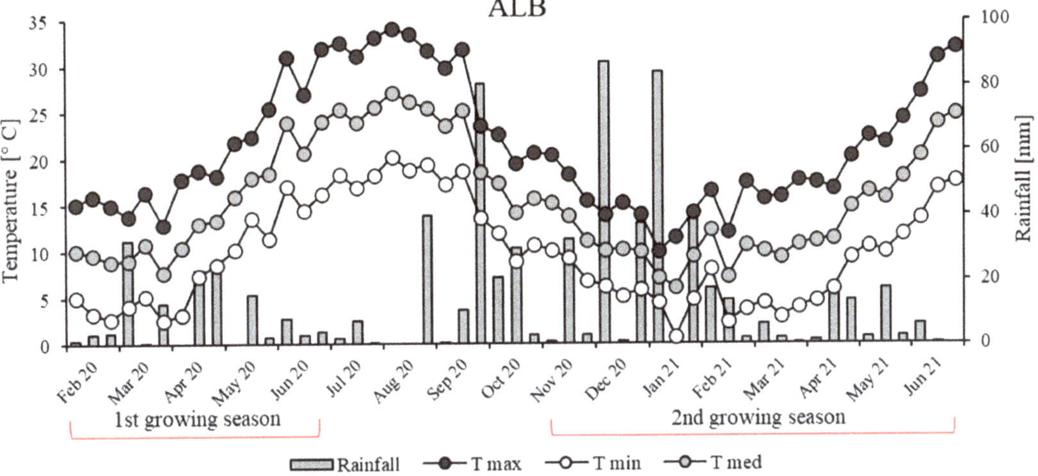

Figure 1. Monthly rainfall and mean, minimum, and maximum temperatures for the entire trial period at Alberese (ALB), Grosseto.

In both locations, the lowest value of SPEI_3 was reached in April 2021 (-1.1 ALB and -1.3 SPG), while for both SPG and ALB, SPEI_6 values lower than -1.0 were never observed (Figures 3 and 4). However, negative values slightly below -1.0 are very frequent in long-term SPEI_3 and SPEI_6 index calculations, so the computed values should be considered in line with the typical meteorological fluctuations of both experimental sites. This suggests that during the two-year trial, there were no periods of extraordinary drought compared to the normal weather condition trend, and the drought level could be considered as the one normally present in the areas. Differences between SPG and ALB, on the absolute values of meteorological drought-related factors (lack of rainfalls and high temperatures), are therefore to be considered those normally present between the two experimental sites. Conversely, positive values of SPEI_6 ($+2.3$; $+2.2$) and SPEI_3 ($+1.9$; $+2.2$) calculated at SPG in January and February 2021, indicate how the rainfall that occurred in the winter 2020–2021 was exceptional, among the highest on record for the previous 30 years.

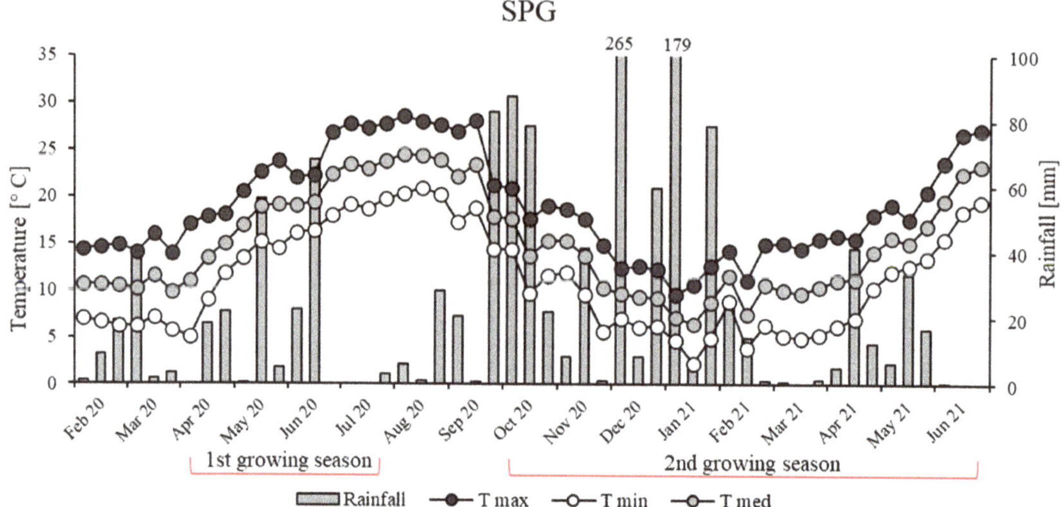

Figure 2. Monthly rainfall and mean, minimum, and maximum temperatures for the entire trial period at San Piero a Grado (SPG), Pisa.

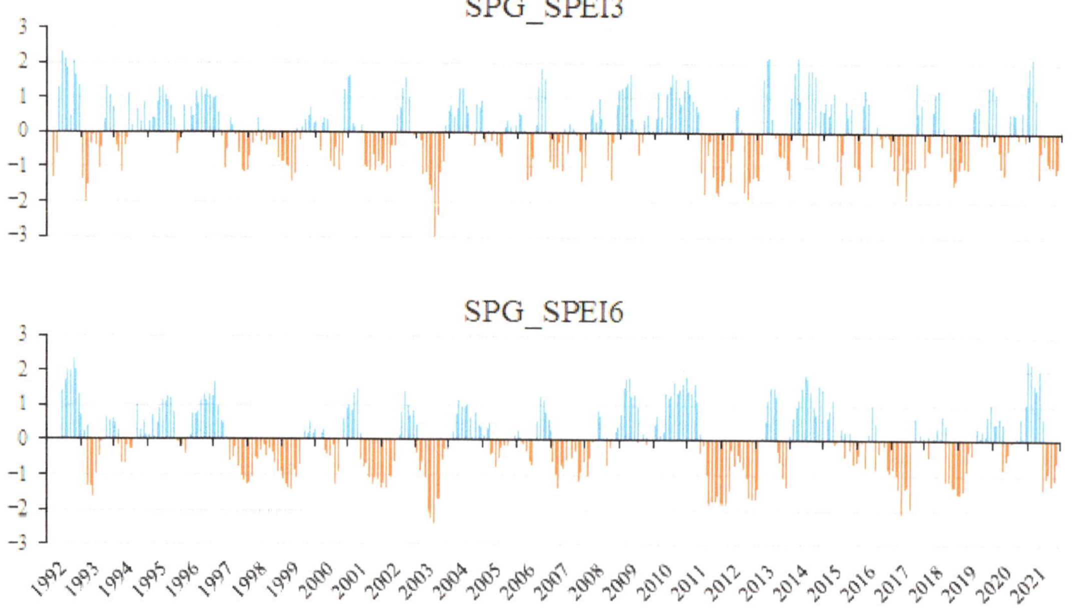

Figure 3. Long-term standardized precipitation-evapotranspiration index (SPEI) calculated at SPG on 3- and 6-month basis. For each year, blue and red bars represent, respectively, monthly positive and negative SPEI values.

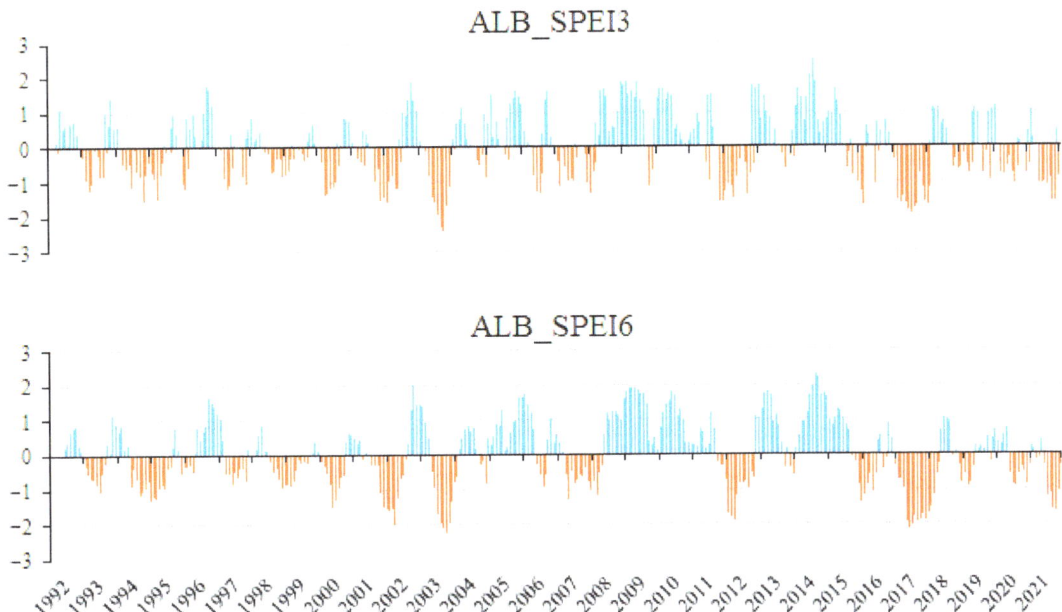

Figure 4. Long term standardized precipitation-evapotranspiration index (SPEI) calculated at ALB on 3- and 6-month basis. For each year, blue and red bars represent, respectively, monthly positive and negative SPEI values.

As expected, linseed varieties showed a shorter growth cycle, from sowing to harvest, when sown in spring compared to the autumn sowing, with an average spring cycle length of 100 and 135 days in SPG and ALB, respectively, and 219 and 223 days for the autumn cycle, in SPG and ALB, respectively (Table 2). At SPG, little difference was detected for the GDD accumulated from sowing to harvest between the two sowing dates (1377 vs. 1463 GDD for spring and autumn sowing, respectively), while more consistent variations have been observed at ALB (1245 vs. 1656 GDD, for spring and autumn sowing, respectively). Regarding varieties, Galaad was the earlier one in both sites, starting flowering after having accumulated 549 and 471 GDD in the spring sowing, at SPG and ALB, respectively, and 727 and 713 GDD in the autumn sowing, at SPG and ALB, respectively. On the other hand, Sideral was characterized by a longer vegetative cycle, in comparison with the other cultivars, in the autumn sowing in both cultivation sites, requiring more days and GDD for flowering (Table 2).

The cumulative rainfall from sowing to harvest differed between the two sowing dates and between the two cultivation sites. Considering the cultivation site, linseed varieties grown in ALB received 34% and 52% less rain than those grown at SPG, in spring and autumn sowing, respectively. At ALB, a strong variation of rain distribution between the vegetative and reproductive phase was observed for both sowing dates, with lower amounts of rainfall between the start of flowering and seed maturity (Table 2). This developmental stage was also characterized by high ET0 levels. On the contrary, very similar values were observed at SPG in both vegetative and reproductive phase, except for rainfall distribution in the autumn sowing.

Table 2. Duration, growing degree days (GDD), cumulative rainfall (mm) and ET0 (mm) evaluated during vegetative (V—from sowing to beginning of flowering) and reproductive (R—from beginning of flowering to maturity) cycles of five linseed varieties in spring (2020) and autumn (2021) sowings.

| | SPG | | | | | | | | ALB | | | | | | | |
| | Duration (Days) | | GDD (°C d) | | Rainfall (mm) | | ET0 (mm) | | Duration (Days) | | GDD (°C d) | | Rainfall (mm) | | ET0 (mm) | |
	V	R	V	R	V	R	V	R	V	R	V	R	V	R	V	R
Spring sowing																
Galaad	49	49	549	782	98	96	157	203	83	52	471	774	97	31	211	270
Libra	57	40	659	659	102	91	193	164	83	52	471	774	97	31	211	270
Sideral	56	42	643	688	103	91	188	172	87	48	515	730	97	31	229	252
Szafir	56	49	643	810	103	91	188	204	87	48	515	730	97	31	229	252
Kaolin	56	47	643	810	103	91	188	195	87	48	515	730	97	31	229	252
Mean ± SD	55 ± 3	45 ± 4	627 ± 44	750 ± 71	102 ± 2	92 ± 2	183 ± 15	188 ± 18	85 ± 2	50 ± 2	497 ± 24	748 ± 24	97 ± 0	31 ± 0	222 ± 10	259 ± 10
Autumn sowing																
Galaad	152	61	727	640	697	101	196	192	141	81	713	943	324	60	202	352
Libra	156	68	755	788	728	70	205	228	144	79	732	924	324	60	209	345
Sideral	166	54	821	668	729	69	235	184	168	55	894	762	354	30	285	269
Szafir	162	56	789	664	729	69	222	188	147	76	753	903	324	60	218	336
Kaolin	158	61	768	694	729	69	210	200	144	79	732	924	324	60	209	345
Mean ± SD	159 ± 5	60 ± 5	772 ± 35	691 ± 58	722 ± 14	76 ± 14	214 ± 15	198 ± 18	149 ± 11	74 ± 11	765 ± 74	891 ± 74	330 ± 13	54 ± 13	225 ± 34	329 ± 34

3.2. Yield and Yields Components

In Tables 3 and 4, F-test results of the effect of location (L), sowing date (SD), genotype (G), and their reciprocal interactions on biometric characteristics, seed yield, yield components and qualitative traits (protein and oil content) are reported. A significant interaction among location, sowing date and genotype was observed for plant height, plant density, number of stems per plant, harvest index, and thousand seed weight. On the contrary, LxSD interaction significantly affected all the analyzed parameters, except given for oil content (Tables 3 and 4). Similarly, significant effects of genotype, sowing date, and location were found. Only the percentage of sterile capsules, oil content and protein content did not vary with the time of sowing, and plant height was the same in both locations (Tables 3 and 4).

Table 3. Main effects of location (L), sowing date (SD) and genotype (G) and their interaction on linseed plant height, plant density, number of stems per plant, number of capsules per stem, sterile capsule percentage, and number of seeds per capsule. Data refer to mean values ± standard deviation.

	Plant Height (cm)	Plant Density (n. m^{-2})	Stems (n. plant^{-1})	Capsule (n. stem^{-1})	Sterile Capsules (%)	Seeds (n. capsule^{-1})
SPG	59.2 ± 7.9 a	341.3 ± 118.9 b	1.3 ± 0.2 a	18.4 ± 9.5 a	9.7 ± 6.6 b	4.0 ± 0.7 b
ALB	58.7 ± 10.3 a	432.0 ± 164.9 a	1.1 ± 0.1 b	9.3 ± 2.5 b	14.5 ± 5.9 a	4.4 ± 1.5 a
Spring	52.6 ± 6.5 b	369.6 ± 100.6 b	1.3 ± 0.2 a	10.0 ± 3.1 b	12.4 ± 5.2 a	3.6 ± 0.9 b
Autumn	65.4 ± 6.4 a	403.7 ± 186.6 a	1.1 ± 0.1 b	17.6 ± 10.0 a	11.8 ± 8.0 a	4.8 ± 1.1 a
Galaad	54.3 ± 6.0 b	324.8 ± 79.4 c	1.2 ± 0.2 b	12.8 ± 7.5 b	10.2 ± 5.1 b	4.4 ± 0.8 a
Libra	62.5 ± 8.5 a	276.0 ± 98.0 c	1.1 ± 0.1 bc	16.7 ± 11.4 a	10.8 ± 7.3 b	4.5 ± 1.5 a
Sideral	54.8 ± 9.7 b	542.5 ± 162.5 a	1.3 ± 0.3 a	10.1 ± 6.0 b	12.6 ± 6.6 ab	3.8 ± 1.2 b
Szafir	63.6 ± 10.1 a	326.0 ± 98.8 c	1.2 ± 0.1 bc	16.7 ± 8.6 a	15.2 ± 7.3 a	4.1 ± 1.2 ab
Kaolin	59.8 ± 7.3 a	463.9 ± 114.2 b	1.1 ± 0.1 c	12.9 ± 5.6 b	11.7 ± 6.6 ab	4.2 ± 1.1 ab
L	n.s.	***	***	***	***	*
SD	***	*	***	***	n.s.	***
G	***	***	***	***	***	**
L × SD	***	***	***	***	***	***
L × G	n.s.	n.s.	***	***	n.s.	***
SD × G	**	***	**	n.s.	*	n.s.
L × SD × G	*	***	***	n.s.	n.s.	n.s.

Means followed by the same letter are not significantly different at $p \leq 0.05$ based on Tukey's HSD test. *** significant at $p \leq 0.001$ level; ** significant at $p \leq 0.01$; * significant at $p \leq 0.05$ level; n.s., not significant.

Plant height was mainly determined by sowing date, as taller plants were obtained adopting autumn sowing compared to spring sowing in both locations (+7.4 cm at SPG and +18.3 cm at ALB) (Table 5). Besides sowing date, plant height was also determined on a genetic basis, with higher values reached on average by Libra, Szafir and Kaolin (+13.5%) compared to Galaad and Sideral (Table 3). Plant height results positively correlated with the number of seeds per capsule, seed yield and crop residues, and negatively correlated with the number of stems per plant, HI, and protein content (Table 6).

At SPG, plants density at maturity was lower during the autumn sowing compared to the spring sowing (−27.5%), while an opposite trend was observed at ALB, where autumn sowing increased the number of plants with almost +177 plants per m^2, compared to the spring sowing (Table 5). However, a clear effect of genotype on this parameter was found, with higher plant density registered for Sideral and Kaolin (Table 3). Plant density was positively correlated with sterile capsule percentage and negatively correlated to number of capsules per stem, and oil content (Table 6).

Table 4. Main effects of location (L), sowing date (SD) and genotype (G) and their interaction on linseed seed yield, vegetative biomass, harvest index (HI), thousand seed weight (TSW), oil and protein concentrations. Data refer to mean values ± standard deviation.

	Seed Yield (Mg ha^{-1})	Crop Residues (Mg ha^{-1})	HI (%)	TSW (g)	Oil Content (g/100 g)	Crude Protein Content (g/100 g)
SPG	2.1 ± 0.6 a	4.0 ± 1.4 a	34.6 ± 5.0 a	7.3 ± 1.0 a	47.2 ± 1.0 a	21.2 ± 1.0 b
ALB	1.3 ± 0.7 b	3.1 ± 2.2 b	32.6 ± 6.7 b	6.0 ± 0.8 b	45.2 ± 1.4 b	22.6 ± 1.5 a
Spring	1.2 ± 0.6 b	2.1 ± 1.2 b	37.4 ± 4.2 a	6.6 ± 0.7 b	46.5 ± 1.7 a	22.6 ± 1.4 a
Autumn	2.1 ± 0.5 a	5.0 ± 1.1 a	29.8 ± 4.9 b	7.5 ± 0.9 a	46.6 ± 2.6 a	21.3 ± 1.3 a
Galaad	1.7 ± 0.8 a	3.1 ± 1.6 b	36.6 ± 7.2 a	8.0 ± 0.7 a	46.1 ± 1.2 b	22.0 ± 1.2 abc
Libra	1.3 ± 0.6 b	3.2 ± 1.8 ab	31.5 ± 4.6 b	6.5 ± 0.6 d	49.0 ± 0.9. a	21.6 ± 1.1 bc
Sideral	1.6 ± 0.8 ab	4.0 ± 2.3 a	30.4 ± 4.9 b	6.0 ± 0.4 e	43.6 ± 1.2 c	23.3 ± 1.1 a
Szafir	1.7 ± 0.7 a	3.5 ± 1.7 ab	34.5 ± 5.4 a	7.5 ± 0.6 b	46.3 ± 1.0 b	22.1 ± 1.1 ab
Kaolin	1.9 ± 0.7 a	3.9 ± 2.0 a	35.1 ± 5.6 a	7.4 ± 0.6 c	47.9 ± 1.2 a	20.6 ± 1.2 c
L	***	***	***	***	***	***
SD	***	***	***	***	n.s.	n.s.
G	***	**	***	***	***	***
L × SD	***	***	***	***	***	n.s.
L × G	**	n.s.	***	***	n.s.	n.s.
SD × G	n.s.	n.s.	n.s.	*	n.s.	n.s.
L × DS × G	n.s.	n.s.	**	*	n.s.	n.s.

Means followed by the same letter are not significantly different at $p \leq 0.05$ based on Tukey's HSD test. *** significant at $p \leq 0.001$ level; ** significant at $p \leq 0.01$; * significant at $p \leq 0.05$ level; n.s., not significant.

Table 5. Interaction effects of location (SPG and ALB) and sowing date (S: spring; A: autumn) on linseed plant height, plant density, number of stems per plant, number of capsules per stem, sterile capsule percentage and number of seeds per capsule. Data refer to mean values ± standard deviation.

Location	Sowing Date	Plant Height (cm)	Plant Density (n. m^{-2})	Stems (n. plant^{-1})	Capsule (n. stem^{-1})	Sterile Capsules (%)	Seeds (n. Capsule^{-1})
SPG	Spring	55.5 ± 7.7 c	395.6 ± 100.6 b	1.4 ± 0.3 a	10.8 ± 3.7 b	14.6 ± 5.4 b	4.0 ± 0.8 b
SPG	Autumn	62.9 ± 6.3 b	286.9 ± 112.5 d	1.1 ± 0.1 c	26.0 ± 7.1 a	4.7 ± 3.1 d	4.1 ± 0.5 b
ALB	Spring	49.6 ± 3.2 d	343.5 ± 96.0 c	1.2 ± 0.1 b	9.3 ± 2.4 b	10.3 ± 4.1 c	3.1 ± 0.7 c
ALB	Autumn	67.9 ± 5.5 a	520.5 ± 173.5 a	1.0 ± 0.0 c	9.3 ± 2.6 b	18.8 ± 4.2 a	5.6 ± 0.9 a

Means followed by the same letter are not significantly different at $p \leq 0.05$ based on Tukey's HSD test.

Regarding the number of stems per plant, no differences were observed among locations with autumn sowing, while with spring sowing, an increase in the number of produced stems was observed at SPG (+17.2%) (Table 5). Little difference among genotypes was highlighted, besides a significant increase (+31%) in the number of stems plant^{-1} produced by Sideral compared to the other varieties (Table 3). No significant correlations have been found among the number of stems per plant and the other investigated parameters (Table 6).

No differences between sowing dates were observed at ALB regarding the number of capsules, with an average value of 9.3 capsule plant^{-1}, while at SPG, the autumn sowing led to an increase of capsules per plant, with the highest values in plants sown in autumn (Table 5). The number of capsules per plant was strongly dependent on genotype, with the highest values reached by Libra and Szafir, followed by the other varieties (Table 3). As expected, the number of capsules plant^{-1} was positively correlated with seed yield, TSW, and oil content, while a negative correlation was observed with the percentage of sterile capsules and protein content (Table 6).

Table 6. Pearson's correlation coefficients and their statistical significance among studied parameters among pairs of variables related to the expression of growth, yield, and qualitative traits in five linseed genotypes under two sowing dates and cultivated in two different locations.

	Plant Height	Plant Density	Stems per Plant	Capsules per Plant	Sterile Capsules	Seeds per Capsule	Seed Yield	Crop Residues	HI	TSW	Oil Content	Protein Content
Plant height	1	0.07 ns	−0.58 **	0.37 ns	0.25 ns	0.69 ***	0.55 *	0.77 ***	−0.71 ***	0.34 ns	0.18 ns	−0.50 *
Plant density		1	0.04 ns	−0.57 **	0.59 **	0.21 ns	0.10 ns	0.33 ns	−0.43 ns	0.35 ns	−0.63 **	0.16 Ns
Stems per plant			1	−0.33 ns	−0.03 ns	−0.28 ns	−0.17 ns	−0.34 ns	0.39 ns	−0.41 ns	−0.20 ns	0.18 ns
Capsules per plant				1	−0.64 **	−0.06 ns	0.53 *	0.37 ns	−0.01 ns	0.53 *	0.49 *	−0.46 *
Sterile capsules					1	0.32 ns	−0.11 ns	0.12 ns	−0.40 ns	−0.24 ns	−0.44 ns	0.20 ns
Seeds per capsule						1	0.47 *	0.61 **	−0.52 *	0.29 ns	0.05 ns	−0.49 *
Seed yield							1	0.86 ***	−0.28 ns	0.62 **	0.13 ns	−0.76 ***
Crop residues								1	−0.72 ***	0.38 ns	−0.05 ns	−0.58 **
HI									1	0.11 ns	0.25 ns	0.08 ns
TSW										1	0.30 ns	−0.58 **
Oil content											1	−0.67 **
Protein content												1

*** significant at $p \leq 0.001$ level; ** significant at $p \leq 0.01$; * significant at $p \leq 0.05$ level; n.s., not significant

167

With spring sowing, sterile capsules were higher at SPG (+4.3%), while with autumn sowing, the percentage of sterile capsules was significantly higher (+14.0%) at ALB compared to SPG (Table 5). This parameter was also genotype-related, with highest percentage of sterile capsules observed in Szafir, while the lowest values were reached by Galaad and Libra (Table 3). No correlation was observed among sterile capsules and the other parameters (Table 6).

The number of seeds per capsule did not vary at SPG depending on sowing date, while at ALB autumn sowing increased this parameter by +79.7% (Table 5). The number of seeds per capsule turned out to be directly correlated with seed yield and crop residues, while a negative correlation was found with HI and protein content (Table 6).

A longer cycle (autumn sowing) increased seed yield in both locations, with higher improvement observed at ALB (+175%) compared to SPG (+37%) (Table 7). As a general trend, the highest yield (2.4 t ha^{-1}) was reached at SPG with autumn sowing. Regarding genotype, in both locations, Libra reached the lowest seeds yield, while no differences among the other four varieties was observed (Table 3). However, for Galaad (−55.9%), Sideral (−45.0%) and Szafir (−31.1%) seeds yield at ALB was significantly lower compared to SPG. Conversely, Kaolin showed the highest grain yield in both locations with no statistically significant differences among ALB and SPG. Seed yield showed a positive correlation with crop residues and TSW and a negative correlation with protein content (Table 6).

Table 7. Interaction effects of location (SPG and ALB) and sowing date (spring and autumn) on seed yield, vegetative biomass, harvest index (HI), thousand seed weight (TSW), oil and protein concentrations of linseed. Data refer to mean values ± standard deviation.

Location	Sowing Date	Seed Yield (Mg ha^{-1})	Crop Residues (Mg ha^{-1})	HI (%)	TSW (g)	Oil Content (g 100 g^{-1})	Crude Protein Content (g 100 g^{-1})
SPG	Spring	1.7 ± 0.4 b	3.1 ± 0.7 b	36.2 ± 4.5 b	6.8 ± 0.8 c	46.6 ± 1.7 b	21.6 ± 0.9 a
	Autumn	2.4 ± 0.5 a	5.0 ± 1.2 a	32.9 ± 4.9 c	7.9 ± 0.8 a	47.8 ± 2.0 a	20.8 ± 1.1 a
ALB	Spring	0.7 ± 0.3 c	1.0 ± 0.3 c	38.6 ± 3.5 a	6.5 ± 0.6 d	46.4 ± 1.9 b	23.5 ± 1.1 a
	Autumn	1.9 ± 0.4 b	5.1 ± 1.0 a	26.7 ± 2.1 d	7.2 ± 0.8 b	45.4 ± 2.0 c	21.8 ± 1.4 a

Means followed by the same letter are not significantly different at $p \leq 0.05$ based on Tukey's HSD test.

Similarly to that observed for seed yield, crop residues reached the highest value in autumn-sown crops in both environments (Table 7). Adopting spring sowing, linseed produced +2 Mg ha^{-1} of residues on a dry matter basis at SPG in comparison with ALB. The lowest amount of crop residues was produced by Galaad, which was found to be significantly different only from Sideral and Kaolin (Table 4). As expected, crop residues were negatively correlated with HI (Table 6). A negative correlation was also found with protein content. According to the trend observed for the seed yield and crop residues, the harvest index significantly decreased with autumn sowing in both locations (Table 7), with a steeper drop (−11.2%) at ALB compared to SPG (−3.3%). On average, Galaad, Szafir, and Kaolin showed the highest HI according to their higher seed production (Table 4). However, a substantial difference was observed in Galaad due to the cultivation site (41.1 vs. 32.0 at SPG and ALB, respectively).

Galaad reached on average the highest values of TSW (Table 4); at SPG this variety showed the highest TSW in both sowing dates (8.0 and 9.0 g in spring and autumn sowing, respectively) while, at ALB comparable values were recorded also for Szafir and Kaolin in spring sowing (7.0 g and 6.8 g, respectively, vs. 7.1 g of Galaad), and for Szafir in autumn sowing (Szafir = 7.9 g; Galaad = 8.0 g). Sideral achieved the lowest TSW in both sowing dates and locations. TSW increased on average with autumn sowing (+16.8% at SPG, +10.7% at ALB) and was higher at SPG compared to ALB (+3.4% with spring sowing, +9.7% with autumn sowing) (Table 7). Moreover, the correlation analysis revealed a negative correlation between TSW and protein content (Table 6).

In spring-sown linseed varieties, oil content did not vary between the two environments, while a significant increase in oil content was recorded at SPG by adopting an autumn sowing (+2.3% at SPG compared to ALB) (Table 7). Higher oil concentration was obtained by Libra and Kaolin, while Sideral expressed the lowest value (Table 4). No differences among sowing dates were highlighted, while, as expected, a negative correlation between oil content and protein content was detected (Table 6). At ALB, a slight increase (+1.4%) of seed protein content was found, while sowing date did not influence this parameter (Table 4). Once again, the genotype played an important role in defining the protein content, with the highest value in the seeds of Sideral and the lowest in Kaolin (Table 4).

3.3. Additive Main Effects and Multiplicative Interaction Analysis: AMMI1 and AMMI2

Figure 5a shows the additive main effects and multiplicative interaction among five genotype and four environments for the grain yield trait. Each environment is given by sowing date–location combination, and they are expressed as SPG_2020 (spring sowing at SPG), SPG_2021 (autumn sowing at SPG), ALB_2020 (spring sowing at ALB) and ALB_2021 (autumn sowing at ALB). In AMMI1, the biplot abscissa and ordinate indicated the first principal component (PC1) term and the trait's significant influence, respectively. Based on genotype mean and interaction with the environment, among the tested genotype, Libra is the lowest-producing variety and the most unstable, even in the most suitable environmental conditions for ALB when sown in spring. Sideral also appeared to be a low-performing variety and, together with Szafir, had the least G × E interaction, their PC1 scores being adjacent to the zero lines of the biplot (Figure 5a). The superior genotypes were Kaolin > Szafir > Galaad. Kaolin and Szafir were more adapted to the specific environment of ALB when sown in autumn (ALB_2021) while Galaad was better suited to the conditions of SPG by adopting a spring sowing (SPG_2020).

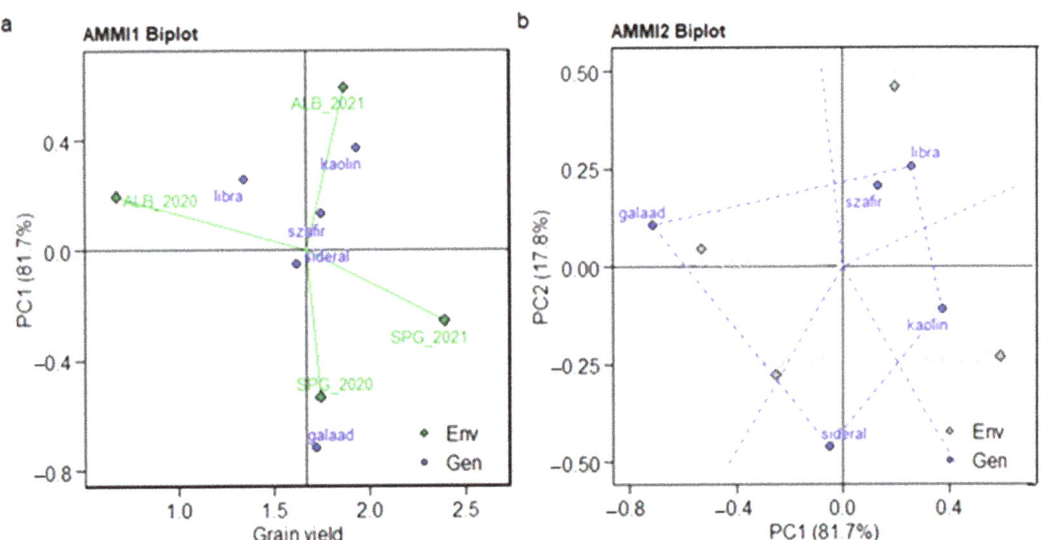

Figure 5. AMMI1 (**a**) and AMMI2 biplot (**b**) analysis for genotype by environment (G × E) interaction in multilocation variety trials for the seed yield trait and 'which-won-where polygon view' of the genotypes and environments.

The AMMI2 biplot (PC1 and PC2; Figure 5b) provides summarized information about genotype x environment interaction: genotypes positioned near the origin had the least interaction, and the genotypes positioned near to the axis had more general stability. Furthermore, any genotypes that are close to each location have specific stability in that environment. In terms of grain yield, our results showed great interaction between genotypes and locations. The presence of a genotype at one of the vertices of the polygon in a sector where the environment markers fall, suggested that this genotype provided a higher seed yield and performed better in that environment. Consequently, our results suggested that Kaolin was suited and more stable for ALB in autumn sowing (ALB_2021); Sideral for SPG_2021; Galaad for SPG_2020, and Libra for ALB_2020. The Szafir genotype placed within the polygon was less respective to the environment than the corner genotypes.

3.4. Genotype Plus Genotype-vs.-Environment Interaction Analysis: GGE Biplot

Our results show the GGE-biplot model in Figure 6a discriminating four environments (sowing date–location combination) with a clear separation in the upper and lower right quadrants, respectively (ALB_2020 and ALB_2021; SPG_2020 and SPG_2021). In Figure 6b, the genotype ranking biplot for the assessment of the best and ideal genotype is reported. Kaolin can be noted as the best leading genotype due to its nearness to the arrowhead in the circle for the seed yield produced. The GGE biplot pattern of 'mean vs. stability' analysis was also performed (Figure 6c). The 'mean vs. stability' helps to simplify the genotype assessment based on the mean performance and stability under a wide range of environments. Line one consists of a single arrow that points towards greater mean grain yield. In our investigation, the 'mean vs. stability' pattern revealed 92.22% for seed yield of G + G × E variation. The arrow sign on the abscissa line directs the ranking of genotypes in increasing order, with a greater value of the productive trait evaluated (Libra > Sideral > Galaad > Szafir > Kaolin). Additionally, the determination of a best-suited environment is crucial for a successful introduction of a new crop variety within a specific cropping system. The two features 'discriminativeness' (the ability of an environment to distinguish genotype) and 'representativeness' (the ability of an environment to represent all other evaluated environments) mean the idealness of the tested environments. In our study, Figure 6d illustrates the 'discriminativeness vs. representativeness', underlining that environment is better than another, but both ALB and SPG in spring (2020) and autumn (2021) gave satisfactory productive results.

3.5. Seed Oil Composition

In Table 8, F-test results of the effects of location (L), sowing date (SD), genotype (G), and their reciprocal interaction on seed oil composition are reported. All the three variability factors and their reciprocal interactions significantly affected the main fatty acid (FA) profile of linseed oil. Linseed oil is characterized by a high degree of unsaturation, with polyunsaturated fatty acids (PUFAs) accounting for about 70%. Among PUFAs, the predominant FAs are represented by α-linolenic (C18:3) and linoleic (C18:2) acids, which represent 56% and 15% of total FAs, respectively. Monounsaturated fatty acids (MUFAs) are around 20%, almost exclusively represented by oleic acid (C18:1), while saturated fatty acids (SFAs) account for 7%, made up mainly of palmitic acid (C16:0). Taking into account the effect of cultivation site, the environmental conditions of SPG favored the accumulation of α-linolenic acid and increased PUFA levels, while seeds obtained by linseed grown at ALB showed the highest levels of MUFA, SFA, oleic acid and linoleic acid (Table 8). According to these profiles, the omega 6 to omega 3 ratio significantly decreased at SPG. These results underlined the key role played by the environmental conditions in defining the FAs profile of linseed oil. Similarly, sowing date also strongly affected the fatty acid composition and oil quality. In general, spring sowing increased the content of the α–linolenic and oleic acids, and, consequently, PUFAs and MUFAs levels as well as PUFAs to SFAs ratio. Accordingly, the lowest omega 6/omega 3 ratio was reached by adopting this sowing time. Finally, genotypic characteristics confirmed to be one of the main factors influencing the FA profile of linseed oil. Among the tested varieties, Libra exhibited simultaneously the highest C18:3 content (61.2% of total FA), PUFA levels (74.5%) as well as PUFAs/SFAs ratio (12.3), but the lowest omega 6/omega 3 ratio. Conversely, an opposite profile was observed for Galaad, which showed the highest levels of C18:1 (22.2%), C18:2 (17.8%), MUFAs (22.3%) and omega 6/omega 3 ratio (0.40), and the lowest content of C18:3 (51.7%). Interestingly, Sazfir was characterized by the highest content of monounsaturated palmitic (8.3%) and heptadecanoic (0.11%) acids.

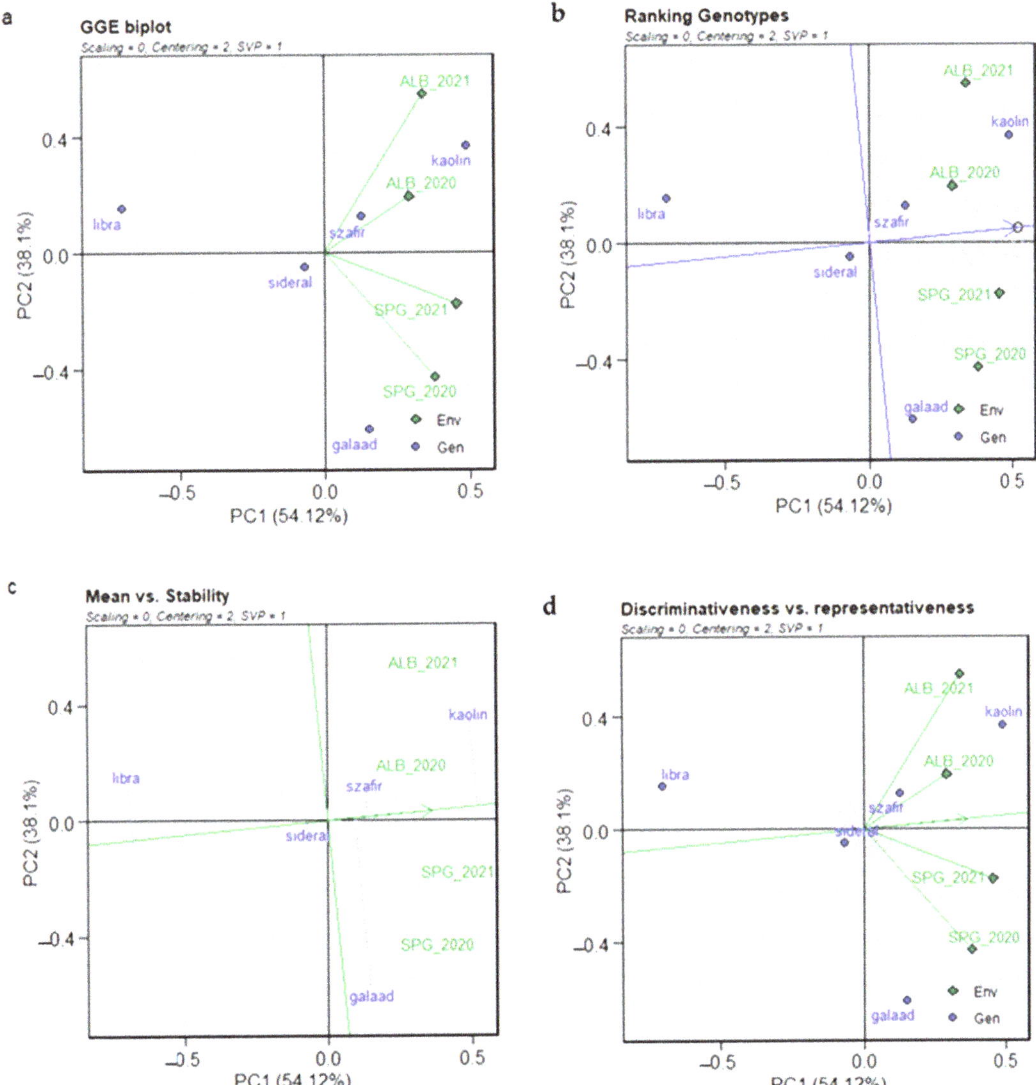

Figure 6. Comprehensive examination of genotypes and environments. GGE Biplot visualization (**a**); ranking genotypes (**b**) within environments and representativeness of test environments; (**c**) comparison of genotype with the 'ideal' genotype for both mean and yield stability; (**d**) comparison of environments based on both discriminating ability and representativeness of the target environment.

Table 8. Fatty acids composition (% of total FA) in linseed in response to location (L), sowing date (SD) and genotype (G). Data refer to mean values.

		C16:0	C16:1	C17:0	C18:1	C18:2	C18:3	SFA	MUFA	PUFA	PUFA/SFA	n6/n3
Location	SPG	6.35 b	0.14	0.084	18.46 b	13.85 b	58.37 a	6.50 b	18.60 b	72.31 a	11.21	0.24 b
	ALB	6.80 a	0.13	0.085	21.11 a	16.80 a	52.94 b	6.95 a	21.25 a	69.74 b	10.94	0.37 a
Sowing	Spring	6.16 b	0.13	0.067 b	20.14 a	13.88 b	57.94 a	6.33 b	20.27 a	71.82 a	11.39 a	0.24 b
date	Autumn	6.95 a	0.13	0.099 a	19.45 b	16.73 a	53.59 b	7.08 a	19.61 b	70.31 b	10.79 b	0.36 a
	Galaad	6.13 bc	0.13 b	0.076 b	22.20 a	17.75 a	51.65 e	6.24 c	22.34 a	69.40 c	11.22 b	0.40 a
	Libra	5.88 c	0.09 c	0.083 b	17.05 d	13.32 c	61.17 a	6.10 c	17.15 d	74.49 a	12.31 a	0.22 d
Genotype	Sideral	6.25 bc	0.11 bc	0.079 b	20.46 b	13.19 c	57.53 b	6.46 b	20.58 b	70.72 b	11.01 b	0.23 c
	Szafir	8.25 a	0.21 a	0.108 a	18.76 c	17.96 a	52.42 d	8.40 a	18.99 c	70.39 bc	9.81 c	0.40 a
	Kaolin	6.38 b	0.11 bc	0.076 b	20.41 b	14.64 b	55.50 c	6.43 b	20.54 b	70.15 bc	11.02 b	0.26 b
	L	**	n.s.	n.s.	**	***	***	***	***	***	n.s.	***
	SD	***	n.s.	***	*	***	***	***	*	**	***	***
	G	***	***	***	***	***	***	***	***	***	***	***
Variability	L × SD	***	***	***	***	***	***	***	***	***	***	***
factors	L × G	***	***	***	***	***	***	***	***	***	***	***
	SD × G	***	***	***	***	***	***	***	***	***	***	***
	L × SD × G	***	***	***	***	***	***	***	***	***	***	***

Means followed by the same letter are not significantly different at $p \leq 0.05$ based on Tukey's HSD test. *** significant at $p \leq 0.001$ level; ** significant at $p \leq 0.01$; * significant at $p \leq 0.05$ level; n.s., not significant.

4. Discussion

Mediterranean cropping systems are mainly based on cereals, often grown in monoculture. Therefore, for increasing crop diversification and cropping system sustainability, organic farmers are looking for alternative crops suited to specific pedoclimatic conditions to introduce in their traditional rotations. So, this study aimed to provide local references about the introduction of linseed under organic management in central Italy, as a suitable and innovative crop, comparing two different environments, five varieties and two sowing dates. As a general trend, the tested linseed varieties demonstrated a good adaptability to both the pedoclimatic conditions of the two contrasting environments (SPG and ALB) and the organic farming system, as confirmed by the good seed and biomass yield, and oil content and quality. However, significant differences in terms of morphological, biometric and yield parameters were detected between the two locations, between the two sowing dates and among varieties, confirming that environmental conditions and agronomic practices can significantly influence linseed performances.

In our study, autumn linseed completed its growing cycle in 226 days, accumulating 1560 GDD (as mean values across locations and varieties); on the other hand, the spring crop required 118 days from sowing to harvest, with a thermal request of 1311 GDD. These findings are in line with those observed by Tavarini et al. [33] who compared two linseed varieties in spring sowing. Čeh et al. [35], in spring experimental trials carried out in four different locations in Slovenia, observed a similar cycle length, ranging from 96 to 120 days depending on location. These authors found that high temperatures and water shortages shortened the growing cycle of the tested linseed varieties. Also, the genetic characteristics were involved in defining the timing and duration of each phenological phase, together with the corresponding growing degree days. In both locations, Galaad was the earlier variety, while Sideral required more days and GDD for flowering. These observations can be extremely useful in the proper variety choice in function of the climatic conditions of a specific environment, preferring early cultivars in environments characterized by high temperatures and very little water availability in the late spring/summer period. It is known as this is a critical period able to significantly influence the duration of seed filling and the translocation of photoassimilates to the seeds, affecting, in turn, seed yield and quality.

Regarding biometric characteristics, no correlation between plant height and plant density was observed, although an increase in the number of plants per unit area generally leads to taller plants due to an increased intraspecific competition for light as reported by Kurtenbach et al. [36]. In any case, shorter plants were obtained in spring sowings for the shorter growing cycle as compared to autumn sowings. Plant height, moving upwards the plants centre of mass, could increase lodging damage [37], so in a windy climatic context, the choice of a small-size genotype might be a determining factor. Our findings are partially in contrast with those reported by Gajardo et al. [38], which suggested that smaller plants with limited vegetative biomass are more efficient in photoassimilate translocation, increasing yield components and seed yield. Conversely, we observed that short genotypes selected for seeds production, such as those used in the present study, if subjected to agronomic practices that favour their vegetative development, such as early sowing, can strongly increase their production level. Moreover, despite the great phenotypic plasticity of linseed, with a great ability to respond to changed spacing, compensating for low plant populations through extensive branching [33,39], our findings showed no significant correlation between plant density and number of stems per plant. On the contrary, plant density negatively affected the number of capsule plant^{-1} according to Benaragama et al. [40]. Lower plants density at SPG in autumn sowing was probably due to the exceptional rainfall which occurred in the 2020–2021 autumn/winter season, which may have caused a significant degree of plant losses in the first development stages. As already discussed, cumulative rainfall at ALB from December 2020 to January 2021 was considerably lower compared to SPG, so the same decline in plant density was not observed. Although to the best of our knowledge no studies on linseed resistance to waterlogging are available, our findings suggested that linseed had some degree of resistance to waterlogging, and that this resistance varied across genotypes. Despite plant density of autumn crop being lower at SPG in comparison to ALB, the highest seed yield was recorded at SPG, confirming that linseed, if no other limiting factors are present, is widely able to compensate the lack of plant m^{-2} with other yields components [41]. The number of stems per plant was strongly dependent on genotype, as previously observed by Kalinina and Lyakh [42] and, together with plant height, is linked with crop interspecific competition for resources [43]. In our study, the negative correlation found together with the genotype-effect involved in the determination of these parameters, can indicate different "strategies" among linseed varieties, for optimal ground cover.

In the Mediterranean climate, it has been observed that linseed was very efficient in photosynthates assimilation, with a high increase in biomass post-anthesis [44], which differs to other species where post-anthesis biomass accumulation can be substantially reduced during grain filling [45]. In any case, the presence of stressor factors after the onset of flowering can negatively affect linseed yields and yield components [46]. In our experiment, water stress was indirectly measured through ET0 computation that indicates the magnitude of the environmental evapotranspiration demand in the two locations, which is particularly important from full flowering to seed maturity. The high evapotranspiration rate in the southern area of Tuscany (ALB) may have been a determining factor in significantly reducing linseed yields at ALB, especially during the first year with spring sowing, when the field trial was carried out on a soil with a predominant sandy component (Table 1) In addition to the water-stress, in the first year the low number of capsules per plant and the high abortion percentage were probably due to a late frost event which occurred on April 8th: at ALB T_{min} reached lower values ($-3.1\ ^\circ$C) compared to SPG ($0\ ^\circ$C). When this event occurred, the tested linseed varieties at ALB had already started flowering, excluding Sideral, which was between 55–57 BBCH phenological stages. At SPG, the earlier-flowering variety (Galaad) started flowering three days later (11th April), and the other varieties began flowering between the 15th and 25th April. This suggests that linseed is particularly sensible to frost events (with $T_{min} < 0\ ^\circ$C) at the beginning of full flowering, with consequent significant and negative effects on important yield components such as the number of capsules and their abortion rate, determining, in turn, a strong reduction in seed yield. Despite the high rusticity and adaptability of linseed to various environments, these findings disclose the important influence of seasonal variation on the different responses of linseed to cultivation site and sowing date due to temperature patterns and to amount and distribution of rainfall. A great variability of seed yield was in fact observed between the two locations as well as between the two sowing dates. The highest seed yield reached at SPG, both in spring and autumn sowing (1.7 and 2.4 Mg ha^{-1}, for spring and autumn-crop, respectively), could be ascribed to milder temperatures which may have alleviated the occurrence of abiotic stress during the critical reproductive phase. These observations confirmed previous reports [41,47], which found low rainfall and high temperatures during the seed-filling period negatively influence seed yield, accelerating maturity, reducing seed size (TSW) and oil content [48]. In addition, in the second year (with autumn sowing) the prolonged vegetative development registered at SPG could have increased the remobilization of photosynthates during the seed-filling stage as confirmed by the highest TSW. Interestingly, linseed's response to organic farming was very satisfactory, reaching yields consistent to those obtained adopting conventional and/or integrated farming techniques, in similar [33,41,49,50] and different environments across Europe [35,51–53]. Among varieties, Kaolin was the most productive with a seed yield of 1.9 Mg ha^{-1}, averaging the yield across environments and sowing dates as confirmed by AMMI ang GGE biplot analyses. Also crop residues were significantly affected by the variability factors considered here, following a similar trend observed for seed yield, with the highest value achieved at SPG in autumn sowing. As observed by Tavarini et al. [49], the C/N ratio of linseed above-ground residues varied from 60 to 100, with a potential nitrogen return to the soil ranging from 20 to 47 kg ha^{-1} per year, depending on growing season, environment, and genotype. All these characteristics are important since the incorporation of crop residues in the soil with tillage, can promote the soil organic matter storage for the subsequent crops in the rotation, although via slow mineralization due the C/N value. Furthermore, location, sowing date and genotype played a key role also in defining harvest index (HI). The observed HI values were consistent with those reported in the literature ranging from 30–34% [49,54]. Our findings point out that HI generally increased with delayed sowing (37.4 vs. 29.8%, in spring and autumn, respectively), despite the lowest seed yield reached in spring-sown crops. This was probably due to the differences in seed and biomass production between the two sowing dates: linseed varieties sown in autumn accumulated larger biomass, thanks to a longer season, but probably translocated less photosynthates to the seeds. Seed oil content and composition represent very important traits in order to evaluate the possibility of the successful introduction of this crop in traditional farming rotations. In addition, the increasing demand of organic vegetable oils and proteins represents a further strength of linseed. In linseed, as in other oilseed crops, climatic conditions and genotypic characteristics are the main factors that influence the oil content and FA profile. Generally, high temperatures during flowering and seed ripening determine a reduction in oil accumulation within the seeds, as evidenced by lower content at ALB in comparison to SPG. An opposite trend was instead observed for the protein content and a negative correlation was found between these two parameters: the higher the oil content, the lower the protein amount [55,56]. Similarly to that observed for oil content, stressor conditions affected the activity of the enzymes responsible for the metabolism of the polyunsaturated fatty acids (PUFAs), decreasing their content with high temperatures and low water availability, with a concomitant increase in MUFA biosynthesis, which is coherent with the observations of Fofana et al. [57]. According to this phenomenon, the oil obtained from SPG cultivation was characterized by higher levels of PUFAs and lower levels of MUFAs. On the contrary, the behavior observed for the two sowing dates was not very clear, with the higher PUFA values and lower MUFAs values recorded in crops sown in spring. Depending on the cultivation site, our results showed an oil content between 47.2% (SPG) and 45.2% (ALB), and PUFAs between 72% (SPG) and 70% (ALB), close to or higher than those previously reported in Mediterranean areas [33,49]. It is known that the FA profile is strongly related to genetic characteristics [58]; in the present study a great variability among the tested varieties was observed, confirming this statement. Alfa-linolenic acid was the most abundant FA in all the varieties, but Libra exhibited the highest level (around 61%). At the same time, this variety had the lowest oleic acid content and, consequently, the lowest MUFAs and major levels of PUFAs. Thanks to the high level of alfa-linolenic acid and PUFAs, linseed oil is characterized by a very high nutritional quality, being an important source of omega 3 fatty acids, with a balanced omega 6/omega 3 ratio. Besides the possibility to use linseed oil as functional food for humans, its chemical composition makes this oil suitable for obtaining a wide range of biobased products.

5. Conclusions

Although a remarkable variability was observed depending on genotype, sowing date and location, linseed showed good agronomic traits that make it suitable to be introduced in the tested environment, contributing to the local production of organic vegetable oils and proteins. In general, autumn sowing coupled with milder and wetter conditions, such as those observed at SPG, seemed to be more favorable for linseed cultivation. Early sowing, in fact, can allow to the crop to escape the drought stress at flowering stage, to which this species is very sensitive. The possibility to organically cultivate linseed as an autumn crop in Mediterranean areas is particularly important due to the limited number of autumn/winter crop options for organic farmers. Crop diversification is one of the pillars of organic agriculture, particularly effective in reducing weed and disease pressure and promoting yields and their stability. From AMMI and GGE multivariate analyses, Kaolin seemed to have the highest production stability, among the tested varieties, while Libra, despite a good oil content and quality, seemed to have poor adaptation to the test areas. Aside from different development length, both "sowing date" and "location" are meteoclimatic-related factors, and the presence of interaction among these with the factor "genotype" reinforces the need for a site-specific approach for choosing the most suitable variety for each pedoclimatic context.

Author Contributions: Conceptualization, L.G.A. and S.T.; Methodology, L.G.A. and S.T.; Formal analysis, A.R., C.C. and S.T.; Investigation, L.G.A. and S.T.; Resources, L.G.A. Data curation, S.T., C.C. and A.R.; Writing—original draft preparation, A.R., S.T. and L.G.A.; Writing—review and editing, A.R., C.C., S.T. and L.G.A.; Visualization A.R. and S.T.; Supervision, L.G.A.; Project administration, L.G.A.; Funding acquisition, L.G.A. and S.T. All authors have read and agreed to the published version of the manuscript.

Funding: This research was funded by the project PIF-SOLEAT—submeasure 16.2 "SIC_OLEAT—Sistemi di Innovazione Colturale per le Oleaginose Toscane" funded by PSR 2014–2020 Regione Toscana.

Data Availability Statement: The data presented in this study are available on request from the corresponding author.

Acknowledgments: The authors wish to express their gratitude to Lara Foschi for her technical assistance.

Conflicts of Interest: The authors declare no conflict of interest. The funders had no role in the design of the study; in the collection, analyses, or interpretation of data; in the writing of the manuscript, or in the decision to publish the results.

References

1. Yang, T.; Siddique, K.H.M.; Liu, K. Cropping systems in agriculture and their impact on soil health—A review. *Global Ecol. Conserv.* **2020**, *23*, e01118. [CrossRef]
2. Sanz-Cobena, A.; Lassaletta, L.; Aguilera, E.; del Prado, A.; Garnier, J.; Billen, G.; Iglesias, A.; Sánchez, B.; Guardia, G.; Abalos, D.; et al. Strategies for greenhouse gas emissions mitigation in Mediterranean agriculture: A review. *Agric. Ecosyst. Environ.* **2017**, *238*, 5–24. [CrossRef]
3. Kehoe, L.; Romero-Muñoz, A.; Polaina, E.; Estes, L.; Kreft, H.; Kuemmerle, T. Biodiversity at risk under future cropland expansion and intensification. *Nat. Ecol. Evol.* **2017**, *1*, 1129–1135. [CrossRef] [PubMed]
4. Mazzocchi, C.; Orsi, L.; Ferrazzi, G.; Corsi, S. The dimension of agricultural diversification: A spatial analysis of Italian municipalities. *Rural Sociol.* **2020**, *85*, 316–345. [CrossRef]
5. Messéan, A.; Viguier, L.; Paresys, L.; Aubertot, J.-N.; Canali, S.; Iannetta, P.; Justes, E.; Karley, A.; Keillor, B.; Kemper, L.; et al. Enabling crop diversification to support transitions towards more sustainable European agrifood systems. *Front. Agric. Sci. Eng.* **2021**, *8*, 474–480. [CrossRef]
6. Vanino, S.; Di Bene, C.; Piccini, C.; Fila, G.; Pennelli, B.; Zornoza, R.; Sanchez-Navarro, V.; Álvaro-Fuentes, J.; Hüppi, R.; Six, J.; et al. A comprehensive assessment of diversified cropping systems on agro-environmental sustainability in three Mediterranean long-term field experiments. *Eur. J. Agron.* **2022**, *140*, 126598. [CrossRef]
7. Chehade, L.A.; Angelini, L.G.; Tavarini, S. Genotype and seasonal variation affect yield and oil quality of safflower (*Carthamus tinctorius* L.) under Mediterranean conditions. *Agronomy* **2022**, *12*, 122. [CrossRef]
8. Hussain, M.I.; Lyra, D.-A.; Farooq, M.; Nikoloudakis, N.; Khalid, N. Salt and drought stresses in safflower: A review. *Agron. Sustain. Dev.* **2016**, *36*, 1–31. [CrossRef]
9. Von Cossel, M.; Lewandowski, I.; Elbersen, B.; Staritsky, I.; Van Eupen, M.; Iqbal, Y.; Mantel, S.; Scordia, D.; Testa, G.; Cosentino, S.L.; et al. Marginal agricultural land low-input systems for biomass production. *Energies* **2019**, *12*, 3123. [CrossRef]
10. Karg, S. New research on the cultural history of the useful plant *Linum usitatissimum* L. (Flax), a resource for food and textiles for 8000 Years. *Veg. Hist. Archaeobot.* **2012**, *21*, 79. [CrossRef]
11. Angelini, L.G.; Tavarini, S.; Antichi, D.; Foschi, L.; Mazzoncini, M. On-farm evaluation of seed yield and oil quality of linseed (*Linum usitatissimum* L.) in inland areas of Tuscany, central Italy. *Ital. J. Agron.* **2016**, *11*, 199–202. [CrossRef]
12. Campos, J.R.; Severino, P.; Ferreira, C.S.; Zielinska, A.; Santini, A.; Souto, S.B.; Souto, E.B. Linseed essential oil—Source of lipids as active ingredients for pharmaceuticals and nutraceuticals. *Curr. Med. Chem.* **2019**, *26*, 4537–4558. [CrossRef] [PubMed]
13. Mazzon, E.; Guigues, P.; Habas, J. Biobased structural epoxy foams derived from plant-oil: Formulation, manufacturing and characterization. *Ind. Crops Prod.* **2020**, *144*, 111994. [CrossRef]
14. Baroncini, E.A.; Yadav, S.K.; Palmese, G.R.; Stanzione, J.F. Recent advances in bio-based epoxy resins and bio-based epoxy agents. *J. Appl. Polym. Sci.* **2016**, *133*. [CrossRef]
15. Xie, Y.; Yan, Z.; Niu, Z.; Coulter, J.A.; Niu, J.; Zhang, J.; Wang, B.; Yan, B.; Zhao, W.; Wang, L. Yield, Oil content, and fatty acid profile of flax (*Linum usitatissimum* L.) as affected by phosphorus rate and seeding rate. *Ind. Crops Prod.* **2020**, *145*, 112087. [CrossRef]

16. Kajla, P.; Sharma, A.; Sood, D.R. Flaxseed—A potential functional food source. *J. Food Sci. Technol.* **2015**, *52*, 1857–1871. [CrossRef]
17. Vlaicu, P.A.; Panaite, T.D.; Turcu, R.P. Enriching laying hens eggs by feeding diets with different fatty acid composition and antioxidants. *Sci. Rep.* **2021**, *11*, 20707. [CrossRef]
18. Vargas-Bello-Pérez, E.; Darabighane, B.; Miccoli, F.E.; Gómez-Cortés, P.; Gonzalez-Ronquillo, M.; Mele, M. Effect of dietary vegetable sources rich in unsaturated fatty acids on milk production, composition, and cheese fatty acid profile in sheep: A meta-analysis. *Front. Vet. Sci.* **2021**, *8*, 641364. [CrossRef]
19. Migliore, G.; Rizzo, G.; Bonanno, A.; Dudinskaya, E.C.; Tóth, J.; Schifani, G. Functional food characteristics in organic food products—the perspectives of Italian consumers on organic eggs enriched with Omega-3 polyunsaturated fatty acids. *Org. Agric.* **2022**, *12*, 149–161. [CrossRef]
20. FAOSTAT. Available online: https://www.fao.org/faostat/en/#data/QCL (accessed on 30 September 2022).
21. Anastasiu, A.E.; Chira, N.A.; Banu, I.; Ionescu, N.; Stan, R.; Rosca, S.I. Oil productivity of seven Romanian linseed varieties as affected by weather conditions. *Ind. Crops Prod.* **2016**, *86*, 219–230. [CrossRef]
22. Green, A.G. Effect of temperature during seed maturation on the oil composition of low-linolenic genotypes of flax. *Crop Sci.* **1986**, *26*, 961–965. [CrossRef]
23. Cross, R.H.; McKay, S.A.B.; McHughen, A.G.; Bonham-Smith, P.C. Heat-stress effects on reproduction and seed set in Linum usitatissimum L. (Flax). *Plant Cell Environ.* **2003**, *26*, 1013–1020. [CrossRef]
24. Bremner, J.M.; Mulvaney, C.S. Nitrogen-total. In *Methods of Soil Analysis, Part 2: Chemical and Microbiological Properties*; Page, A.L., Miller, R.H., Keeney, D., Eds.; American Society of Agronomy, Inc.; Soil Science Society of America, Inc.: Madison, WI, USA, 1982; pp. 595–624.
25. Nelson, D.W.; Sommers, L.E. Total carbon. In *Methods of Soil Analysis: Part 2 Chemical and Microbiological Properties*, 2nd ed.; Page, A.L., Miller, R.H., Keeney, D.R., Eds.; American Society of Agronomy Inc.; Soil Science Society of America, Inc.: Madison, WI, USA, 1982; pp. 539–579.
26. Gardin, L.; Chiesi, M.; Fibbi, L.; Maselli, F. Mapping soil organic carbon in Tuscany through the statistical combination of ground observations with ancillary and remote sensing data. *Geoderma* **2021**, *404*, 115386. [CrossRef]
27. Vicente-Serrano, S.M.; Beguería, S.; López-Moreno, J.I. A multiscalar drought index sensitive to global warming: The Standardized Precipitation Evapotranspiration Index. *J. Clim.* **2010**, *23*, 1696–1718. [CrossRef]
28. Peña-Gallardo, M.; Martín Vicente-Serrano, S.; Domínguez-Castro, F.; Beguería, S. The impact of drought on the productivity of two rainfed crops in Spain. *Nat. Hazards Earth Syst. Sci.* **2019**, *19*, 1215–1234. [CrossRef]
29. Vergni, L.; Vinci, A.; Todisco, F. Effectiveness of the new standardized deficit distance index and other meteorological indices in the assessment of agricultural drought impacts in Central Italy. *J. Hydrol.* **2021**, *603*, 126986. [CrossRef]
30. Allen, R.G.; Pereira, L.S.; Raes, D.; Smith, M. *Crop Evapotranspiration—Guidelines for Computing Crop Water Requirements*; United Nations Food and Agriculture Organization, Irrigation and Drain: Rome, Italy, 1998.
31. Smith, J.M.; Froment, M.A. A growth stage key for winter linseed (*Linum usitatissimum*). *Ann. Appl. Biol.* **1998**, *133*, 297–306. [CrossRef]
32. *International Rules for Seed Testing*; ISTA: Wallisellen, Switzerland, 2004; ISBN 3-906549-38-0.
33. Tavarini, S.; Castagna, A.; Conte, G.; Foschi, L.; Sanmartin, C.; Incrocci, L.; Ranieri, A.; Serra, A.; Angelini, L.G. Evaluation of Chemical Composition of Two Linseed Varieties as Sources of Health-Beneficial Substances. *Molecules* **2019**, *24*, 3729. [CrossRef]
34. Yan, W.; Kang, M.S.; Ma, B.; Woods, S.; Cornelius, P.L. GGE biplot vs. AMMI analysis of genotype-by-environment data. *Crop Sci.* **2007**, *47*, 643–653. [CrossRef]
35. Čeh, B.; Štraus, S.; Hladnik, A.; Kušar, A. Impact of linseed variety, location and production year on seed yield, oil content and its composition. *Agronomy* **2020**, *10*, 1770. [CrossRef]
36. Kurtenbach, M.E.; Johnson, E.N.; Gulden, R.H.; Duguid, S.; Dyck, M.F.; Willenborg, C.J. Integrating cultural practices with herbicides augments weed management in flax. *Agron. J.* **2019**, *111*, 1904–1912. [CrossRef]
37. Dey, P.; Mahapatra, B.S.; Pramanick, B.; Pyne, S.; Pandit, P. Optimization of seed rate and nutrient management levels can reduce lodging damage and improve yield, quality and energetics of subtropical flax. *Biomass Bioenergy* **2022**, *157*, 106355. [CrossRef]
38. Gajardo, H.A.; Quian, R.; Soto-Cerda, B. Agronomic and quality assessment of linseed advanced breeding lines varying in seed mucilage content and their use for food and feed. *Crop Sci.* **2017**, *57*, 2979–2990. [CrossRef]
39. Hassan, F.U.; Leitch, M.H. Influence of seeding density on contents and uptake of N, P and K in linseed (*Linum usitatissimum* L.). *J. Agron. Crop Sci.* **2000**, *185*, 193–199. [CrossRef]
40. Benaragama, D.I.; Johnson, E.N.; Gulden, R.H.; Willenborg, C.J. Integrated agronomy for high yield and stable flax production in Canada. *Agron. J.* **2022**, *114*, 2230–2242. [CrossRef]
41. Casa, R.; Russell, G.; Lo Cascio, B.; Rossini, F. Environmental effects on linseed (Linum usitatissimum L.) yield and growth of flax at different stand densities. *Eur. J. Agron.* **1999**, *11*, 267–278. [CrossRef]
42. Kalinina, E.Y.; Lyakh, V.A. Combining ability according to the traits of stem branching and plant height in linseed lines. *Cytol. Genet.* **2011**, *45*, 293–297. [CrossRef]
43. Korres, N.E.; Froud-Williams, R.J. Effects of winter wheat cultivars and seed rate on the biological characteristics of naturally occurring weed flora. *Weed Res.* **2002**, *42*, 417–428. [CrossRef]
44. Sulas, L.; Re, G.A.; Sanna, F.; Bullitta, S.; Piluzza, G. Fatty acid composition and antioxidant capacity in linseed grown as forage in Mediterranean environment. *Ital. J. Agron.* **2019**, *14*, 50–58. [CrossRef]

45. Ercoli, L.; Arduini, I.; Mariotti, M.; Masoni, A. Post-anthesis dry matter and nitrogen dynamics in durum wheat as affected by nitrogen and temperature during grain filling. *Cereal Res. Commun.* **2010**, *38*, 294–303. [CrossRef]
46. Hocking, P.J.; Pinkerton, A. Response of growth and yield components of linseed to the onset or relief of nitrogen stress at several stages of crop development. *Field Crops Res.* **1991**, *27*, 83–102. [CrossRef]
47. Dordas, C.A. Variation of physiological determinants of yield in linseed in response to nitrogen fertilization. *Ind. Crops Prod.* **2010**, *31*, 455–465. [CrossRef]
48. Luhs, W.; Friedt, W. Major oil crops. In *Designer Oil Crops: Breeding, Processing and Biotechnology*; Murphy, D.J., Ed.; VCH Press: Weinheim, Germany, 1994; pp. 5–71.
49. Tavarini, S.; Angelini, L.G.; Casadei, N.; Spugnoli, P.; Lazzeri, L. Agronomical evaluation and chemical characterization of Linum usitatissimum L. as oilseed crop for bio-based products in two environments of Central and Nothern Italy. *Ital. J. Agron.* **2016**, *11*, 122–132. [CrossRef]
50. D'Antuono, L.F.; Rossini, F. Yield potential and ecophysiological traits of the Altamurano linseed (*Linum usitatissimum* L.), a landrace of southern Italy. *Genet. Resour. Crop Evol.* **2006**, *53*, 65–75. [CrossRef]
51. Diepenbrock, W.A.; Leon, J.; Clasen, K. Yielding ability and yield stability of linseed in Central Europe. *Agron. J.* **1995**, *87*, 84–88. [CrossRef]
52. Dordas, C.A. Nitrogen nutrition index and its relationship to N use efficiency in linseed. *Eur. J. Agron.* **2011**, *34*, 124–132. [CrossRef]
53. Klein, J.; Zikeli, S.; Claupen, W.; Gruber, S. Linseed (*Linum usitatissimum*) as an oil crop in organic farming: Abiotic impacts on seed ingredients and yield. *Org. Agric.* **2017**, *7*, 1–19. [CrossRef]
54. Berti, M.; Fischer, S.; Wilckens, R.; Hevia, F.; Johnson, B. Adaptation and genotype × environment interaction of flaxseed (*Linum usitatissimum* L.) genotypes in South Central Chile. *Chil. J. Agric. Res.* **2010**, *70*, 345–356.
55. Bhatty, R.S.; Cherdkiatgumchai, P. Compositional analysis of laboratory- prepared and commercial samples of linseed meal and of hull isolated from flax. *J. Am. Oil Chem. Soc.* **1990**, *67*, 79–84. [CrossRef]
56. Saastamoinen, M.; Pihlava, J.M.; Eurola, M.; Klemola, A.; Jauhiainen, L.; Hietaniemi, V. Yield, SDG lignan, cadmium, lead, oil and protein contents of linseed (*Linum usitatissimum* L.) cultivated in trials and at different farm conditions in the southwestern part of Finland. *Agric. Food Sci.* **2013**, *22*, 296–306. [CrossRef]
57. Fofana, B.; Cloutier, S.; Duguid, S.; Ching, J.; Rampitsch, C.G. Gene expression of stearoyl-ACP desaturase and Δ12 fatty acid desaturase 2 is modulated during seed development of flax (*Linum usitatissimum*). *Lipids* **2006**, *41*, 705–712. [CrossRef] [PubMed]
58. Ataii, E.; Mirlohi, A.; Sabzalian, M.R.; Goli, S.A.H.; Sadri, N.; Sharif-Moghaddam, N.; Gheysari, M. Genetic variability of seed yield and oil nutritional attributes in linseed dominated by biennial variation. *Crop Pasture Sci.* **2021**, *72*, 443–457. [CrossRef]

Article

Influence of Microalgae *Planktochlorella nurekis* Clones on Seed Germination

Małgorzata Karbarz [1,*], Magdalena Piziak [1], Janusz Żuczek [2] and Magdalena Duda [2]

[1] Department of Biology, University of Rzeszow, Pigonia 1, 35-310 Rzeszow, Poland
[2] Bioorganic Technologies sp. z o.o., Sielec 1A, 39-120 Sedziszow Malopolski, Poland
* Correspondence: mkarbarz@ur.edu.pl

Abstract: Microalgae are a rich source of plant hormones, vitamins, and other substances that can influence plant physiological metabolism, which in turn affects plant development, biotic and abiotic stress resistance, and yield. This study aimed at testing microalgae *Planktochlorella nurekis* clones obtained by co-treatment with colchicine and cytochalasin on four plant species to check their potential use as biostimulators in agriculture. The results are valuable for breeders, farmers, and microgreen producers. Eleven clone extracts in 1%, 5%, and 10% concentration were tested on four plant species: lettuce, wheat, broccoli, and radish. Germination and seedling characteristics (leaf and root length, fresh weight) were measured for each species. *P. nurekis* extracts show both a stimulating and inhibitory effect on tested plants, depending on the tested concentration, plant species, and algal clone tested. Co-treatment with colchicine and cytochalasin may be a good source of clones for potential use in agriculture as biostimulators and herbicides.

Keywords: microalgae; *Planktochlorella nurekis*; seed germination; sprouts

1. Introduction

Continuous population growth leads to an increase in the demand for food [1]. In addition, modern agriculture is increasingly looking for products of natural origin in order to reduce the use of chemicals [2]. The biotechnology of microalgae, which is used for the production of natural fertilizers, is helpful in this respect. Microalgae are increasingly attracting the attention of scientists around the world [3,4]. In Roman times, plant seedlings were wrapped with algae to improve their growth. Nowadays, algal extracts are regaining popularity, and a few products based on algal extracts are already commercially available [5].

Microalgae are single-celled, autotrophic and photosynthetic organisms. Their natural habitats are freshwater and saltwater. Microalgae can be divided into four groups: cynaobacteria, rhodophytes, chlorophytes, and chromophytes [6]. Microalgae have many uses, ranging from use as food additives for humans and animals to applications in the cosmetics industry. They are also used for the production of such molecules as, for example, dyes or fatty acids as well as biofuels [7].

Plant germination is initiated by the activation of gibberellins (GA). The activation of GA inhibits abscisic acid (ABA) activity as it is active during seed dormancy and plays a role in germination inhibition [8]. Apart from germination, GA are involved in a few developmental pathways in the plant, namely stress regulation and cell division, which affect plant growth and development [9]. Auxin (IAA) is involved in the cell cycle, growth, and development of plant parts, including pollen, embryo, leaves, and roots [10]. It also plays a role in seed dormancy [11]. Cytokinins boost cell division in the plant and also during seed germination. Together with auxins, they play a role in shoot and root elongation, respectively [12]. It is worth mentioning that the individual application of the phytohormones BAP, NAA, or GA3 did not improve the germination rate compared to the water baseline control [13]. Microalgae are a great source of phytohormones, and in

Citation: Karbarz, M.; Piziak, M.; Żuczek, J.; Duda, M. Influence of Microalgae *Planktochlorella nurekis* Clones on Seed Germination. *Agronomy* **2023**, *13*, 9. https://doi.org/10.3390/agronomy13010009

Academic Editors: Christian Frasconi, Marco Fontanelli and Daniele Antichi

Received: 18 November 2022
Revised: 14 December 2022
Accepted: 15 December 2022
Published: 20 December 2022

inner, natural composition with other substances, they can boost germination, which is important not only in crop plant cultivation but also in sprout and microgreen production. Nowadays, consumers are seeking food that provides health benefits, hence the increasing popularity of sprouts [14]. The results suggest that extracts from algae are suitable as an alternative to synthetic biostimulants used in the production of sprouts [15,16].

Biostimulants are defined as materials other than fertilizers that promote plant growth when used in small amounts [17]. Plant growth regulators (PGRs) can promote germination and plant development, speed up the production cycle, and increase production efficiency, which is significant when it comes to high-value crops, not only in field environments but especially in controlled environments [18]. Phytohormones are produced by plants as well as other organisms, including microalgae [19]. In recent years, the use of algae extracts as fertilizers in organic farming has been increasing [20–22]. Algae extracts contain a rich set of phytohormones, amino acids, fatty acids, and microelements responsible for controlling the growth and development of plants and increasing resistance to pathogens. Data confirming the positive effect of algae and algal extracts on the growth of vegetables, fruits, and other crops are available in the literature. Algae extracts are used for seed conditioning as well as soil or foliar application during the growing and flowering period. They stimulate seed germination, growth, and yields of various crops. Algae components such as macro- and micronutrients, amino acids, vitamins, cytokinins, auxins, and abscisic acid-like hormones (ABA) influence metabolism in treated plants, leading to increased growth and yield [3,20]. They also improve soil fertility as well as pest and disease control, and they contribute to the removal of toxins. Despite the analysis of various chemical components of algae extracts, their mechanism of action on plants is not fully understood. We are probably dealing with a synergistic effect of many substances, both known and unknown [23].

Planktochlorella nurekis Skaloud & Nemcová is a new species of microalgae reported in 2014. This species has mononuclear, spherical cells surrounded by a two-layer cell wall. *Planktochlorella nurekis* is characterized by an asexual mode of reproduction using autospores [7]. Recently, this species was characterized in terms of biochemistry, and 11 clones were obtained by co-treatment with colchicine and cytochalasin. Clones showed increased antioxidant activity, increased content of B vitamins, higher lipid levels, and antimicrobial activity [24–26]. This study investigated the influence of the 11 clones on the germination capacity of selected crop plants and the morphological features of the tested plant sprouts.

2. Materials and Methods

2.1. Materials

Research material, from the company Bioorganic Technologies Sp. z o.o. (Sielec, Poland), was the dry mass of 11 selected algae clones and was compared to the wild type (WT). A detailed description of the research material procured and its characteristics is presented in the publication by E. Szpyrka et al., 2020 [25].

Studies were conducted on selected crops:

- *Brassica oleracea* L.—cultivar: Cezar (W. Legutko, Przedsiębiorstwo Handlowo-Nasienne Sp. z o.o.)
- *Lactuca sativa* L.—cultivar: May King—(W. Legutko, Przedsiębiorstwo Handlowo-Nasienne Sp. z o.o.)
- *Triticum aestivum* L.—spring wheat, cultivar: Serenada, (Hodowla Roślin Strzelce)
- *Raphanis sativus* L.—cultivar: Mino Early (W. Legutko, Przdsiębiorstwo Handlowo-Nasienne Sp. z o.o.)

2.2. Preparation of Seaweed Extract

The dry mass of 12 strains of algae was ground with laboratory grinder WŻ-1S (Gadkiewicz Instruments) for 5 s, then 15 g of each were selected. The prepared samples were flooded with 150 mL distilled water, incubated 12 h, and constantly shaken at room temperature. The extracts obtained were diluted in distilled water to obtain test concentrations of 1%, 5%, and 10% (related to dry mass) (Figure 1) The control test was distilled water (0%).

Figure 1. Extracts obtained from clone number 8 (1%, 5%, and 10%).

2.3. Germination Test

The seeds of the test plant were sterilized in 4% sodium hypochlorite solution for 10 min and then rinsed three times in sterile water. Plant cultivation was carried out in Petri dishes (diameter 9 cm) lined with two discs of filter paper (medium quality filters, weight 80, ash content 0.1). Each pan was lined with ten seeds of the test plant. The studies were conducted in three biological repetitions for each concentration of the extract and control. The seeds were exposed to different concentrations of the algae strains tested by watering them with 5 mL of the appropriate concentration extract and 5 mL of distilled water (control). The plants were grown in a breeding room at $24 \pm 2\,^\circ\text{C}$ day/night (16/8) for 5 days. During this time, seed germination was observed, and the number of germinated plants was counted each day of the experiment. The length of the longest roots, the length of the leaves, and fresh mass were measured after 5 days (Table 1) [27]. The following parameters were determined from the results obtained:

- Germination energy (%GE) [28] after 3 days: %GE = (number of germinated seeds/total number of seeds) $\times$ 100
- Growth parameters—length of root and leaves (cm)
- Fresh plant weight (g).

Table 1. Plant seedlings measurements.

Observation	Method	Unit	Day of Measurement
Germination energy	%GE formula	%	at day 3
Length of root and leaves	Ruler	cm	at day 5
Fresh plant weight	Scale (Radwag PS 210.R2, precision 1 mg)	g	at day 5

2.4. Statistical Analysis

Mean values $\pm$ SD were calculated on the basis of at least three independent experiments. All data were analyzed for significant differences by analysis of variance—ANOVA. Statistical significance was evaluated using GraphPad Prism 8 using one-way ANOVA and Dunnett's test. A *p*-value < 0.05 was considered as a statistically significant.

3. Results

3.1. Germination Energy

Seed germination for all tested plants differed depending on the algae clone used and the concentration of extracts; there were also visible differences in seed germination in relation to the control sample. Differences in the germination energy of lettuce (*Lactuca sativa* L.), wheat (*Triticum aestivum* L.), and radish (*Raphanis sativus* L.) seeds treated with 1%, 5%, and 10% concentration of extracts reached statistical significance as compared to the control. In the case of broccoli (*Brassica oleracea* L.), statistically significant differences in germination energy appeared in relation to the control when the extract with 1% concentration was used.

The conducted broccoli seed germination test showed that treatment with 1% concentration of clone 5 and 11 extracts decreased the germination energy as compared to the control (Figure 2a, Table 2). The remaining clones did not significantly affect seed germination. The biggest differences between the action of the extracts are visible in the germination energy of lettuce seeds. Six algae clones decreased germination energy (Figure 2b, Table 2). Clones 1 and 9 had the greatest impact. For wheat, clone 8 decreased the seed germination rate (Figure 2c, Table 2). On the other hand, clone 7, used in radish seeds, increased the germination energy (Figure 2d, Table 2).

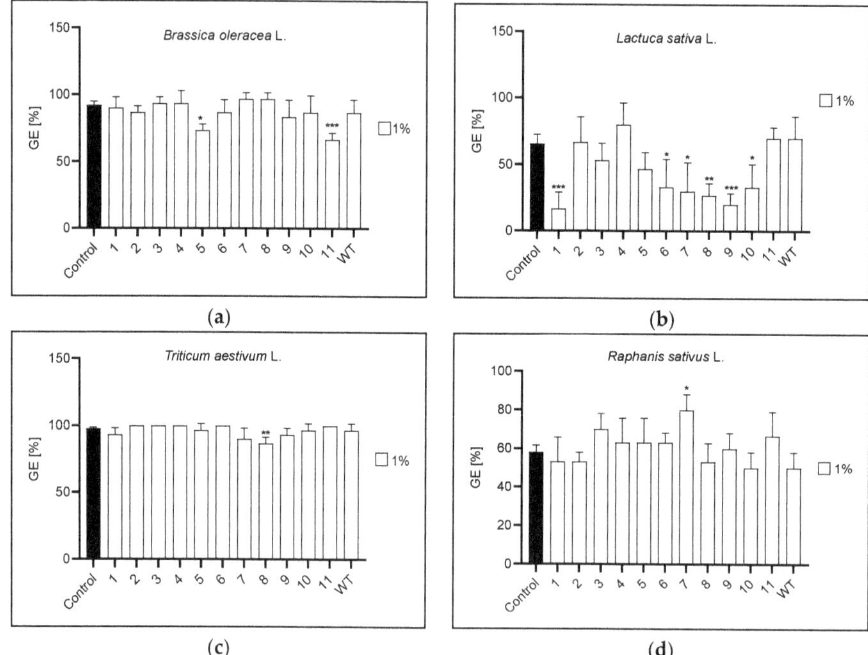

Figure 2. Percentage of germination of broccoli (**a**), lettuce (**b**), wheat (**c**), and radish (**d**) seeds treated with 1% concentration for all algae clones tested (1-11, W1) and control (Control). Values represent mean ± SD, n = 10 plants, * $p < 0.1$, ** $p < 0.01$, *** $p < 0.001$ compared to control (ANOVA and Dunnett's a posteriori test).

The changes in the germination energy of broccoli seeds resulting from the use of algae extracts at a concentration of 5% did not reach a statistically significant level (Figure 3a, Table 2). Therefore, it can be concluded that this concentration does not affect the germination energy of broccoli seeds. Germination of lettuce seeds differed depending on the algae clone used at 5% concentration (Figure 3b, Table 2). Clone 11 caused a slight increase in the germination energy, while clones 1, 2, 3, 6, 7, 8, 9, and WT caused a decrease in the germination energy of lettuce seeds. Clones 1, 3, 6, 7, 8, and 9 had the greatest impact on seed germination ($p < 0.0001$). In the case of wheat seeds, no significant changes in the germination energy were observed; only clone 10 caused a slight decrease in the germination energy (Figure 3c, Table 2). The application of 5% concentrations of algae clones to radish seeds caused a decrease in the germination energy for clones 2 and 3 but at a small level ($p < 0.1$) (Figure 3d, Table 2).

Table 2. Influence of different algal clones concentrations (conc.) on four tested parameters in four plant species. Each number represents the clone. In green are positive effects, and in red are negative effects. Neutral effects are represented by n. Significant effects are displayed in bold for *** $p < 0.001$ or **** $p < 0.0001$ or without bold for * $p < 0.1$ or ** $p < 0.01$.

Parameter	Conc.	Broccoli	Lettuce	Wheat	Radish
Germination energy	1%	5, **11**	1,6,7,8,**9**,10	8	7
	5%	n	1,2,3,6,7,8,9,11,WT	10	2,3
	10%	n	1,2,3,4,5,6,7,8,9,10,11,**WT**	4	2,3,4,7,10,WT
Root length	1%	1,3	1,3,4,5,6,7,	1,5,6,7,8,**WT**	9,WT
	5%	3,9,10	7,10,11,WT	1,3,5,7,8,9,**10**,WT	3,6,**11**
	10%	n	10	1,2,3,5,7,8,9,**10**,11,WT	7,11
Leaf length	1%	n	10	2,8,9,11	2,4,8,9,11,**WT**
	5%	11	1,WT	1,2,**4**,6,7,9,11	1,3,**11**
	10%	n	9	4,6,8	1,2,3,4,5,6,7,8,9,10,11,**WT**
Fresh weight	1%	n	6,8	**WT**	7,9,10,WT
	5%	10	1,3,5,6,7,8,9,11,**WT**	3	1,3,4,8,11
	10%	n	2,3,4,5,6,7,9,WT	4	1,4,5,7,9,10,11

For broccoli seeds, the use of a 10% concentration of the studied algae clones did not cause statistically significant changes (Figure 4a, Table 2). This concentration had a great influence on the germination energy of lettuce seeds. Each of the used algae clones caused a significant decrease in the germination energy at the level of $p < 0.0001$ (Figure 4b, Table 2). Clone 1 completely inhibited the sprouting of lettuce seeds. Germination of wheat seeds after applying a 10% concentration of clone 4 slightly decreased the germination energy (Figure 4c, Table 2). The remaining clones did not affect seed germination. In the case of radish, clones 2, 3, 4, 7, 10, and WT slightly decreased germination energy ($p < 0.1$) (Figure 4d, Table 2).

Analysis of the influence of algae clones at different concentrations on the germination of selected seeds showed that lettuce seeds were the most susceptible to their effects. With an increase in the concentration of the extract used, the germination energy of seeds decreased. The exception was clone 11, which, when used at 5% concentration, caused a slight increase in the germination energy. For all seeds, the use of algae extracts was observed to cause a decrease in germination energy, with a few exceptions (clone 7–1%— radish, clone 11–5%—lettuce), but these changes were small.

3.2. Root and Leaf Length

The length of the roots and leaves of the sprouts of the tested plants also depended on the algae clone used and the concentration of its extract. Differences in root and leaf length with respect to control for each plant reached statistical significance under the influence for all test concentrations.

3.2.1. Root Length

The root length of broccoli sprouts and lettuce increased under the influence of some algae clones at a concentration of 1%. For broccoli, clones 1 and 3 caused increased root growth (Figure 5a, Table 2); in the case of lettuce, it was clones 1, 3, 4, 5, 6, and 7 (Figure 5b, Table 2). Algae clones at a concentration of 1% caused slight wheat root growth. The remaining algae clones (1, 6, 7, 8, and WT) reduced the root length of wheat germ (Figure 5c,

Table 2). In the case of radish sprouts, clone 9 and WT slightly decreased root length (Figure 5d, Table 2).

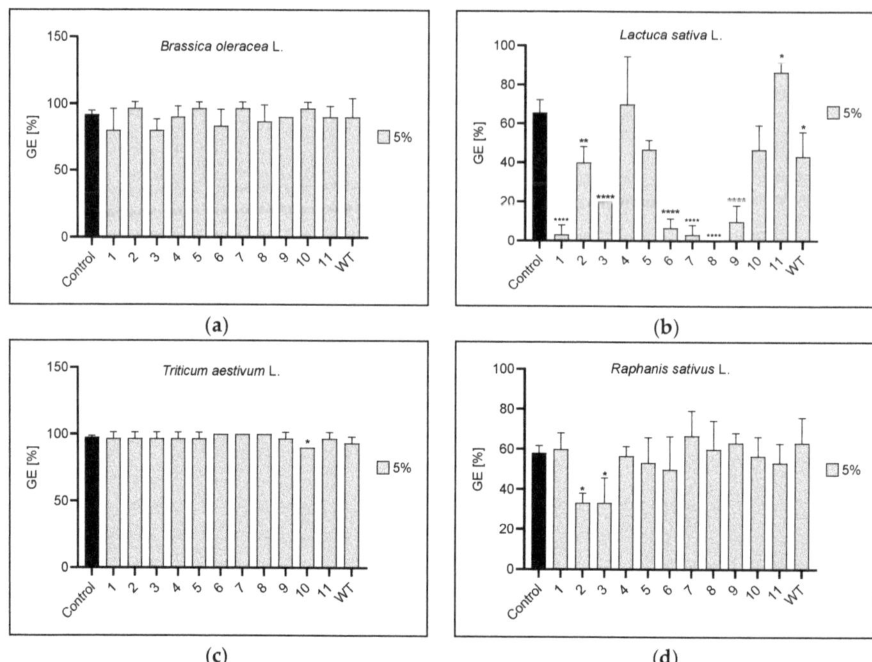

Figure 3. Percentage of germination of broccoli (**a**), lettuce (**b**), wheat (**c**), and radish (**d**) seeds treated with 5% concentration for all algae clones tested (1-11, WT) and control (Control). Values represent mean ± SD, n = 10 plants, * $p < 0.1$, ** $p < 0.01$, **** $p < 0.0001$ compared to control (ANOVA and Dunnett's a posteriori test).

The use of algae clones at a concentration of 5% had an effect on the root length for all the sprouts of the tested plants. In the case of broccoli sprouts, three algae clones (3, 9, and 10) slightly increased the root length (Figure 6a, Table 2). Similarly, in the case of lettuce sprouts, algae clone 10, 11, and WT increased the length of the roots, while clone 7 caused a decrease in the length of the roots of the lettuce sprouts (Figure 6b, Table 2). The greatest changes in root length were seen in wheat germ. Clones 3 and 5 caused an increase in root length, while clones 1, 7, 8, 9, 10, and WT caused a decrease in the root length of wheat germ (Figure 6c, Table 2). In the case of radish sprouts, three clones caused changes in root length (Figure 6d, Table 2). The third and eleventh clones reduced the length of the roots, while the sixth significantly contributed to the growth of the length of the radish sprouts.

The use of a 10% concentration of algae clones did not affect the root length of broccoli sprouts (Figure 7a, Table 2). Clone 10 caused an increase in the length of lettuce sprouts to a small extent (Figure 7b). In the case of wheat germ, clones 4 and 6 did not affect the size of the roots. Clone 3 and 5 caused a slight growth of sprout roots. The remaining algae clones reduced the root length of wheat germ (Figure 7c, Table 2). Two clones (7 and 11) influenced the length of radish sprouts, causing a reduction in the length of the sprouts' roots (Figure 7d, Table 2)

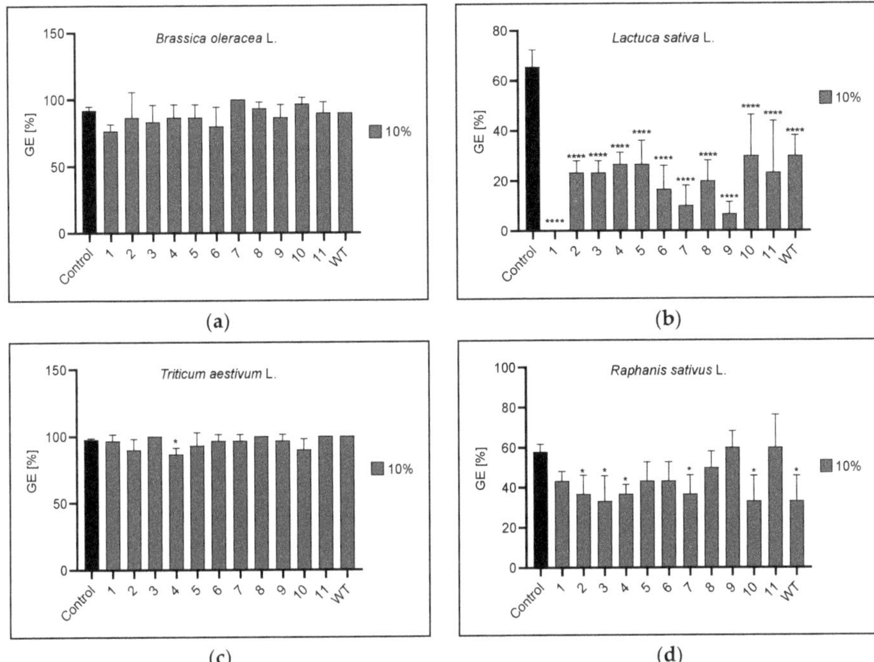

(a)   (b)

(c)   (d)

Figure 4. Percentage of germination of broccoli (**a**), lettuce (**b**), wheat (**c**), and radish (**d**) seeds treated with 10% concentration for all algae clones tested (1-11, WT) and control (Control). Values represent mean ± SD, n = 10 plants, * $p < 0.1$, **** $p < 0.0001$ compared to control (ANOVA and Dunnett's a posteriori test).

3.2.2. Leaf Length

Algae clones applied at 1% concentration did not change the length of broccoli sprouts (Figure 8a, Table 2). Clone 10 caused a decrease in leaf length of the lettuce sprouts (Figure 8b, Table 2). In the case of wheat germ, the leaf length increased when four algae clones were applied (Figure 8c, Table 2). Clone 2 had the greatest impact at $p < 0.001$. The greatest changes were seen in the size of the leaves of the radish sprouts. Six algae clones reduced the leaf length of the sprouts (Figure 8d, Table 2).

The use of algae extract at a concentration of 5% caused a slight increase in the length of broccoli sprouts treated with clone 11 (Figure 9a, Table 2). The leaf length of lettuce sprouts decreased under the influence of the first clone and WT (Figure 9b, Table 2). Seven algae clones had an effect on the leaf length of wheat germ (Figure 9c, Table 2). Clone 4 caused a significant reduction in leaf length, while the remaining ones elongated the leaves. Radish sprout leaf length was reduced under the influence of clones 1, 3, and 11 (Figure 9d, Table 2).

The leaf length of broccoli sprouts did not change under the influence of a concentration of 10% (Figure 10a, Table 2). In the case of lettuce sprouts, clone 9 caused a reduction in the length of sprout leaves (Figure 10b, Table 2). The length of wheat leaves was reduced under the influence of two strains (clones 4 and 6); meanwhile, clones 2 and 8 caused an increase in the length of wheat germ leaves (Figure 10c, Table 2). The concentration of 10% for all algae clones caused a significant reduction in the length of the leaves of radish sprouts (at the level of $p < 0.0001$) (Figure 10d, Table 2).

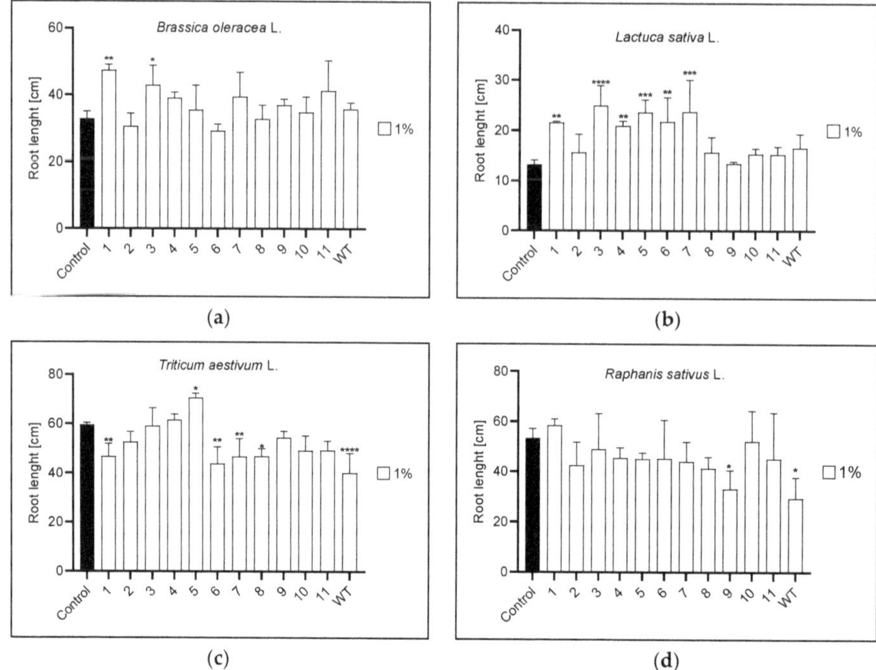

(a)　　　　　　　　　　　　　　　　(b)

(c)　　　　　　　　　　　　　　　　(d)

Figure 5. Root length of broccoli (**a**), lettuce (**b**), wheat (**c**), and radish (**d**) sprouts treated with 1% concentration for all algae clones tested (1-11, WT) and control (Control). Values represent mean $\pm$ SD, n = 10 plants, * $p < 0.1$, ** $p < 0.01$, *** $p < 0.001$, **** $p < 0.0001$ compared to control (ANOVA and Dunnett's a posteriori test).

3.3. Fresh Plant Weight

The influence of selected algae clones on the plants' fresh mass was also investigated. Apart from the concentration of 1% in the case of broccoli sprouts, the remaining concentrations reached statistically significant differences in the weight of the plants compared to the control.

The changes that occurred in the weight of sprouts of the tested plants under the influence of algae extracts at a concentration of 1% are presented in the graphs (Figure 11a–d, Table 2). The changes in the weight of broccoli sprouts were not significant (Figure 11a, Table 2). In the case of lettuce sprouts, the sixth clone caused a significant decrease in weight, while clone 8 was marked by a significant increase in the weight of the lettuce sprouts (Figure 11b, Table 2). The WT clone caused a decrease in the weight of wheat germ (Figure 11c, Table 2). The weight of radish sprouts slightly increased under the influence of clone 10, and clone 7 caused a slight decrease in weight, while clone 9 and WT had the greatest effect on the weight loss of radish sprouts (Figure 11d).

Under the influence of algae extracts at a concentration of 5%, the weight of the tested plants' sprouts either decreased or increased. Clone 10 caused an increase in the weight of broccoli sprouts (Figure 12a, Table 2). Nine out of twelve analyzed algae clones caused a decrease in the weight of lettuce sprouts to a greater or lesser extent (Figure 12b, Table 2). In the case of wheat germ, the weight increased as a result of the action of the third algae clone (Figure 12c, Table 2). The weight of radish sprouts decreased under the influence of clones 1, 3, 4, 8, and 11 (Figure 12d, Table 2).

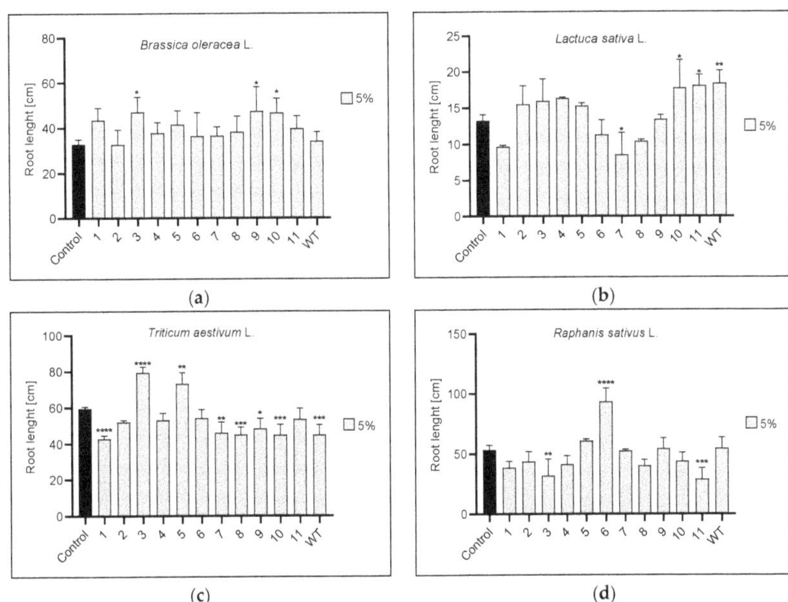

Figure 6. Root length of broccoli (**a**), lettuce (**b**), wheat (**c**), and radish (**d**) sprouts treated with 5% concentration for all algae clones tested (1-11, WT) and control (Control). Values represent mean ± SD, n = 10 plants, * $p < 0.1$, ** $p < 0.01$, *** $p < 0.001$, **** $p < 0.0001$ compared to control (ANOVA and Dunnett's a posteriori test).

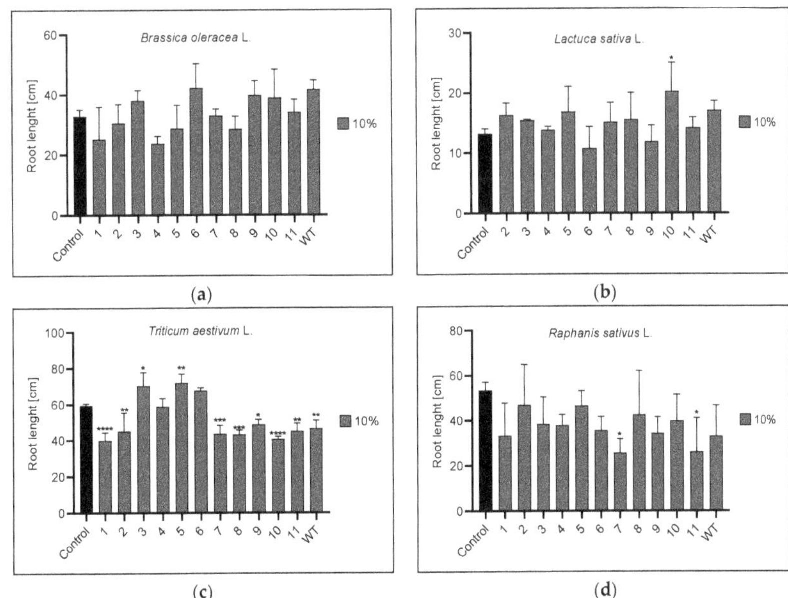

Figure 7. Root length of broccoli (**a**), lettuce (**b**), wheat (**c**), and radish (**d**) sprouts treated with 10% concentration for all tested algae clones (1-11, WT) and control (Control). Values represent mean ± SD, n = 10 plants, * $p < 0.1$, ** $p < 0.01$, *** $p < 0.001$, **** $p < 0.0001$ compared to control (ANOVA and Dunnett's a posteriori test).

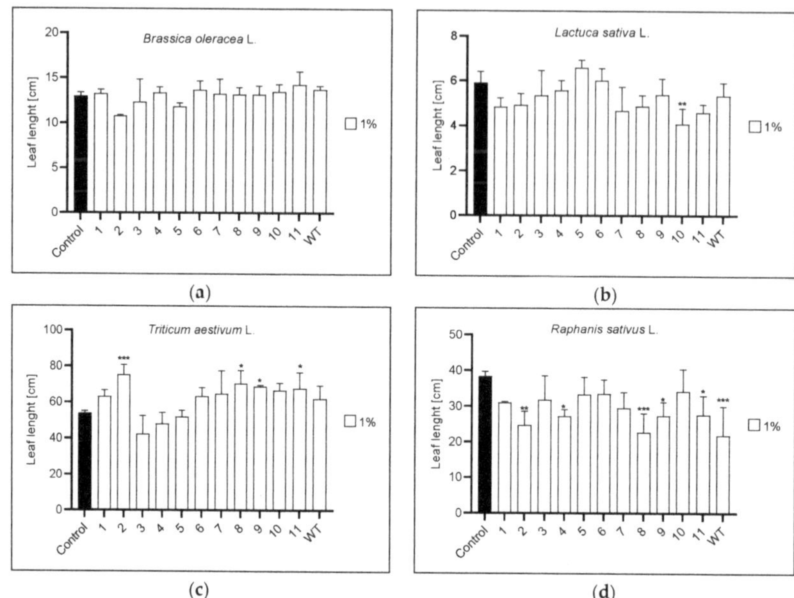

Figure 8. Leaf length of broccoli (**a**), lettuce (**b**), wheat (**c**), and radish (**d**) sprouts treated with 1% concentration for all algae clones tested (1-11, WT) and control (Control). Values represent mean ± SD, n = 10 plants, * $p < 0.1$, ** $p < 0.01$, *** $p < 0.001$ compared to control (ANOVA and Dunnett's a posteriori test).

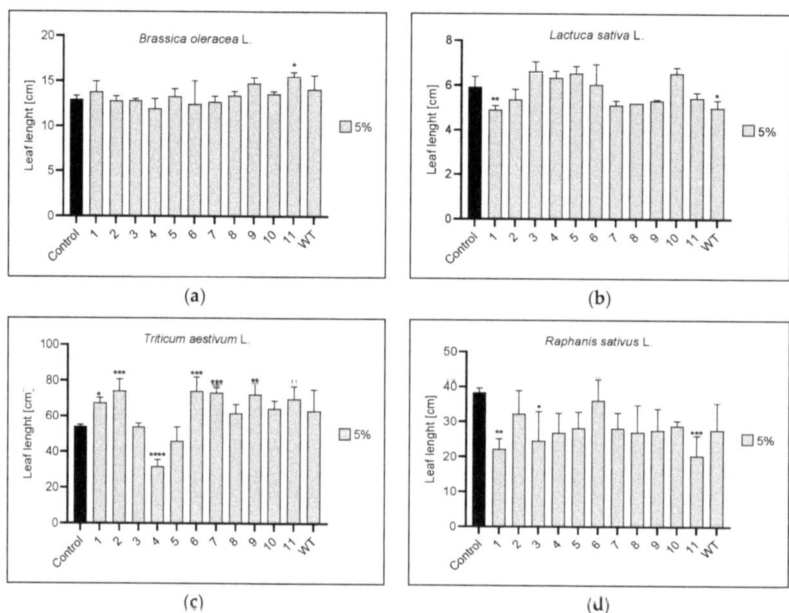

Figure 9. Leaf length of broccoli (**a**), lettuce (**b**), wheat (**c**), and radish (**d**) sprouts treated with 5% concentration for all algae clones tested (1-11, WT) and control (Control). Values represent mean ± SD, n = 10 plants, * $p < 0.1$, ** $p < 0.01$, *** $p < 0.001$, **** $p < 0.0001$ compared to control (ANOVA and Dunnett's a posteriori test).

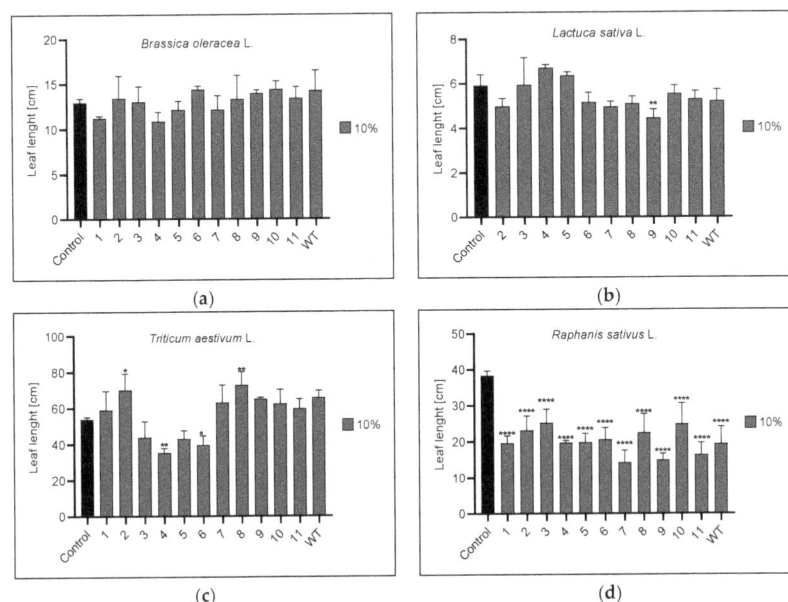

(a)

(b)

(c)

(d)

Figure 10. Leaf length of broccoli (**a**), lettuce (**b**), wheat (**c**), and radish (**d**) sprouts treated with 10% concentration for all algae clones tested (1-11, WT) and control (Control). Values represent mean ± SD, n = 10 plants, * $p < 0.1$, ** $p < 0.01$, **** $p < 0.0001$ compared to control (ANOVA and Dunnett's a posteriori test).

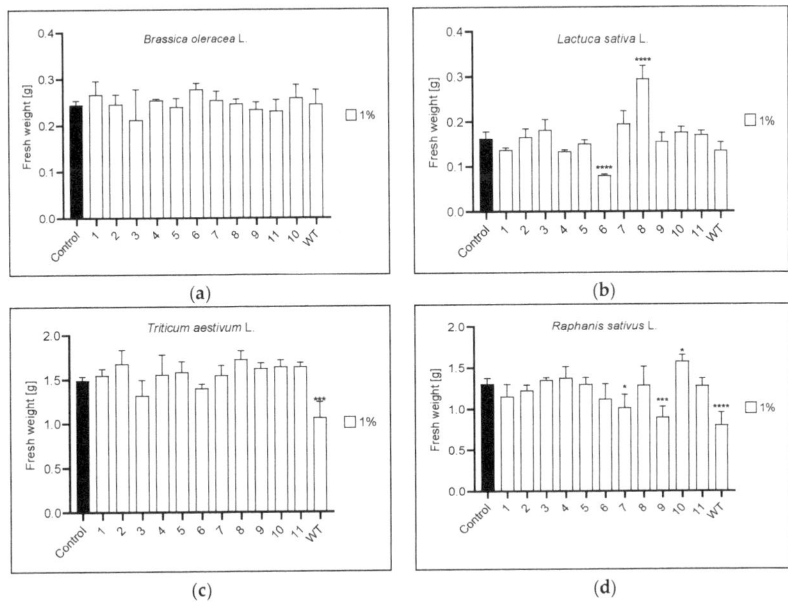

(a)

(b)

(c)

(d)

Figure 11. Fresh weight of broccoli (**a**), lettuce (**b**), wheat (**c**), and radish (**d**) sprouts treated with 1% concentration for all algae clones tested (1-11, WT) and control (Control). Values represent mean ± SD, n = 10 plants, * $p < 0.1$, *** $p < 0.001$, **** $p < 0.0001$ compared to control (ANOVA and Dunnett's a posteriori test).

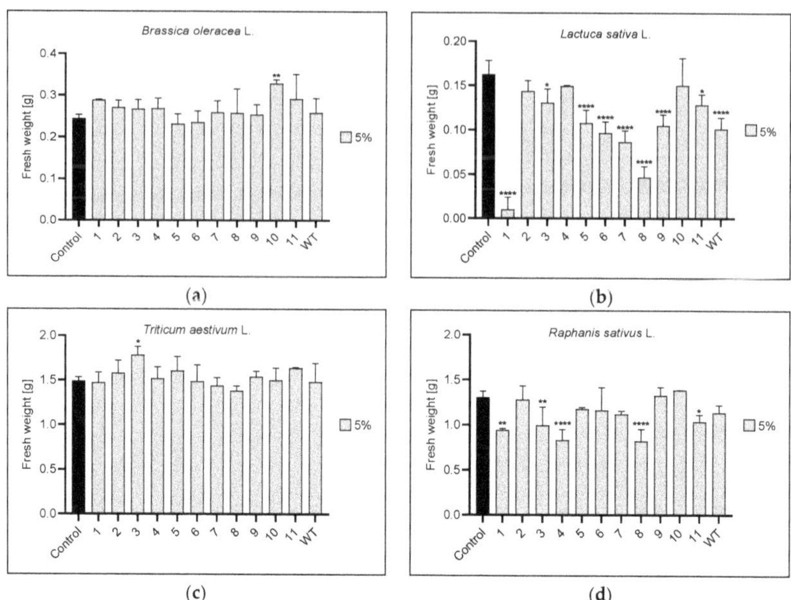

Figure 12. Fresh weight of broccoli (**a**), lettuce (**b**), wheat (**c**), and radish (**d**) sprouts treated with 5% concentration for all algae clones tested (1-11, WT) and control (Control). Values represent mean $\pm$ SD, n = 10 plants, * $p < 0.1$, ** $p < 0.01$, **** $p < 0.0001$ compared to control (ANOVA and Dunnett's a posteriori test).

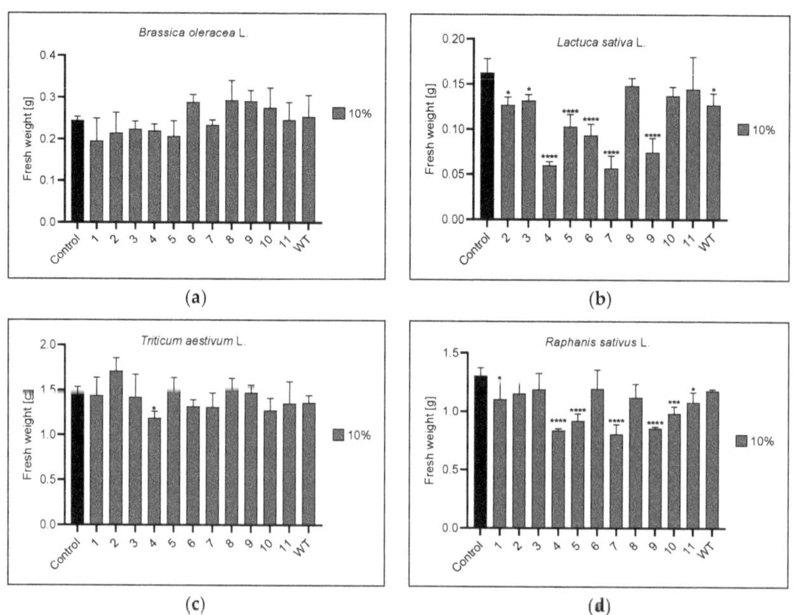

Figure 13. Fresh weight of broccoli (**a**), lettuce (**b**), wheat (**c**), and radish (**d**) sprouts treated with a 10% concentration for all algae clones tested (1-11, WT) and control (Control). Values represent mean $\pm$ SD, n = 10 plants, * $p < 0.1$, *** $p < 0.001$, **** $p < 0.0001$ compared to control (ANOVA and Dunnett's a posteriori test).

Algae extracts at a concentration of 10% did not cause changes in the weight of broccoli sprouts compared to the control (Figure 13a, Table 2). In the case of lettuce sprouts, the weight decreased under the influence of the action of eight algae clones. Clones 4, 5, 6, 7, and 9 had the greatest impact (Figure 13b, Table 2). The weight of wheat germ decreased slightly under the influence of clone 4 (Figure 13c, Table 2). The radish sprouts also showed a decrease in weight under the action of seven algae clones (Figure 13d, Table 2). Four clones (4, 5, 7 and 9) caused a significant decrease in the weight of radish sprouts.

4. Discussion

Algae extracts have been shown to contain substances that positively affect plant growth and development. Algae extracts are rich in vitamins, polysaccharides, amino acids, lipids, polyphenols, and plant hormones such as cytokinins, gibberellins, or auxins; they are also a source of macro- and microelements. The use of algae extracts results in improved seed germination, root and leaf development, as well as plant resistance to a variety of pathogens [3,28].

Many algal strains improve soil properties, crop growth, and fruit quality and have therefore been used as fertilizers [29]. However, no study has been performed to date on *Planktochlorella nurekis*. This is the first time that extracts obtained from the biomass of eleven *Planktochlorella nurekis* clones were tested on plants and resulted in improved biochemical characteristics compared to their unmodified counterpart. Environmental factors (light intensity, temperature) were constant during all treatments, so the differences between treatments are due to the different concentrations of extracts and different clones tested.

As a result of the analysis of the effect of algae extracts on the seedlings of four plant species (lettuce, wheat, radish and broccoli), it can be concluded that, in most cases, they have a negative or neutral effect on the species tested. In some cases, however, the effect was positive at low concentrations (1% and 5%). The overall germination efficiency was affected only in lettuce, for which a decrease was observed. Studies by Kumar et al. show that diluted algae extracts promote seed germination more than concentrated ones; growth of seedlings appears delayed at higher concentrations (15 and 20%) [30]. Our results show a similar trend in lettuce. In 1% concentration, three clones show an inhibitory effect on germination efficiency, in 5%—six, and in a concentration of 10%, it is exhibited by all clones and WT. Priming wheat seeds with low concentrations of seaweed extracts from *U. linza* and *C. officinalis* (20%) resulted in high percentages of germination [31]. We did not observe this effect in any of the tested clones or concentrations. Extracts seemed to be neutral to tested wheat seedlings, maybe because our concentrations were lower. Low concentrations (1%, 5%) of *P. nureckis* extracts stimulated root length positively in lettuce and negatively in wheat (1%, 5%, 10%). *U. linza* 20% extract effectively increased the total plant length of wheat. The highest shoot length, root length, and fresh weight were observed in 20% concentration of *C. tomentosum* liquid fertilizer in wheat plants [32]. Leaf length decreased in radish. For all of the clones, the strongest effect was observed with 10% extract. In the rest of the tested plant species, the effect was positive or negative only for single clones. According to Kasim et al. the imposition of salinity stress inhibits the shoot length of radish. However, presoaking of seeds in seaweed extracts significantly improved this parameter under salinity stress [33]. Of the different doses of microalgae, 5% resulted in the greatest improvement in growth components in broccoli (shoot FW and DW, length, and leaf area per plant) under drought stress [34]. In our study, 5% extract did not affect all tested parameters. The fresh weight of seedlings was negatively affected by extracts in concentrations of 5 and 10% in lettuce and radish. Data on the effect of *Chlorella vulgaris* extract on lettuce sprouting can be found in the literature [35]. The authors report an increase of about 60% in the fresh weight of lettuce seedlings (3- and 6-day-old seedlings). The cultivation conditions were very similar, but the extract was added to the soil, and the plants were not watered in Petri dishes, as in our case. We observed mostly a negative effect on the fresh weight of lettuce, but one clone caused fresh weight to increase at a similar level.

In low concentrations, algal extracts efficiently stimulate growth and plant yield, which can be explained by the corresponding low concentration of phytohormones. The concentration of phytohormones depends on botanical origin, time and place of collection, and method of biomass extraction [5]. Germination is mainly regulated by phytohormones: ABA, cytokinins, brassinosteroids, ethylene, and gibberellic acid (GA3) [36]. Phytohormones are believed to be very active components of algal extracts. They are derived from groups with different chemical structures and can influence plant metabolisms in many ways. Apart from phytohormones, algae contain a variety of other substances, such as polysaccharides, fatty acids, phenolic compounds, betaines, and a diverse assortment of other biologically active components. This great diversity of active compounds may affect plant growth synergistically or antagonistically. They can have a stimulatory or inhibitory effect on plants. Our results show both effects, depending on the tested extract, its concentration, and the tested plant. For example: root length was affected positively in lettuce and negatively in wheat. Fresh weight was negatively affected in lettuce and radish. Overall, the effect was usually neutral or negative, which may be explained by the excessive concentration of phytohormones. In higher doses, phytohormones can inhibit important processes occurring in plants [37]. Additionally, recent studies on *Lemna minor* have shown that algal methanol extract can act similarly to herbicide [38]. Rupawalla et al. quantified 15 known phytohormones in microalgae and did not obtain a clear correlation between phytohormone profile and seed germination [13]. They suggest that more different phytohormones and other substances are present in microalgae and that more complex interactions are taking place during plant germination, both among them as well as between them and the plant. In microalgae, a range of different substances, including auxins, cytokinins, giberellins, abscisic acid, ethyle, brassinosteroids, salicylates, signal peptiades, jasmonic acid, and polyamines, were found [39–41]. These can influence seed germination and seed development.

Microalgae extracts were tested on spinach seedlings, and Rupawalla et al. concluded that the most important biostimulators were located inside microalgae cells, and their availability for plant seedlings could be increased through cell disruption. [13] Their potential for use in drip irrigation was identified. These results also suggested that the insoluble fraction, which contains cell debris and precipitants, showed inhibitory effects on seedling development. In our case, all extract (not fractioned) was applied to seeds, and the neutral or inhibitory effects observed may be due to the presence of the insoluble fraction. It would be important to identify the soluble fractions of the 11 tested clones on plant seedlings.

Apart from their effect on growth and development parameters, algal extracts can also play a role in the stimulation of defense responses to stress factors such as high salinity, drought, low temperature, or pathogen attack [42], but sometimes the effect is not stimulating. Algal extracts enriched in amino acids demonstrated a biostimulating effect on plant physiology characterized by an increase of photosynthetic pigments and anthocyanin. No positive effect was observed in terms of plant growth or yield. Additionally, algal extracts enriched in amino acids increased the susceptibility of lettuce to powdery mildew [43].

Here we present the results of screening tests of 11 microalgae clones on four cultivated plant species. However, further studies need to be conducted to optimize the dose and type of fraction added to seedlings. Microalgae are a potential source of plant growth regulators, but whole extracts can have a neutral or inhibitory effect as other substances are also present in the extract. It would be interesting for future research to test different fractions of the clones on plants to choose the clone and fraction that would best improve germination, root and shoot development, and have the strongest potential application in agriculture. Our results are valuable not only for breeders and farmers but also for sprout and microgreen producers. Current studies show the positive effect of algal extracts on Amaranth sprouts. Aqueous seaweed extracts from *P. durvillei* and *U. lactuca* exhibited biostimulant activity when applied to amaranth seeds, improving the growth, yield, and functional quality of amaranth sprouts [16]. The plant species we tested are all mentioned in a recent paper by Ebert [44]. Broccoli is one of six commonly grown and consumed microgreen species [45]. Together with radish and four other species, it is a member of the *Brassicaceae* family, which dominated the global sprout market in 2019 [46]. Lettuce and wheat sprouts are less popular but nevertheless of interest to consumers.

Apart from having an overall negative or neutral effect, some single clones seem to increase activity in both stimulating and inhibitory ways, which proves that co-treatment with colchicine and cytochalasin is a good tool to obtain *P. nurekis* clones with great value to agriculture and a potential use as both biostimulators and herbicides. Interest in bio-based chemicals is increasing. There is a demand for plant biostimulators that increase crop yield and quality and reduce usage of synthetic agrochemicals in plant production. Further tests need to be performed to carefully establish suitable concentrations of extracts for each plant species. Such tests will undoubtedly become important given the ongoing growth of the global market for plant biostimulators, predicted to be worth more than USD 4000 million by 2023 [47]. Biostimulators based on microalgae are a great potential source of active substances, both for seedlings and older plants. They would have a great impact not only on plant yield and environmental protection but also on the quality of the sprouts, vegetables, and grains we consume. Microalgae offer a promising platform for biostimulant production, and tests such as ours have an important role to play providing proof of function and enabling further screening to be planned.

5. Conclusions

Our results show the importance of testing algal extracts in different concentrations on different plant species. The study presented here is preliminary. The response of the clones tested is highly variable in some species and regular in others. It would be interesting for further studies to establish why. Additionally, in some cases, WT is very similar to control, which suggests that co-treatment with colchicine and cytochalasin may be a good source of clones with properties that are desired in agriculture and are different than those of WT. Some clones can have a negative effect, while others have a positive effect. Our conclusion is that both effects are desired in agriculture, as both natural herbicides and biostimulators can have a great impact on the safe production of healthy food.

Author Contributions: Conceptualization, M.K.; Methodology, M.K. and M.P.; Investigation, M.K. and M.P.; Writing—Original Draft Preparation, M.P. and M.K.; Writing—Review and Editing, M.K.; Resources, J.Ż.; Data Curation, M.D. and J.Ż.; Supervision, J.Ż.; Project Administration, M.D. and J.Ż.; Funding Acquisition, J.Ż. All authors have read and agreed to the published version of the manuscript.

Funding: The study was supported by the project no. POIR.01.01.01-00-0405/17 from the National Centre for Research and Development. The project was co-financed by the European Union as part of the Smart Growth Operational Programme—Measure: R&D projects of enterprises; Sub-measure: Industrial research and development work implemented by enterprises.

Data Availability Statement: The data presented in this study are available on request from the corresponding author.

Conflicts of Interest: The authors declare no conflict of interest. The funders had no role in the design of the study; in the collection, analyses, or interpretation of data; in the writing of the manuscript; or in the decision to publish the results.

Sample Availability: Samples of the of algal biomasses are not available from the authors.

References

1. FAO. Global Report on Food Crises. Available online: https://www.fao.org/3/cb9997en/cb9997en.pdf (accessed on 11 December 2022).
2. FAO. The International Code of Conduct for the Sustainable Use and Management of Fertilizers. Available online: https://www.fao.org/3/ca5253en/ca5253en.pdf (accessed on 7 December 2022).
3. Khan, W.; Rayirath, U.P.; Subramanian, S.; Jithesh, M.N.; Rayorath, P.; Hodges, D.M.; Critchley, A.T.; Craigie, J.S.; Norrie, J.; Prithiviraj, B. Seaweed Extracts as Biostimulants of Plant Growth and Development. *J. Plant Growth Regul.* **2009**, *28*, 386–399. [CrossRef]

4. Tang, D.Y.Y.; Khoo, K.S.; Chew, K.W.; Tao, Y.; Ho, S.H.; Show, P.L. Potential utilization of bioproducts from microalgae for the quality enhancement of natural products. *Bioresour. Technol.* **2020**, *304*, 122997. [CrossRef] [PubMed]

5. Górka, B.; Korzeniowska, K.; Lipok, J.; Wieczorek, P.P. The Biomass of Algae and Algal Extracts in Agricultural Production | SpringerLink. In *Developments in Applied Phycology*; SpringerLink: Berlin/Heidelberg, Germany, 2018; pp. 103–114.

6. Mobin, S.; Alam, F. Some Promising Microalgal Species for Commercial Applications: A review. *Energy Procedia* **2017**, *110*, 510–517. [CrossRef]

7. Škaloud, P.; Němcová, Y.; Pytela, J.; Bogdanov, I.N.; Bock, C.; Pickinpaugh, H.S. Planktochlorella nurekis gen. et sp. nov. (Trebouxiophyceae, Chlorophyta), a novel coccoid green alga carrying significant biotechnological potential. *Fottea* **2014**, *14*, 53–62. [CrossRef]

8. Moore, T.C. *Biochemistry and Physiology of Plant Hormones*; Springer Science & Business Media: Berlin/Heidelberg, Germany, 2012.

9. vander Knaap, E.; Kim, J.H.; Kende, H. A Novel Gibberellin-Induced Gene from Rice and Its Potential Regulatory Role in Stem Growth. *Plant Physiol.* **2000**, *122*, 695–704. [CrossRef]

10. Popko, J.; Hansch, R.; Mendel, R.R.; Polle, A.; Teichmann, T. The role of abscisic acid and auxin in the response of poplar to abiotic stress-Popko-2010-Plant Biology-Wiley Online Library. *Plant Biol.* **2010**, *12*, 242–258. [CrossRef]

11. Liu, X.; Zhang, H.; Zhao, Y.; Feng, Z.; Li, Q.; Yang, H.-Q.; Luan, S.; Li, J.; He, Z.-H. Auxin controls seed dormancy through stimulation of abscisic acid signaling by inducing ARF-mediated ABI3 activation in Arabidopsis. *Proc. Natl. Acad. Sci. USA* **2013**, *110*, 15485–15490. [CrossRef]

12. Hussain, S.; Nanda, S.; Zhang, J.; Rehmani, M.I.A.; Suleman, M.; Li, G.; Hou, H. Auxin and Cytokinin Interplay during Leaf Morphogenesis and Phyllotaxy. *Plants* **2021**, *10*, 1732. [CrossRef]

13. Rupawalla, Z.; Shaw, L.; Ross, I.L.; Schmidt, S.; Hankamer, B.; Wolf, J. Germination screen for microalgae-generated plant growth biostimulants. *Algal Res.* **2022**, *66*, 102784. [CrossRef]

14. Rouphael, Y.; Colla, G.; Pascale, S.D. Sprouts, Microgreens and Edible Flowers as Novel Functional Foods. *Agronomy* **2021**, *11*, 2568. [CrossRef]

15. Liu, H.; Kang, Y.; Zhao, X.; Liu, Y.; Zhang, X.; Zhang, S. Effects of elicitation on bioactive compounds and biological activities of sprouts. *J. Funct. Foods* **2019**, *53*, 136–145. [CrossRef]

16. Osuna-Ruíz, I.; Ledezma, A.K.D.; Martínez-Montaño, E.; Salazar-Leyva, J.A.; Tirado, V.A.R.; García, I.B. Enhancement of in-vitro antioxidant properties and growth of amaranth seed sprouts treated with seaweed extracts. *J. Appl. Phycol.* **2022**, 1–11. [CrossRef]

17. Zhang, X.; Schmidt, R.E. The impact of growth regulators on the α-tocopherol status in water-stressed Poa pratensis. *Int. Turfgrass Soc. Res. J.* **1997**, *8*, 1364–1371.

18. González-Pérez, B.K.; Rivas-Castillo, A.M.; Valdez-Calderón, A.; Gayosso-Morales, M.A. Microalgae as biostimulants: A new approach in agriculture. *World J. Microbiol. Biotechnol.* **2021**, *38*, 1–12. [CrossRef] [PubMed]

19. Wang, C.; Qi, M.; Guo, J.; Zhou, C.; Yan, X.; Ruan, R.; Cheng, P. The Active Phytohormone in Microalgae: The Characteristics, Efficient Detection, and Their Adversity Resistance Applications. *Molecules* **2021**, *27*, 46. [CrossRef]

20. Michalak, I.; Chojnacka, K.; Saeid, A. Plant Growth Biostimulants, Dietary Feed Supplements and Cosmetics Formulated with Supercritical CO2 Algal Extracts. *Molecules* **2017**, *22*, 66. [CrossRef]

21. Pan, S.; Jeevanandam, J.; Danquah, M.K. Benefits of Algal Extracts in Sustainable Agriculture. In *Grand Challenges in Algae Biotechnology*; Hallmann, A., Rampelotto, P., Eds.; SpringerLink: Berlin/Heidelberg, Germany, 2020; pp. 501–534.

22. Das, P.; Khan, S.; Chaudhary, A.K.; AbdulQuadir, M.; Thaher, M.I.; Al-Jabri, H. Potential Applications of Algae-Based Bio-fertilizer. In *Biofertilizers for Sustainable Agriculture and Environment. Soil Biology*; Giri, B., Prasad, R., Wu, Q.S., Varma, A., Eds.; SpringerLink: Berlin/Heidelberg, Germany, 2019.

23. Fornes, F.; Sánchez-Perales, M.; Guardiola, J.L. Effect of a Seaweed Extract on the Productivity of 'de Nules' Clementine Mandarin and Navelina Orange. *Botanica Marina* **2002**, *45*, 486–489. [CrossRef]

24. Adamczyk-Grochala, J.; Wnuk, M.; Duda, M.; Zuczek, J.; Lewinska, A. Treatment with Modified Extracts of the Microalga Planktochlorella nurekis Attenuates the Development of Stress-Induced Senescence in Human Skin Cells. *Nutrients* **2020**, *12*, 1005. [CrossRef]

25. Szpyrka, E.; Broda, D.; Oklejewicz, B.; Podbielska, M.; Slowik Borowiec, M.; Jagusztyn, B.; Chrzanowski, G.; Kus-Liskiewicz, M.; Duda, M.; Zuczek, J.; et al. A Non-Vector Approach to Increase Lipid Levels in the Microalga Planktochlorella nurekis. *Molecules* **2020**, *25*, 270. [CrossRef]

26. Potocki, L.; Oklejewicz, B.; Kuna, E.; Szpyrka, E.; Duda, M.; Zuczek, J. Application of Green Algal Planktochlorella nurekis Biomasses to Modulate Growth of Selected Microbial Species. *Molecules* **2021**, *26*, 4038. [CrossRef]

27. Czyczyło-Mysza, I.; Marcińska, I.; Skrzypek, E.; Cyganek, K.; Juzoń, K.; Karbarz, M. QTL mapping for germination of seeds obtained from previous wheat generation under drought. *Cent. Eur. J. Biol.* **2014**, *9*, 374–382. [CrossRef]

28. Hernández-Herrera, R.M.; Santacruz-Ruvalcaba, F.; Ruiz-López, M.A.; Norrie, J.; Hernández-Carmona, G. Effect of liquid seaweed extracts on growth of tomato seedlings (*Solanum lycopersicum* L.). *J. Appl. Phycol.* **2013**, *26*, 619–628. [CrossRef]

29. Lu, Q.; Yu, X. From manure to high-value fertilizer: The employment of microalgae as a nutrient carrier for sustainable agriculture. *Algal Res.* **2022**, *67*, 102855. [CrossRef]

30. Kumar, N.A.; Vanlalzarzova, B.; Sridhar, S.; Baluswami, M. Effect of liquid seaweed fertilizer of Sargassum wightii grev. on the growth and biochemical content of green gram (*Vigna radiata* (L.) R. wilczek). *Recent Res. Sci. Technol.* **2015**, *4*, 40–45.

31. Hamouda, M.M.; Saad-Allah, K.M.; Gad, D. Potential of Seaweed Extract on Growth, Physiological, Cytological and Biochemical Parameters of Wheat (Triticum aestivum L.) Seedlings. *J. Soil Sci. Plant Nutr.* **2022**, *22*, 1818–1831. [CrossRef]
32. Mohy El-Din, S.M. Utilization of seaweed extracts as bio-fertilizers to stimulate the growth of wheat seedlings. *Egypt. J. Exp. Biol.* **2015**, *11*, 31–39.
33. Kasim, W.A.E.A.; Saad-Allah, K.M.; Hamouda, M. Seed priming with extracts of two seaweeds alleviates the physiological and molecular impacts of salinity stress on radish (*Raphanus sativus*). *Int. J. Agric. Biol.* **2016**, *18*, 653–660. [CrossRef]
34. Kusvuran, S. Microalgae (Chlorella vulgaris Beijerinck) alleviates drought stress of broccoli plants by improving nutrient uptake, secondary metabolites, and antioxidative defense system. *Hortic. Plant J.* **2021**, *7*, 221–231. [CrossRef]
35. Faheed, F.A.; Fattah, Z.A. Effect of Chlorella vulgaris as Bio-fertilizer on Growth Parameters and Metabolic Aspects of Lettuce Plant. *Plant J. Agric. Soc. Sci.* **2008**, *4*, 165–169. [CrossRef]
36. Finkelstein, R.R.; Gampala, S.S.L.; Rock, C.D. Abscisic Acid Signaling in Seeds and Seedlings. *Plant Cell* **2002**, *14*. [CrossRef]
37. Asif, R.; Yasmin, R.; Mustafa, M.; Ambreen, A.; Mazhar, M.; Rehman, A.; Umbreen, S.; Ahmad, M.; Asif, R.; Yasmin, R.; et al. Chapter 7: Phytohormones as Plant Growth Regulators and Safe Protectors against Biotic and Abiotic Stress. In *Plant Hormones: Recent Advances, New Perspectives and Applications*; Hano, C., Ed.; IntechOpen: London, UK, 2022. [CrossRef]
38. Asimakis, E.; Shehata, A.A.; Eisenreich, W.; Acheuk, F.; Lasram, S.; Basiouni, S.; Emekci, M.; Ntougias, S.; Taner, G.; May-Simera, H.; et al. Algae and Their Metabolites as Potential Bio-Pesticides. *Microorganisms* **2022**, *10*, 307. [CrossRef]
39. Han, X.; Zeng, H.; Bartocci, P.; Fantozzi, F.; Yan, Y. Phytohormones and Effects on Growth and Metabolites of Microalgae: A Review. *Fermentation* **2018**, *4*, 25. [CrossRef]
40. Kapoore, R.V.; Wood, E.E.; Llewellyn, C.A. Algae biostimulants: A critical look at microalgal biostimulants for sustainable agricultural practices. *Biotechnol. Adv.* **2021**, *49*, 107754. [CrossRef]
41. Du, K.; Tao, H.; Wen, X.; Geng, Y.; Li, Y. Enhanced growth and lipid production of Chlorella pyrenoidosa by plant growth regulator GA3. *Fresenius Environ. Bull.* **2015**, *24*, 3414–3419.
42. Tarakhovskaya, E.R.; Maslov, Y.I.; Shishova, M.F. Phytohormones in algae. *Russ. J. Plant Physiol.* **2007**, *54*, 163–170. [CrossRef]
43. Rover, S.; de Freitas, M.B.; Barcelos-Oliveira, J.L.; Stadnik, M.J. An algal extract enriched with amino acids increases the content of leaf pigments but also the susceptibility to the powdery mildew of lettuce. *Phytoparasitica* **2022**, 1–12. [CrossRef]
44. Ebert, A.W. Sprouts and Microgreens—Novel Food Sources for Healthy Diets. *Plants* **2022**, *11*, 571. [CrossRef]
45. Michell, K.A.; Isweiri, H.; Newman, S.E.; Bunning, M.; Bellows, L.L.; Dinges, M.M.; Grabos, L.E.; Rao, S.; Foster, M.T.; Heuberger, A.L.; et al. Microgreens: Consumer sensory perception and acceptance of an emerging functional food crop. *J. Food Sci.* **2020**, *85*, 926–935. [CrossRef] [PubMed]
46. Microgreens Market by Type (Broccoli, Cabbage, Cauliflower, Arugula, Peas, Basil, Radish, Cress, Others), by Farming (Indoor vertical farming, Commercial greenhouse, Others), by End User (Retail, Food Service, Others): Global Opportunity Analysis and Industry Forecast, 2019–2028. Available online: https://www.alliedmarketresearch.com/microgreens-market-A08733 (accessed on 10 December 2022).
47. Caracciolo, F.; El-Nakhel, C.; Raimondo, M.; Kyriacou, M.C.; Cembalo, L.; Pascale, S.D.; Rouphael, Y. Sensory Attributes and Consumer Acceptability of 12 Microgreens Species. *Agronomy* **2020**, *10*, 1043. [CrossRef]

Article

Complete Mitochondrial Genome Sequence, Characteristics, and Phylogenetic Analysis of *Oenanthe javanica*

Xiaoyan Li [1,†], Qiuju Han [1,†], Mengyao Li [1], Qing Luo [1], Shunhua Zhu [2], Yangxia Zheng [1,*] and Guofei Tan [2,*]

1   College of Horticulture, Sichuan Agricultural University, Chengdu 611130, China; lxy2324804342@163.com (X.L.); hanqiuju0816@163.com (Q.H.); limy@sicau.edu.cn (M.L.); lqing985@foxmail.com (Q.L.)
2   Institute of Horticulture, Guizhou Academy of Agricultural Sciences, Guiyang 550006, China; zsh2801@163.com
*   Correspondence: zhengyx13520@sicau.edu.cn (Y.Z.); tgfei@foxmail.com (G.T.)
†   These authors contributed equally to this work.

Abstract: The plant mitochondria play a crucial role in various cellular energy synthesis and conversion processes and are essential for plant growth. Watercress (*Oenanthe javanica*) is a fast-growing vegetable with strong adaptability and wide cultivation range, and it possesses high nutritional value. In our study, we assembled the *O. javanica* mitochondrial genome using the Illumina and Nanopore sequencing platforms. The results revealed that the mitochondrial genome map of watercress has a circular structure of 384,074 bp, containing 28 tRNA genes, 3 rRNA genes, and 34 protein-coding genes. A total of 87 SSR (simple sequence repeat) loci were detected, with 99% composed of palindrome repeats and forward repeats, while no complementary repeats were identified. Codon preference analysis indicated that watercress prefers to use codons encoding leucine, isoleucine, and serine with a preference for A/U-ending codons. Phylogenetic analysis showed that watercress is closely related to species of *Bupleurum*, *Apium*, *Angelica*, and *Daucus*, with the closest evolutionary relationship observed with *Saposhnikovia divaricata* and *Apium graveolens*. This study provides a valuable resource for the study of the evolution and molecular breeding of watercress.

Keywords: watercress; mitochondrial genome; repetitive sequence; phylogenetic analysis

Citation: Li, X.; Han, Q.; Li, M.; Luo, Q.; Zhu, S.; Zheng, Y.; Tan, G. Complete Mitochondrial Genome Sequence, Characteristics, and Phylogenetic Analysis of *Oenanthe javanica*. *Agronomy* **2023**, *13*, 2103. https://doi.org/10.3390/agronomy13082103

Academic Editors: David Edwards, Daniele Antichi, Christian Frasconi and Marco Fontanelli

Received: 3 July 2023
Revised: 1 August 2023
Accepted: 7 August 2023
Published: 10 August 2023

1. Introduction

Plant genomes consist of nuclear, chloroplast, and mitochondrial components, with each exhibiting different genetic and evolutionary patterns [1]. Among them, mitochondria are essential organelles in most eukaryotes and play a crucial role in cellular energy production [2]. Plant mitochondrial genomes, known as the largest organelle genomes, exhibit low synonymous substitution rates, relatively frequent rearrangements, and high levels of inversions and recombinations [3]. Horizontal transfer of mitochondrial genes was observed in early stages of plant evolution [4], and most transferred mitochondrial genes are ribosomal-protein-coding genes [5]. The transferred genetic information includes not only complete functional genes but also noncoding sequences and gene fragments [6]. Repeat sequences play a vital role in maintaining the structure of noncoding regions in the plant mitochondrial genome through their involvement in genome rearrangements [7], inversions, insertions, and deletions [8,9]. Plant mitochondrial genomes are known to vary significantly between species, and scattered repeat sequences can give rise to multipartite structures within a species [10]. In addition, plant mitochondria have lower mutation rates than their nuclear or chloroplast genomes with high variation in genome size, gene order, and intergenic sequences but high conservation in protein-coding sequences [7]. Hence, the highly conserved features of the mitochondrial genome can provide valuable information for the study of plant evolution and phylogenetics [11].

Plant mitochondrial genomes have generally received less attention than plastid or nuclear genomes. This is mainly due to the larger size of plant mitochondrial genomes (typically around 400 kb) compared to animal mitochondrial genomes (usually 10–39 kb) [12] as well as the substantial differences in genome size, structure, and rearrangement patterns, leading to extensive variation in structural dynamics and repetitive DNA content [13]. Consequently, the complexity of the structure presents a challenging task for mitochondrial genome sequencing [14]. However, with the rapid development of sequencing platforms and assembly programs, a growing number of mitochondrial genomes have been completely sequenced [2], providing a rich source of data for molecular breeding of plant genetic resources, such as *Coptis chinensis* [15], *Physcomitrella patens* [16], *Nymphaea colorata* [17], *Piper betle* [18], *Actinidia chinensis* [19], *Apium graveolens* [20], etc. Against this research background, exploring the intrinsic information of mitochondrial genomes at a broader taxonomic scale will provide a theoretical basis for the study of plant genetics, evolution, and genetic resources.

Watercress (*Oenanthe javanica* (Blume) DC.) is a perennial aquatic herb of the *Oenanthe* genus in the Apiaceae family [21], cultivated in tropical and temperate regions of Asia for thousands of years [22]. Watercress is highly nutritious, containing fiber, vitamins, minerals, flavonoids, and other substances in its edible stems and leaves [23,24]. It aids digestion and has medicinal properties such as anti-inflammatory, antioxidant, and antihypertensive effects, making it a vegetable with both culinary and medicinal value [25]. In addition, watercress is appreciated for its unique texture and flavor, which makes it popular with people [26]. With its fast growth, wide geographical distribution, high adaptability, and extensive growing range, watercress is a desirable off-season vegetable and a favorite with consumers. Currently, research on watercress has mainly focused on its nutritional composition and cultivation physiology, and the lack of comprehensive genomic data has limited in-depth studies on watercress. Although some phylogenetic studies have reported the chloroplast genome sequence, transcriptome, and miRNA data of watercress [27,28], a complete mitochondrial genome sequence is still lacking. Therefore, analyzing the sequence structure characteristics of watercress mitochondria will help to enrich its genomic data and elucidate the phylogenetic relationships between watercress and other species in the Apiaceae family.

In this study, we assembled the complete mitochondrial genome of watercress and analyzed its gene content, repetitive sequences, codon preference, RNA editing sites, Pi nucleotide diversity, and phylogenetic relationships. In addition, we studied the structural analysis of the mitochondria and the gene transfer between the chloroplast and the mitochondrial genome. These findings will improve our understanding of the organellar genome and genetic diversity of watercress and provide opportunities for further genomic breeding research in watercress.

2. Materials and Methods

2.1. Plant Materials

Watercress samples were collected from Xiongshi Village, Qixingguan District, Bijie City, Guizhou Province. The plants were transplanted to the greenhouse of the Institute of Horticulture Research, Guizhou Academy of Agricultural Sciences, in January 2022. During the stage of vigorous leaf growth in April 2022, the leaves were harvested and rapidly frozen in liquid nitrogen and stored at −80 °C for mitochondrial genome sequencing. High-quality total genomic DNA was extracted from fresh leaves using a plant DNA extraction kit (TransGene, Beijing, China), and the DNA quality and concentration were checked using 1% agarose gel electrophoresis and NanoDrop ND 2000 (ThermoFischer, Waltham, MA, USA).

2.2. Mitogenome Assembly and Annotation

To obtain the full-length mitochondrial genome with high accuracy, short-read and long-read sequencing technologies were combined in this study. The short-read sequencing platform was Illumina Novaseq 6000 (Illumina, San Diego, CA, USA), and the paired-end sequencing (PE) read length was 150 bp. The fastp software (v0.20.0, https://github.com/OpenGene/fastp, accessed on 15 June 2022) was used to filter the original data and obtain high-quality reads [29]. The long-read sequencing platform was Nanopore PromethION (Nanopore, Oxford, UK), and the sequencing data were filtered by filtlong (v0.2.1) software. Plant mitochondrial genes (CDS and rRNA) are highly conserved. Taking advantage of this characteristic, the alignment software Minimap2 (v2.1) was used to align the raw long-read sequencing data with the reference gene sequences (plant mitochondrial core genes, https://github.com/xul962464/plant_mt_ref_gene, accessed on 15 June 2022) in order to obtain all the long-read sequencing data of the mitochondrial genome. Then, the assembly software canu [30] was used to correct the long-read sequencing data obtained, and bowtie2 (v2.3.5.1) was used to align the short-read sequencing data to the corrected sequence. Then, the default parameter Unicycler (v0.4.8) was used to compare the above short-read sequencing data and the corrected long-read sequencing data for concatenation. Finally, the complete mitochondrial genome of watercress was submitted to the NCBI database under the accession number OR209169.

2.3. Mitochondrial Gene Annotation and Analysis

Protein-coding genes and rRNA genes were annotated by comparing them with published plant mitochondrial sequences and further adjusted based on closely related species. Related species included *Apium graveolens* (MZ328722.1), *Apium leptophyllum* (MZ328723.1), *Ferula sinkiangensis* (OK585063.1), *Bupleurum chinense* (OK166971.1), *Saposhnikovia divaricate* (MZ128146.1), and *Daucus carota* (NC_017855.1). tRNA annotation was performed using tRNAscan-SE software (http://lowelab.ucsc.edu/tRNAscan-SE/, accessed on 17 June 2022) [31]. ORFs were predicted using the NCBI Open Reading Frame Finder (https://www.ncbi.nlm.nih.gov/orffinder/, accessed on 17 June 2022) with a minimum length of 102 bp. Redundant sequences and sequences overlapping with known genes were excluded. Sequences longer than 300 bp were annotated by searching against the nonredundant database (nr) using BLASTn software (v2.10.1). The mitochondrial genome map was generated using OGDRAW (https://chlorobox.mpimp-golm.mpg.de/OGDraw.html, accessed on 17 June 2022).

2.4. Repeat Sequence Analysis and Chloroplast-to-Mitochondrial DNA Transfer Analysis

SSRs were identified using MISA v1.0 software. Tandem repeats were identified using TRF software (trf409.linux64), and dispersed repeats were identified using BLASTn software (v2.10.1) with the redundant and tandem repeats removed. The identified repeats were visualized using Circos (v0.69-5). Chloroplast data of watercress with the GenBank accession number MT622521 were downloaded from the NCBI. BLAST search was performed to find homologous sequences between the chloroplast and mitochondrial genomes, with the similarity threshold set at 70% and the E-value set at 10-5. The results of homologous sequences were visualized using Circos (v0.69-5).

2.5. Mitochondrial Genome Feature Analysis

RNA editing sites were predicted using the online software PREP Suite (http://prep.unl.edu/, accessed on 18 June 2022). Codon usage bias analysis was performed using a Perl script developed in-house following the method described by Li et al. [32] for analyzing codon usage bias in protein-coding genes. Multiple sequence alignment of homologous gene sequences from different species was performed using MAFFT v7.427 software, and the pi value for each gene was calculated using DnaSP. The software CGVIEW (RELEASE-2017_09_19, http://stothard.afns.ualberta.ca/cgview_server/, accessed on 18 June 2022) was used with default parameters for comparative analysis of mitochondrial genome structures among closely related species.

2.6. Genome Comparative Analysis

Sequences from six species in the Apiaceae family, including *Saposhnikovia divaricate* (MZ128146.1), *Apium leptophyllum* (MZ328723.1), *Daucus carota* (JQ248574.1), *Bupleurum chinense* (KX887330.1), and *Ferula sinkiangensis* (OK585063.1), were download for comparative analysis of mitochondrial genomes. The software nucmer (4.0.0beta2) was used for genome alignment of other sequences and assembled sequences to generate dot plot graphs. The assembled species were compared with the selected species using the blastn (2.10.1+) algorithm, setting the -word_size to 7 and the E-value to 1×10^{-5} and selecting fragments with a comparison length greater than 300 bp to construct the multiplexed synteny map.

2.7. Phylogenetic Analysis

Mitochondrial genome sequences of 29 species in the family Apiaceae, including *Bupleurum chinense* (OK166971.1), *Saposhnikovia divaricata* (MZ128146.1), and *Lactuca sativa* (MK642355.1), were selected from the NCBI database for collinearity analysis with *Oenanthe*. Multiple sequence alignment of shared CDS sequences was performed using MAFFT in auto mode, and the aligned sequences were concatenated and trimmed using trimAl (v1.4.rev15, parameter: -gt 0.7). The model was predicted using jmodeltest-2.1.10 software, and the GTR model type was determined. Finally, the maximum likelihood phylogenetic tree was constructed using RAxML (v8.2.10) with the GTRGAMMA model and 1000 bootstrap replicates.

3. Results

3.1. Assembly and Basic Features of the O. javanica Mitochondrial Genome

The mitochondrial genome of *O. javanica* was sequenced, resulting in a total of 7,359,287,400 base pairs (bp) of gene sequences. The mitochondrial genome showed a complex multibranched structure. After excluding repetitive regions, a high quality clean read dataset of 24,530,958 bp was obtained. Finally, the mitochondrial genome of watercress was assembled and mapped to a circular structure with 384,074 bp (Figure 1 and Table S1). The GC content was 44.97%, and the base composition showed an AT bias. Only three tRNA genes contained one intron each, indicating that the watercress mitochondrial genome is compact. The watercress mitochondrial genome was annotated, revealing a total of 65 genes, including 28 tRNA genes, 3 rRNA genes, and 34 protein-coding genes (Table 1 and Table S2). The protein-coding genes (PCGs) included 25 unique core mitochondrial genes: five ATP synthase genes, nine NADH dehydrogenase genes, one cytochrome c biogenesis gene, four cytochrome c oxidase genes, two transporter membrane protein genes, one mature enzyme gene, and three cytochrome c oxidase genes. In addition, nine noncore mitochondrial genes were identified, including four large subunit ribosomal protein genes and five small subunit ribosomal protein genes. Among the tRNA genes identified, *trnI-GAT* and *trnP-CGG* contained one intron each. Among all the mitochondrial genes, five were found to be multicopy genes, including the membrane transport protein gene (*mttB*), the ribosomal large subunit gene (*rpl10*), and three tRNA genes (*trnM-CAT, trnP-TGG,* and *trnS-TGA*). The *trnM-CAT* gene had up to seven copies, while the other genes (*mttB, rpl10, trnP-TGG,* and *trnS-TGA*) had two copies each.

3.2. Repeat Sequence Analysis

Microsatellites, also known as simple sequence repeats (SSRs), consist of single, double, triple, quadruple, or quintuple repeated DNA motifs [33]. They exhibit dominant inheritance and are important for genetic and evolutionary studies of plant mitochondrial genomes [34]. A total of 87 SSRs were identified in the mitochondrial genome of watercress. Mononucleotide and dinucleotide SSRs accounted for 35.63% of the SSRs, with adenine (A) mononucleotide repeats accounting for 60.00% of the mononucleotide SSRs. AG and CT repeats were the most common types of dinucleotide SSRs, accounting for 62.50% of the dinucleotide SSRs. Hexanucleotide SSRs are widespread in eukaryotic and prokaryotic genomes, but only two were identified in the mitochondrial genome of watercress, representing the lowest proportion (Figure 2A). In addition, the basic composition of SSR motifs in the watercress mitochondrial genome showed a strong bias towards AT-rich sequences. Research has shown that plant mitochondrial genomes may undergo recombination at the locations of scattered repetitive sequences, especially longer ones [18]. The distribution of scattered repetitive sequences in the mitochondrial genome was analyzed and a distribution map was generated (Figure 2B and Tables S3 and S4). A total of 585 pairs of repetitive sequences with a length $\geq$ 30 were observed, including 278 pairs of palindromic repeats, 301 pairs of forward repeats, and 6 pairs of reverse repeats. No complementary repeats were found. The longest palindromic repeat and forward repeat were 4018 and 4801 bp, respectively.

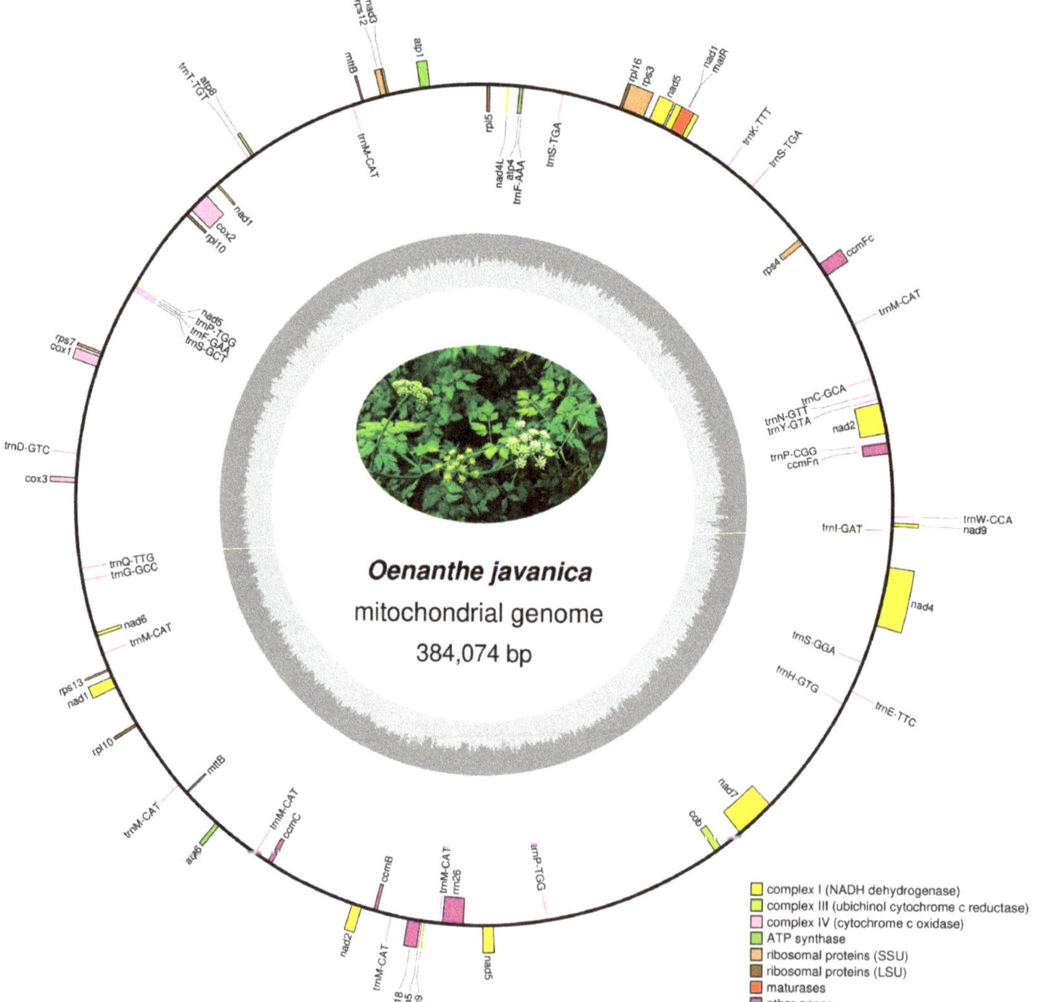

Figure 1. Gene map of the mitochondrial genome of *O. javanica*. Forward coding genes are on the outside of the circle, reverse-coding genes are on the inside of the circle, and the inner gray circle represents the GC content.

Table 1. Gene composition of the mitogenome of *O. javanica*.

Types	Group of Genes	Name of Genes
Protein coding genes (PCGs)	ATP synthase	*atp1, atp4, atp6, atp8, atp9*
	NADH dehydrogenase	*nad1, nad2, nad3, nad4, nad4L, nad5, nad6, nad7, nad9*
	Cytochrome c biogenesis	*cob*
	Ubiquinol cytochrome c reductase	*ccmB, ccmC, ccmFC, ccmFN*
	Cytochrome c oxidase	*cox1, cox2, cox3*
	Maturases	*matR*
	Transport membrane protein	*mttB(×2)*
	Ribosomal proteins (LSU)	*rpl5, rpl10(×2), rpl16*
	Ribosomal proteins (SSU)	*rps3, rps4, rps7, rps12, rps13*
Ribosomal RNAs		*rrn5, rrn18, rrn26*
Transfer RNAs		*trnC-GCA, trnD-GTC, trnE-TTC, trnF-AAA, trnF-GAA, trnG-GCC, trnH-GTG, trnI-GAT *, trnK-TTT, trnM-CAT(×7), trnN-GTT, trnP-CGG *, trnP-TGG(×2), trnQ-TTG, trnS-GCT, trnS-GGA, trnS-TGA(×2), trnT-TGT *, trnW-CCA, trnY-GTA*

Notes: * indicates that the gene contains one intron; the number in parentheses represents the number of copies of the gene (×2 means having two copies).

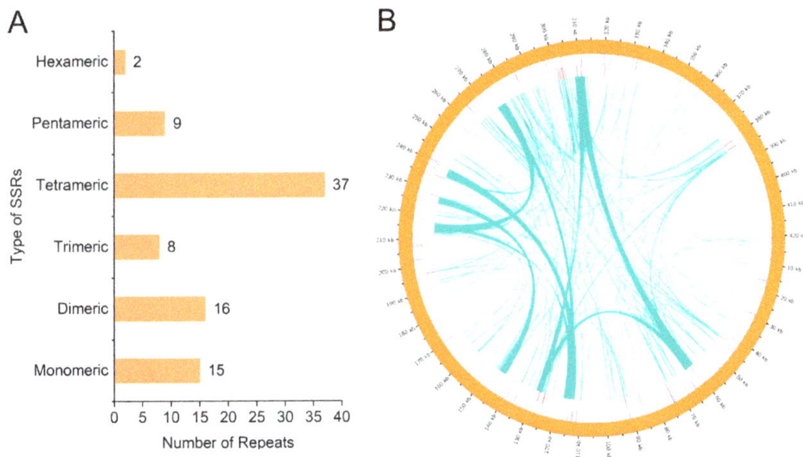

Figure 2. Distribution of tandem repeats in the watercress mitochondrial genome. (**A**) Type and number of tandem repeats. (**B**) Distribution of repeats on the genome. The yellow color is the genome scale, with simple repeats in the outermost circle (blue), followed (red) by tandem repeats, and scattered repeats in the innermost concatenation.

3.3. Analysis of Codon Usage Bias

Research has shown that there are differences in the usage of codons in the mitochondrial genomes of different species. This preference provides a basis for studying species evolution [35]. Codon usage bias analysis was performed on the watercress mitochondrial genome, and codon usage for each amino acid is shown in Table S5. A total of 9900 codons were detected, with leucine, isoleucine, and serine having the highest frequency of use with 1022, 918, and 764 occurrences, respectively. Cysteine and tryptophan had the lowest number of codons used with 133 and 145 occurrences, respectively. These results are consistent with observations in other plant mitochondrial genomes [36,37], indicating that the watercress mitochondrial genome is relatively conserved. Generally, codons with a relative synonymous codon usage (RSCU) value greater than 1 are considered to be preferred by amino acids. As shown in Figure 3, most mitochondrial protein-coding genes (PCGs) show a general preference for certain codons. For example, histidine (His) has a high preference for the AUG codon, with the highest RSCU value of 3.00 among mitochondrial PCGs, followed by alanine (Ala) with a preference for the GCU codon and an RSCU value of 1.60. It is worth noting that phenylalanine (Phe), valine (Val), and tryptophan (Trp) have maximum RSCU values less than 1.2, indicating a weak codon usage bias. There are 6485 codons with RSCU values greater than 1.00, of which 614 codons end with G (UUG, UGG, and AUG), and as many as 5871 codons end in A/U, suggesting a preference for A/U-ending codons in the protein-coding genes of the watercress mitochondrial genome.

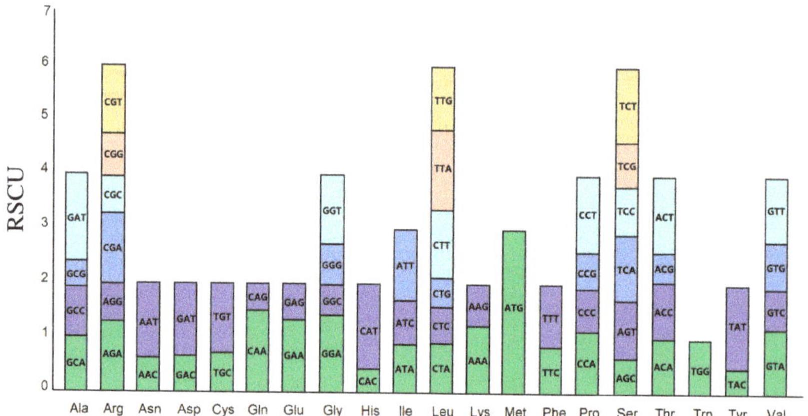

Figure 3. RSCU of the watercress mitochondrial genome.

3.4. Analysis of Pi Nucleotide Diversity

Pi values can reveal variation in the nucleic acid sequences of different plants, and regions of higher variability can provide potential molecular markers for population genetics [38]. Pi values were calculated for 35 genes in watercress and ranged from 0.00217 to 0.03741 (Figure 4). It is worth noting that rpl5 had the highest Pi value (0.03714), followed by *atp8*, *nad1*, and *rps3* with corresponding Pi values of 0.037, 0.02838, and 0.02251, respectively. We speculate that these four hotspots may contain information about evolutionary sites and could potentially serve as molecular markers [32]. Furthermore, in terms of the total number of mutations, *rps3* has the highest number of mutations, reaching 97, with a region length of 1896 bp, followed by *nad1* and *matR* with 72 and 71 mutations, respectively (Table S6). These three genes with the highest mutation rates may be the main sources of variation in Apiaceae plants.

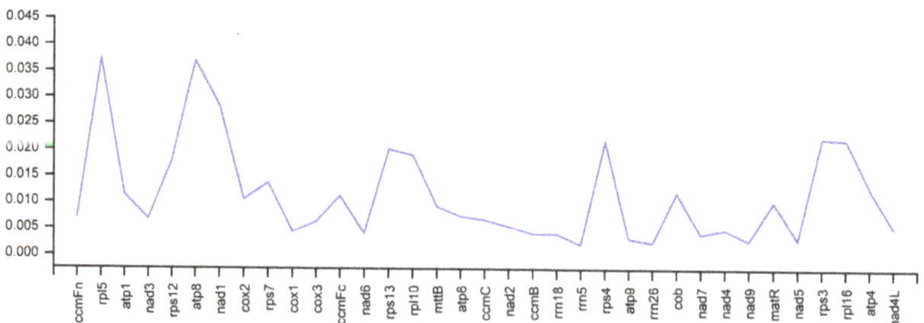

Figure 4. The nucleotide variability (Pi) value of genes in watercress mitochondrial genome.

3.5. Analysis of Chloroplast-to-Mitochondria Genomic Transfers in Watercress

We identified 24 fragments of the chloroplast genome that were transferred to the mitochondrial genome of watercress. These fragments included both genes and intergenic regions (Table S7), and their homologous segments are shown in Figure 5. The total length of homologous sequences between the watercress mitochondrial and chloroplast genomes was 18,959 bp, which accounts for approximately 4.94% of the assembled mitochondrial genome. The lengths of these transferred fragments ranged from 30 to 4333 bp. Among the fragments with more than 95% sequence identity, we found 17 fragments, with the smallest being *psaA* and *trnS-GCU* with 34 and 30 bp, respectively. Furthermore, by aligning the chloroplast nucleotide sequences with the mitochondrial genome, we discovered sequences with 100% identity in *trnS-GGA*, *trnH-GUG*, *ycf2*, *psaA*, and *trnS-GCU*. The longest gene fragment among these was *trnS-GGA*, which spanned 459 bp.

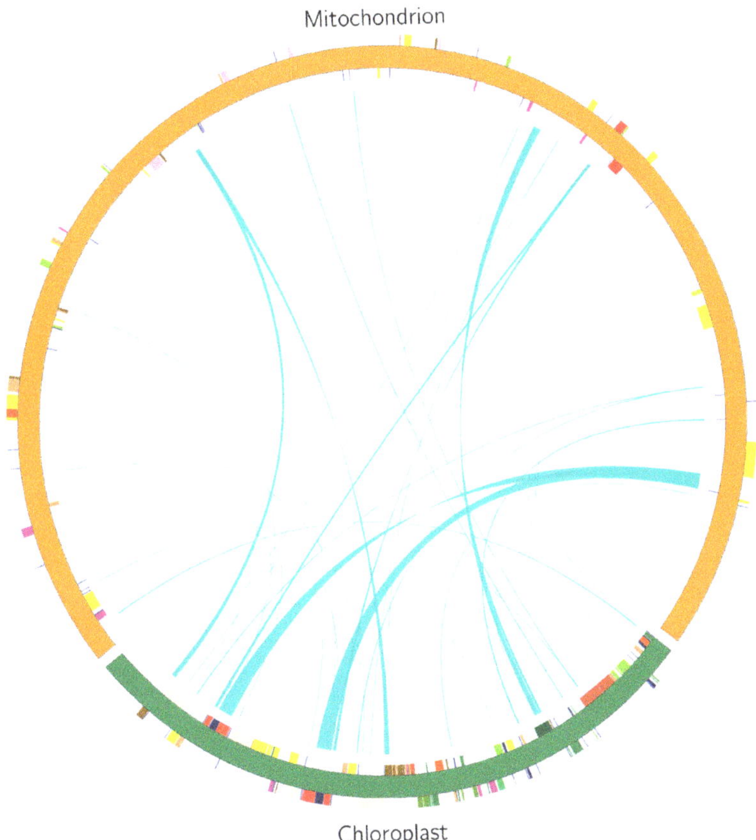

Figure 5. Chloroplast and mitochondrial sequence homologous fragments from watercress. Chloroplast is the chloroplast sequence, and the others are mitochondrial sequences. Genes from the same complex are labeled with the same color, and the middle line connection indicates homologous sequences.

3.6. RNA Editing Site Analysis

We identified RNA editing sites in the protein-coding genes encoded by the watercress mitochondrial genome. The results revealed a total of 512 RNA editing sites in watercress (Figure 6A,B). The genes with the highest number of RNA editing sites were the NADH dehydrogenase genes *nad4* and *nad2* with 45 and 30 substitution sites, respectively. This was followed by the cytochrome C biogenesis genes *ccmFn*, *ccmB*, and *ccmC*, which had 36, 34, and 30 substitution sites, respectively. In contrast, ribosomal protein genes (*rps3*, *rps4*, *rps7*, *rps12,* and *rps13*) and ribosomal large subunit genes (*rpl5, rpl10,* and *rpl16*) had relatively few RNA-editing-induced substitutions (2–14 sites). These results indicate that RNA editing substitutions in Oenanthe are mainly distributed in core genes, whereas fewer substitutions are found in noncore genes (Figure 6A). To date, the most common type of

RNA editing discovered is C→T, and in watercress, all RNA editing types were C→T, which is consistent with other species. RNA editing leads to changes in the hydrophilic and hydrophobic properties of amino acids. Among the altered properties, hydrophilic to hydrophobic editing accounted for the highest percentage at 47.85%, followed by hydrophobic to hydrophilic at 30.08%. The lowest score corresponding to GCT (A) ≥ GTT (V) in the *rpl5* was only 0.22 (Figure 6B and Table S8).

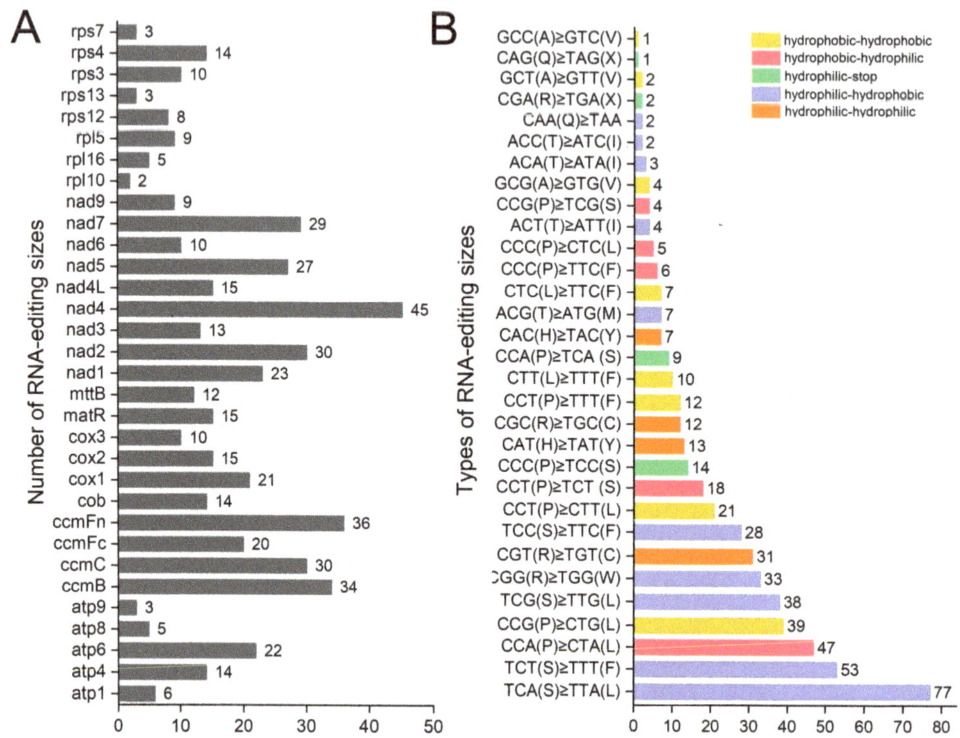

Figure 6. Statistics of the number of RNA editing sites in watercress. (**A**) Statistics of the number of RNA editing sites per gene. (**B**) Statistics of the change in hydrophilicity of amino acids caused by RNA editing.

3.7. Comparative Analysis of Mitochondrial Structures

Graphical genome maps have been widely used to assess genome features and sequence characteristics [39]. In this study, we used CGView to compare and analyze the sequence similarity between the mitochondrial genomes of watercress and other closely related species within the watercress genus using the annotated mitochondrial genome sequence of watercress as a reference. The results indicate a high degree of similarity in the rRNA and tRNA coding regions in watercress, with a relatively high GC content in the regions corresponding to rRNA genes (Figure 7). Homologous collinear blocks were detected between watercress and its closely related species (Figure 8A). However, the arrangement order of collinear blocks differed within individual mitochondrial genomes. The absence of longer diagonal segments in Figure 8B suggests structural variations in watercress during the process of evolution compared to other closely related species. In conclusion, the mitochondrial genome of watercress has undergone extensive genomic rearrangements with closely related species, indicating a highly nonconservative structure in the mitochondrial genome.

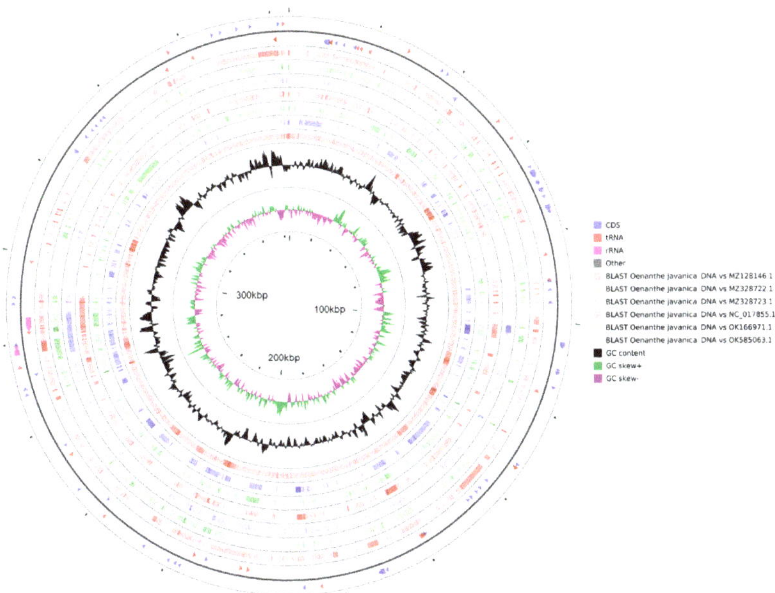

Figure 7. Comparative analysis of mitochondrial structure. The two outermost circles represent the gene lengths and orientations of the mitochondrial genome, the inner circle represents the similarity results between comparisons with other genomes, and the black circle represents the GC content.

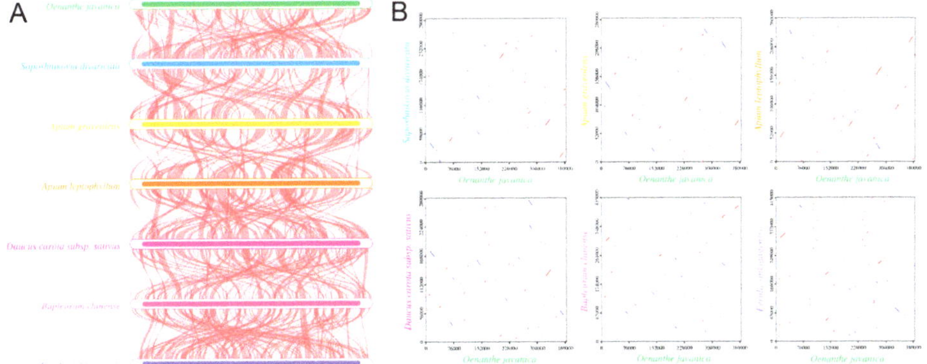

Figure 8. Mitochondrial sequence covariance analysis of watercress. (**A**) Multiple synteny plot of watercress with closely related species. The boxes in each row indicate a genome, and the connecting lines in the middle indicate regions of homology. (**B**) Dot plot of watercress with closely related species. The horizontal coordinate in each box indicates the assembled sequence, the vertical coordinate indicates the other sequences, the red line in the box indicates the forward comparison, and the blue line indicates the reverse complementary comparison.

3.8. Phylogenetic Analysis

In this study, a maximum likelihood method was used to construct a phylogenetic tree of 29 plants, including watercress. The results showed that of the 27 nodes in the generated tree, 21 nodes had bootstrap support values greater than 70%, with 14 nodes having 100% support (Figure 9). The phylogenetic tree showed that watercress forms a clade with the genera *Ferula*, *Apium*, *Angelica*, and *Daucus*. *Bupleurum* formed another clade, while *Nymphaea* formed a distinct major clade, with most nodes showing 100% support. The closest evolutionary relationships of watercress were observed with *S. divaricata* (MZ128146.1) and *A. graveolens* (MZ328722.1), which is consistent with the collinearity analysis results (Figure 8A). It is worth noting that watercress emerged as a monophyletic lineage independently without sister relationships with other genera. *Ferula* and *Apium* appeared as sister groups with 100% support, while *Angelica* and *Daucus* formed a sister group with 100% support.

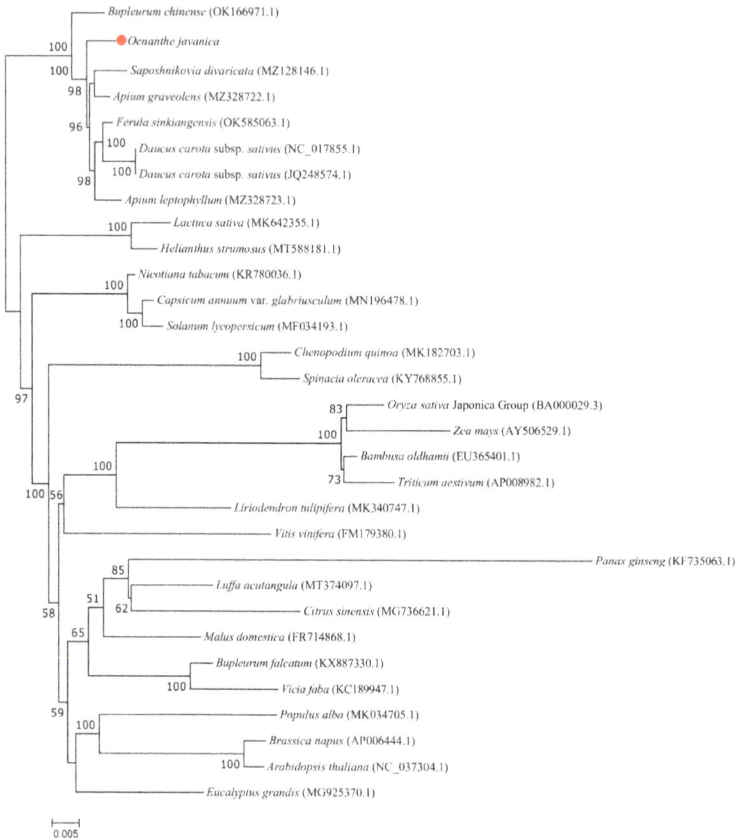

Figure 9. Phylogenetic tree based on protein-coding genes in mitochondrial genomes of 29 species. The numbers at each node are bootstrap support values (expressed as a percentage of 1000 replicates). The text corresponding to the red dots in the figure indicates watercress (*Oenanthe javanica*).

4. Discussion

The emergence of high-throughput sequencing technologies has greatly facilitated the study of plant mitochondrial genomes [17]. In this study, we successfully assembled the high-quality mitochondrial genome of watercress, which could serve as a valuable resource for future investigations into the evolution of watercress mitochondrial genomes. Sequence assembly revealed that the watercress mitochondrial genome is a circular molecule of 384,074 bp with a typical AT bias in base composition. It contains 28 tRNA genes (including 3 with introns), 3 rRNA genes, and 34 protein-coding genes. Among them, the *trnM CAT* gene has up to seven copies. Studies have shown that the variation in size among plant mitochondrial genomes is not determined by the number of genes or introns in these coding genes [19]. Instead, most plant mitochondrial genomes contain noncoding sequences of varying sizes [40]. As products of biological evolution, repetitive sequences are often used to reveal the historical imprints of long-term exchange and recombination of genetic material between species [41]. Dispersed repetitive sequences are associated with the generation of genetic diversity and make a significant contribution to genome evolution. In our study, the watercress mitochondrial genome contained a total of 87 SSRs, the majority of which were mononucleotide or dinucleotide repeats, suggesting that the size variation in the watercress mitochondrial genome may be due to repetitive short sequences. Dispersed repetitive sequences typically include forward repeats, reverse repeats, complementary repeats, or palindromic repeats [42]. In the watercress mitochondrial genome, 99% of the repeats were composed of palindromic repeats and forward repeats, with only six pairs of reverse repeats detected and no complementary repeats identified, which is consistent with the findings in other species [43].

Codon usage bias refers to the differential frequency of synonymous codons used in coding sequences by an organism [44]. RCUS, based on the hypothesis that codon usage affects translation dynamics, has been proposed to regulate translation efficiency, accuracy, and protein folding, which provides a basis for the study of species evolution [45]. The highest codon usage frequencies in the watercress mitochondrial genome were observed for leucine, isoleucine, and serine. Screening for codons with RSCU values greater than 1.00 revealed that 90.53% of protein-coding genes in the watercress mitochondrial genome preferentially use A/U-ending codons, while the remaining 9.47% use A/G ending codons (UUG, UGG, and AUG). Nucleotide diversity analysis is commonly used to detect hotspots in the genome, and Pi values reflect the variation of nucleotide sequences in plants [38]. The Pi values of the watercress genes ranged from 0.00217 to 0.03741, with the *rpl5* gene having the highest Pi value among the 35 genes. In addition, the *rps33*, *nad1*, and *matR* genes had the highest mutation levels in watercress. We also identified 24 plastid genome fragments in the watercress mitochondrial genome, representing 4.94% of the assembled mitochondrial genome. Chloroplast genome transfer to the mitochondria is associated with potential molecular functions and physiological processes in plants [46]. The genes with the most RNA editing in the watercress mitochondrial genome were NADH dehydrogenase genes and cytochrome c biogenesis genes. So far, the majority of RNA editing types discovered in plant species are C→T, and all RNA editing types in watercress were also C→T, showing a high degree of consistency with other species.

In recent years, with the increasing abundance of genomic and related genomic information, the convenience and novelty of graphical genome maps have been widely applied in the assessment of genomic features [39]. We utilized CGView to analyze the collinearity among mitochondrial genomes of closely related species in the Apiaceae family to investigate DNA rearrangement events within the mitochondrial genome. The mitochondrial genome sequences of seven species in the watercress genus exhibited certain collinearity but with significant structural variations, indicating a highly nonconservative structure in the watercress mitochondrial genome. Using the annotated mitochondrial genome sequence of watercress as a reference, we compared it with the assembled sequences of the other six species and no significant structural variations were found. In previous studies, the methods used to construct phylogenetic trees were based on one or more relatively short sequences [47]. However, due to horizontal gene transfer between populations and differences in rates of genetic evolution, phylogenetic trees based on single or few genes cannot fully represent the phylogenetic relationships. As DNA sequencing techniques mature, whole genome sequencing is increasingly being used in plant phylogenetics and population genetics [48,49]. In this study, the phylogenetic relationship of watercress was further analyzed based on mitochondrial genome information, and a sequence-based phylogenetic tree was constructed using protein-coding genes. The results indicate a close relationship between watercress, *S. divaricata*, and *A. graveolens*. Therefore, under the assumption of mitochondrial genome evolution and diversification mechanisms, the watercress mitochondrial genome may provide a potential for studying the evolution and development of the watercress genus and provide an important theoretical basis for improving watercress production, developing varietal resources, and maintaining stable morphology.

5. Conclusions

The watercress mitochondrial genome is a circular molecule of 384,074 bp. The nucleotide composition has a typical AT bias and contains 28 tRNA genes, 3 rRNA genes, and 34 protein-coding genes. A total of 87 SSR sites were detected, with 99% consisting of palindromic and forward repeats, only 6 pairs of reverse repeats, and no complementary repeats detected. Codon preference analysis revealed that watercress shares have the same codon preferences as most plant species, preferring to use codons encoding leucine, isoleucine, and serine and showing a preference for A/U-ending codons. In addition, the phylogenetic tree based on the mitochondrial genomes of 29 species helps in the scientific classification of watercress. In conclusion, this study provides information on the genetic characteristics, phylogenetic relationships, and evolution of watercress and provides a basis for species identification and biological studies of watercress and other species of Apiaceae.

Supplementary Materials: The following supporting information can be downloaded at: https://www.mdpi.com/article/10.3390/agronomy13082103/s1, Table S1: Sequencing statistics of watercress mitochondrial genome; Table S2: Specific information on the mitogenome composition of watercress; Table S3: Tandem repeat sequence information in the mitogenome of watercress; Table S4: Information on the distribution of repetitive sequences on the mitogenome of watercress; Table S5: The use of amino acid pairs of codons in the mitogenome of watercress; Table S6: Analysis of Pi nucleic acid diversity in watercress; Table S7: Chloroplast nucleic acid sequences compared to mitogenome information. Table S8: Statistical table of amino acid hydrophilicity changes caused by RNA editing.

Author Contributions: Conceptualization, Q.H. and G.T.; data processing, X.L.; formal analysis, X.L. and Q.H.; investigation, Q.L.; software, S.Z.; writing—original draft preparation, X.L. and Q.H.; writing—review and editing, M.L., Y.Z. and G.T.; funding acquisition, Y.Z. and G.T. All authors have read and agreed to the published version of the manuscript.

Funding: This research was funded by the Project of Guizhou Provincial Department of Science and Technology "Construction and Utilization of Horticultural Platform of Plant GenBank Creation" (No. Qiankehe Fuqi [2022] 005, No. Qiankehe Support [2022] Key 019); Guizhou Modern Agriculuture Research System (GZMARS)—Plateau Characteristic Vegetable Industry; Shaanxi Provincial Department of Science and Technology: 2022ZY1-CGZY-07; and the Project of Research and Demonstration of Key Technologies for Quality Improvement and Efficiency Enhancement of Vegetables in Bazhong City.

Institutional Review Board Statement: No application.

Informed Consent Statement: No application.

Data Availability Statement: The mitochondrial genome sequence of *Oenanthe javanica* that were generated are deposited at NCBI under the accession number OR209169.

Conflicts of Interest: The authors declare no conflict of interest.

References

1. Grosser, M.R.; Sites, S.K.; Murata, M.M.; Lopez, Y.; Chamusco, K.C.; Love Harriage, K.; Grosser, J.W.; Graham, J.H.; Gmitter, F.; Chase, C.D. Plant mitochondrial introns as genetic markers-conservation and variation. *Front. Plant Sci.* **2023**, *14*, 1116851.
2. Jang, W.; Lee, H.O.; Kim, J.-U.; Lee, J.-W.; Hong, C.-E.; Bang, K.-H.; Chung, J.-W.; Jo, I.-H. Complete mitochondrial genome and a set of 10 novel kompetitive allele-specific PCR markers in ginseng (*Panax ginseng* C. A. Mey.). *Agronomy* **2020**, *10*, 1868.
3. Xiong, Y.; Yu, Q.; Xiong, Y.; Zhao, J.; Lei, X.; Liu, L.; Liu, W.; Peng, Y.; Zhang, J.; Li, D.; et al. The complete mitogenome of *Elymus sibiricus* and insights into its evolutionary pattern based on simple repeat sequences of seed plant mitogenomes. *Front. Plant Sci.* **2022**, *12*, 802321.
4. Maréchal, A.; Brisson, N. Recombination and the maintenance of plant organelle genome stability. *New Phytol.* **2010**, *186*, 299–317.
5. Adams, K.L.; Palmer, J.D. Evolution of mitochondrial gene content: Gene loss and transfer to the nucleus. *Mol. Phylogenet. Evol.* **2003**, *29*, 380–395.
6. Beier, S.; Thiel, T.; Münch, T.; Scholz, U.; Mascher, M. MISA-web: A web server for microsatellite prediction. *Bioinformatics* **2017**, *33*, 2583–2585.
7. Alverson, A.J.; Rice, D.W.; Dickinson, S.; Barry, K.; Palmer, J.D. Origins and recombination of the bacterial-sized multichromosomal mitochondrial genome of cucumber. *Plant Cell* **2011**, *23*, 2499–2513.
8. Palmer, J.D.; Adams, K.L.; Cho, Y.; Parkinson, C.L.; Qiu, Y.L.; Song, K. Dynamic evolution of plant mitochondrial genomes: Mobile genes and introns and highly variable mutation rates. *Proc. Natl. Acad. Sci. USA* **2000**, *97*, 6960–6966.

9. Cheng, N.; Lo, Y.S.; Ansari, M.I.; Ho, K.C.; Jeng, S.T.; Lin, N.S.; Dai, H. Correlation between mtDNA complexity and mtDNA replication mode in developing cotyledon mitochondria during mung bean seed germination. *New Phytol.* **2017**, *213*, 751–763.

10. Varré, J.-S.; D'Agostino, N.; Touzet, P.; Gallina, S.; Tamburino, R.; Cantarella, C.; Ubrig, E.; Cardi, T.; Drouard, L.; Gualberto, J.M.; et al. Complete Sequence, Multichromosomal Architecture and Transcriptome Analysis of the Solanum tuberosum Mitochondrial Genome. *Int. J. Mol. Sci.* **2019**, *20*, 4788. [CrossRef]

11. Bi, C.; Paterson, A.H.; Wang, X.; Xu, Y.; Wu, D.; Qu, Y.; Jiang, A.; Ye, Q.; Ye, N. Analysis of the complete mitochondrial genome sequence of the diploid cotton *Gossypium raimondii* by comparative genomics approaches. *BioMed. Res. Int.* **2016**, *2016*, 5040598.

12. Mustafa, S.I.; Heslop-Harrison, J.S.; Schwarzacher, T. The complete mitochondrial genome from Iraqi Meriz goats and the maternal lineage using whole genome sequencing data. *Iran. J. Appl. Anim. Sci.* **2021**, *12*, 321–328.

13. Kozik, A.; Rowan, B.A.; Lavelle, D.; Berke, L.; Schranz, M.E.; Michelmore, R.W.; Christensen, A.C. The alternative reality of plant mitochondrial DNA: One ring does not rule them all. *PLoS Genet.* **2019**, *15*, e1008373.

14. Alverson, A.J.; Wei, X.; Rice, D.W.; Stern, D.B.; Barry, K.; Palmer, J.D. Insights into the evolution of mitochondrial genome size from complete sequences of *Citrullus lanatus* and *Cucurbita pepo* (Cucurbitaceae). *Mol. Biol. Evol.* **2017**, *27*, 1436–1448.

15. Zhong, F.; Ke, W.; Li, Y.; Chen, X.; Zhou, T.; Xu, B.; Qi, L.; Yan, Z.; Ma, Y. Comprehensive analysis of the complete mitochondrial genomes of three *Coptis* species (*C. chinensis, C. deltoidea* and *C. omeiensis*): The important medicinal plants in China. *Front. Plant Sci.* **2023**, *29*, 1166420.

16. Wang, B.; Xue, J.; Li, L.; Liu, Y.; Qiu, Y.L. The complete mitochondrial genome sequence of the liverwort *Pleurozia purpurea* reveals extremely conservative mitochondrial genome evolution in liverworts. *Curr. Genet.* **2009**, *55*, 601–609.

17. Dong, S.; Zhao, C.; Chen, F.; Liu, Y.; Zhang, S.; Wu, H.; Zhang, L.; Liu, Y. The complete mitochondrial genome of the early flowering plant *Nymphaea colorata* is highly repetitive with low recombination. *BMC Genom.* **2018**, *19*, 614.

18. Yu, R.; Chen, X.; Long, L.; Jost, M.; Zhao, R.; Liu, L.; Mower, J.P.; de Pamphilis, C.W.; Wanke, S.; Jiao, Y. De novo assembly and comparative analyses of mitochondrial genomes in *Piperales*. *Genom. Biol. Evol.* **2023**, *15*, evad041.

19. Ren, W.; Wang, L.; Feng, G.; Tao, C.; Liu, Y.; Yang, J. High-quality assembly and comparative analysis of *Actinidia latifolia* and *A. valvata* mitogenomes. *Genes* **2023**, *14*, 863.

20. Cheng, Q.; Wang, P.; Li, T.; Liu, J.; Zhang, Y.; Wang, Y.; Sun, L.; Shen, H. Complete mitochondrial genome sequence and identification of a candidate gene responsible for cytoplasmic male sterility in celery (*Apium graveolens* L.). *Int. J. Mol. Sci.* **2021**, *22*, 8584.

21. Chen, C.M.; Chen, I.C.; Chen, Y.L.; Lin, T.H.; Chen, W.L.; Chao, C.Y.; Wu, Y.R.; Lu, Y.T.; Lee, C.Y.; Chien, H.C.; et al. Medicinal herbs *Oenanthe javanica* (Blume) DC., *Casuarina equisetifolia* L. and *Sorghum bicolor* (L.) Moench protect human cells from MPP⁺ damage via inducing *FBXO7* expression. *Phytomedicine* **2016**, *23*, 1422–1433.

22. Lu, C.L.; Li, X.F. A review of *Oenanthe javanica* (Blume) DC. as traditional medicinal plant and its therapeutic potential. *Evid. Based Complement. Altern. Med.* **2019**, *2019*, 6495819.

23. Feng, K.; Kan, X.-Y.; Liu, Q.; Yan, Y.-J.; Sun, N.; Yang, Z.-Y.; Zhao, S.-P.; Wu, P.; Li, L.-J. Metabolomics Analysis Reveals Metabolites and Metabolic Pathways Involved in the Growth and Quality of Water Dropwort [*Oenanthe javanica* (Blume) DC.] under Nutrient Solution Culture. *Plants* **2023**, *12*, 1459. [CrossRef]

24. Kumar, S.; Huang, X.; Ji, Q.; Qayyum, A.; Zhou, K.; Ke, W.; Zhu, H.; Zhu, G. Influence of Blanching on the Gene Expression Profile of Phenylpropanoid, Flavonoid and Vitamin Biosynthesis, and Their Accumulation in *Oenanthe javanica*. *Antioxidants* **2022**, *11*, 470. [CrossRef]

25. Kumar, S.; Li, G.; Yang, J.; Huang, X.; Ji, Q.; Liu, Z.; Ke, W.; Hou, H. Effect of salt stress on growth, physiological parameters, and ionic concentration of water dropwort (*Oenanthe javanica*) cultivars. *Front. Plant Sci.* **2021**, *12*, 660409.

26. Seo, W.H.; Baek, H.H. Identification of characteristic aroma-active compounds from water dropwort (*Oenanthe javanica* DC.). *J. Agric. Food Chem.* **2005**, *53*, 6766–6770.

27. Zhang, Z.; Dong, H.; Yuan, M.; Yu, Y. The complete chloroplast genome of *Oenanthe javanica*. *Mitochondrial DNA B Resour.* **2020**, *5*, 3151–3153.

28. Jiang, Q.; Wang, F.; Tan, H.W.; Li, M.Y.; Xu, Z.S.; Tan, G.F.; Xiong, A.S. De novo transcriptome assembly, gene annotation, marker development, and miRNA potential target genes validation under abiotic stresses in *Oenanthe javanica*. *Mol. Genet. Genom.* **2015**, *290*, 671–683.

29. Yao, J.; Zhao, F.; Xu, Y.; Zhao, K.; Quan, H.; Su, Y.; Hao, P.; Liu, J.; Yu, B.; Yao, M.; et al. Complete chloroplast genome sequencing and phylogenetic analysis of two *Dracocephalum* plants. *BioMed Res. Int.* **2020**, *2020*, 4374801.

30. Koren, S.; Walenz, B.P.; Berlin, K.; Miller, J.R.; Phillippy, A.M. Canu: Scalable and accurate long-read assembly via adaptive k-mer weighting and repeat separation. *Genom. Res.* **2017**, *27*, 722–736.

31. Chan, P.P.; Lowe, T.M. tRNAscan-SE: Searching for tRNA genes in genomic sequences. *Methods Mol. Biol.* **2019**, *1962*, 1–14.

32. Li, M.Y.; Zhang, R.; Li, R.; Zheng, K.M.; Xiao, J.C.; Zheng, Y.X. Analyses of chloroplast genome of *Eutrema japonicum* provide new Insights into the evolution of *Eutrema* Species. *Agronomy* **2021**, *11*, 2546.

33. Powell, W.; Machray, G.C.; Provan, J. Polymorphism revealed by simple sequence repeats. *Trends Plant Sci.* **1996**, *1*, 215–222.

34. Avvaru, A.K.; Saxena, S.; Sowpati, D.T.; Mishra, R.K. MSDB: A comprehensive database of simple sequence repeats. *Genom. Biol. Evol.* **2017**, *9*, 1797–1802.

35. Gu, C.; Dong, B.; Xu, L.; Tembrock, L.R.; Zheng, S.; Wu, Z. The complete chloroplast genome of *Heimia myrtifolia* and comparative analysis within myrtales. *Molecules* **2018**, *23*, 846.

36. Dong, S.; Chen, L.; Liu, Y.; Wang, Y.; Zhang, S.; Yang, L.; Lang, X.; Zhang, S. The draft mitochondrial genome of *Magnolia biondii* and mitochondrial phylogenomics of angiosperms. *PLoS ONE* **2020**, *15*, e0231020.
37. Shi, Y.; Liu, Y.; Zhang, S.; Zou, R.; Tang, J.; Mu, W.; Peng, Y.; Dong, S. Assembly and comparative analysis of the complete mitochondrial genome sequence of *Sophora japonica* 'JinhuaiJ2'. *PLoS ONE* **2018**, *13*, e0202485.
38. Sun, Y.; Li, M.Y.; Ma, J.Y.; He, M.L.; Zheng, Y.X. Complete chloroplast genome sequence of a new variety of *Brasenia schreberi*: Genome characteristics, comparative analysis, and phylogenetic relationships. *Agronomy* **2022**, *12*, 2972.
39. Stothard, P.; Grant, J.R.; Van Domselaar, G. Visualizing and comparing circular genomes using the CGView family of tools. *Brief. Bioinform.* **2019**, *20*, 1576–1582.
40. Spinelli, J.B.; Haigis, M.C. The multifaceted contributions of mitochondria to cellular metabolism. *Nat. Cell Biol.* **2018**, *20*, 745–754.
41. Wynn, E.L.; Christensen, A.C. Repeats of unusual size in plant mitochondrial genomes: Identification, incidence and evolution. *G3* **2019**, *9*, 549–559.
42. Ni, Y.; Li, J.; Chen, H.; Yue, J.; Chen, P.; Liu, C. Comparative analysis of the chloroplast and mitochondrial genomes of *Saposhnikovia divaricata* revealed the possible transfer of plastome repeat regions into the mitogenome. *BMC Genom.* **2022**, *23*, 570. [CrossRef] [PubMed]
43. Li, Z., Bai, D.; Zhong, Y.; Abid, M.; Qi, X.; Hu, C.; Fang, J. Physiological responses of two contrasting kiwifruit (*Actinidia* spp.) rootstocks against waterlogging stress. *Plants* **2021**, *10*, 2586. [CrossRef] [PubMed]
44. Hanson, G.; Coller, J. Codon optimality, bias and usage in translation and mRNA decay. *Nat. Rev. Mol. Cell Biol.* **2018**, *19*, 20–30.
45. Tang, D.; Huang, S.; Quan, C.; Huang, Y.; Miao, J.; Wei, F. Mitochondrial genome characteristics and phylogenetic analysis of the medicinal and edible plant *Mesona chinensis* Benth. *Front. Genet.* **2023**, *13*, 1056389. [CrossRef] [PubMed]
46. Yu, C.H.; Dang, Y.; Zhou, Z.; Wu, C.; Zhao, F.; Sachs, M.S.; Liu, Y. Codon usage influences the local rate of translation elongation to regulate co-translational protein folding. *Mol. Cell* **2015**, *59*, 744–754. [CrossRef] [PubMed]
47. He, P.; Xiao, G.; Liu, H.; Zhang, L.; Zhao, L.; Tang, M.; Huang, S.; An, Y.; Yu, J. Two pivotal RNA editing sites in the mitochondrial *atp1* mRNA are required for ATP synthase to produce sufficient ATP for cotton fiber cell elongation. *New Phytol.* **2018**, *218*, 167–182. [CrossRef]
48. Zhuang, Y.; Tripp, E.A. The draft genome of *Ruellia speciosa* (Beautiful Wild Petunia: Acanthaceae). *DNA Res.* **2017**, *24*, 179–192.
49. Feng, Y.; Xiang, X.; Akhter, D.; Pan, R.; Fu, Z.; Jin, X. Mitochondrial phylogenomics of fagales provides insights into plant mitogenome mosaic evolution. *Front. Plant Sci.* **2021**, *12*, 762195. [CrossRef]

MDPI
St. Alban-Anlage 66
4052 Basel
Switzerland
www.mdpi.com

Agronomy Editorial Office
E-mail: agronomy@mdpi.com
www.mdpi.com/journal/agronomy